TEACHING
ENGINEERING

TEACHING ENGINEERING

Phillip C. Wankat
Frank S. Oreovicz
Purdue University

McGraw-Hill, Inc.
New York   St. Louis   San Francisco   Auckland   Bogotá   Caracas
Lisbon   London   Madrid   Mexico   Milan   Montreal   New Delhi
Paris   San Juan   Singapore   Sydney   Tokyo   Toronto

TEACHING ENGINEERING

1 2 3 4 5 6 7 8 9 0 DOC DOC 9 0 9 8 7 6 5 4 3 2

ISBN 0-07-068154-6

The editor was B.J. Clark;
the production supervisor was Leroy A. Young.
R. R. Donnelley & Sons Company was printer and binder.

Library of Congress Cataloging-in-Publication Data

Wankat, Phillip C., (date).
Teaching engineering / Phillip C. Wankat, Frank S. Oreovicz.
    p.   cm.
Includes index.
ISBN 0-07-068154-6
1. Engineering—Study and teaching (Higher)—United States.
I. Oreovicz, Frank S.   II. Title.
T73.W18   1993
620'.0071'173—dc20   92-37689

CONTENTS

PREFACE

With his characteristic cleverness, George Bernard Shaw armed several generations of cynics with his statement "Those who can, do; those who can't, teach." But in today's world, engineering professors have to be able to do engineering *and* to teach engineering. How they prepare for this task is the subject of this book, which grew out of our conviction that new faculty are entering the university well prepared and well mentored in doing research, but almost totally at sea when it comes to the day-to-day requirements of teaching. At best, graduate students obtain only a second-hand knowledge of teaching, rarely having the opportunity to conduct an entire class for an extended period of time. If their role models are good or, better yet, master teachers, then some of the luster may wear off and they may gain valuable exposure to the craft. More often than not, the opposite occurs. An individual with a desire to teach has to rely on his or her own interest in teaching, and later discovers, with the mounting pressure of producing publications and research, that he or she can give only minimal attention to the classroom. This is a risky way to ensure the future of our discipline.

In 1983 we developed and taught for the first time a graduate course, Educational Methods for Engineers, geared toward Ph.D. candidates who were interested in an academic career. Our sources came from a variety of disciplines, journals, and books because we immediately noticed that no textbook was available which focused solely on engineering. Classic texts such as Highet's and McKeachie's became starting points and we scoured the literature for what was available in engineering. With a grant from the National Science Foundation in 1990 we expanded the course to include all of engineering, conducted a summer workshop, and began this book much earlier than we otherwise could have. Although the writing of this book was supported by NSF, all of the views in this book are the authors' and do not represent the views of either the National Science Foundation or Purdue University.

Many people have helped us, often unknowingly, in developing the ideas presented in this book. The writings and lectures of the following engineering professors have helped to shape our thinking: Richard Culver, Raymond Fahien, Richard Felder, Scott Fogler, Gordon Flammer, Lee Harrisberger, Billy Koen, Richard Noble, Helen Plants, John Sears, Bill Schowalter, Dendy Sloan, Karl Smith, Jim Stice, Charles Wales, Patricia Whiting, Don Woods and Charles Yokomoto.

At Purdue, Ron Andres suggested the partnership of W & O; others influential include Ron Barile, Kent Davis, Alden Emery, John Feldhusen, Dick Hackney, Neal Houze, Lowell Koppel, John Lindenlaub, Dick McDowell, Dave Meyer, Cheryl Oreovicz, Sam Postlethwait, Bob Squires, and Henry Yang, plus many other faculty members. Our students in classes and workshops tested the manuscript, and their comments have been extremely helpful. Professor John Wiest audited the entire class and his discussion and comments helped to mold this book. Professor Felder's critique of the book led us to reorganize the order of presentation. Professor Phil Swain was extremely helpful in polishing Chapter 8. Without question, the work of Mary McCaulley in extending and explicating the ideas of Katherine Briggs and Isabel Briggs-Myers formed our thinking on psychological type and its relevance to engineering education. Catherine Fitzgerald and John DiTiberio provided first-hand exposure to Type theory in action.

In the early formatting stages, Margaret Hunt provided invaluable assistance; Stephen Carlin drew the final figures and did the final formatting of the text. Betty Delgass provided the index as well as helpful suggestions and comments on both style and substance. We also wish to acknowledge the careful and helpful close reading by the McGraw-Hill copy editors, as well as the patient guidance through the publishing process provided by editors B.J. Clark and John Morriss. Through it all, our secretaries, Karen Parsons and Paula Pfaff, tirelessly dealt with two authors who often made changes independently.

Finally, we dedicate this book to our families in appreciation for their patience and support: To our wives, Dot and Sherry, for listening to our complaints; and to our children—Charles and Jennifer, and John and Mary-Kate: with their future in mind we wrote this book.

INTRODUCTION:
TEACHING ENGINEERING

It is possible to learn how to teach well. That is the thesis of this book. We want to help new professors get started toward effective, efficient teaching so that they can avoid the "new professor horror show" in the first class they teach. And by exposing them to a variety of theories and methods, we want to open the door for their growth as educators. Since one goal is immediate and the second is long-term, we have included both immediate how-to procedures and more theoretical or philosophical sections. Written mainly for Ph.D. students and professors in all areas of engineering, the book may be used as a text for a graduate-level class or by professionals who wish to read it on their own. Although our focus is engineering, much of this book should be useful to teachers in other technical disciplines. Teaching is a complex human activity, so it's impossible to develop a formula which guarantees that it will be excellent. But by becoming more efficient, professors can learn to do a good job and end up with more time to do other things such as research.

1.1. WHY TEACH TEACHING NOW?

The majority of engineering professors have never had a formal course in education, and some can even produce a variety of challenging rationalizations why such a course is unnecessary:

1 I didn't need a teaching course.
2 I learned how to teach by watching my teachers.
3 Good teachers are born and not made.
4 Teaching is unimportant.

1

5 Teaching courses have not improved the teaching in high schools and grade schools.
6 Engineers need more technical courses.
7 If I am a good researcher, I will automatically be a good teacher.
8 Even if a teaching course might be a good idea, none is available.

1 The first criticism can be answered in several ways. Just because someone did not need a teaching course does not logically imply that he or she would not have benefited from one. What is more important, times have changed. In the past, young assistant professors received a good deal of on-the-job training in how to teach. New assistant professors were mentored in teaching and were expected to teach several classes a semester. Now, mentoring is in research, and an assistant professor in engineering at a research university may teach only one course a semester. In the past the major topic of discussion with older professors was teaching; now it is research and grantsmanship. Because of these changes, formal training in teaching methods is now much more important. Van Ness (1989) has presented a detailed description of the changes in chemical engineering education which closely match changes in other areas of engineering education.

The problems facing engineering education have also changed. According to demographic studies, the number of traditional engineering students—white, male eighteen-year olds—is expected to go through a minimum from 1992 to 1994 and then increase very slowly (Hodgkinson, 1985; Reynolds and Oaxaca, 1988). In order to have enough engineers to remain internationally competitive, we must recruit, teach, and retain nontraditional students such as women and underrepresented minorities. There is also a moral imperative for reaching out to these nontraditional students. They offer different challenges and require different educational methods. A related problem is how to encourage enough U.S. citizens, particularly women and minorities, to earn a Ph.D. and then become educators. Many students see the workloads of assistant professors as oppressive and do not want the sword of the tenure decision hanging over their heads. A course on efficient, effective teaching would reduce the trauma of starting an academic career and help these students to see the joys of teaching.

2 You undoubtedly learned something about teaching from your teachers, but what if they were bad teachers? Even if you did have good teachers, this method at best gives the new professor a limited repertoire and does not provide for any of the necessary practice. This approach also does not help you incorporate new educational technology into the classroom unless you have had the rare opportunity to take a course from one of the pioneers in these areas. An opinion contrary to this is given by Highet (1976, p. 112), who argues that a course on education during graduate study is not needed since students can learn by watching good and bad teachers.

3 Some of the characteristics of good teachers may well be inborn and not made, but the same can be said for engineers. We expect engineers to undergo rigorous training to become proficient. It is logical to require similar rigorous training in the teaching methods of engineering professors. Experience in teaching engineering students how to teach shows that everyone can improve her or his teaching (e.g., see Wankat and Oreovicz, 1984; Stice, 1991). Even those born with an innate affinity for teaching or research can improve by study and practice. Finally, in its extreme, this argument removes all responsibility and all possibility for change from an individual.

4 There is no doubt that teaching is very important to students, parents, alumni, accreditation boards, and state legislatures. Unfortunately, at many universities research is more important than teaching in the promotion process. When assistant professors are denied tenure, it is because of lack of research, not because they have not been good teachers. An efficient teacher can do a good job teaching in the same amount of time an inefficient teacher spends doing a poor job. New professors who study educational methods will likely be better prepared to teach and will be more efficient during their first years in academia.

5 There is a general trend toward reducing the number of courses in pedagogy and increasing the number of content courses for both grade school and high school teachers. However, there is no trend toward zero courses or no practice in how to teach. The optimum number of courses in teaching methods undoubtedly lies between the large number required of elementary school teachers and the zero number taken by most engineering professors.

6 The demand for more and more technical courses is frequently heard at both the undergraduate and graduate levels. At the graduate level some of the most prestigious universities require the fewest number of courses. Thus, arguments that instructors must cover more technical content lack conviction at the graduate level. Courses on teaching can be very challenging and can open up entirely new vistas to the student. A course on teaching methods will be useful to all students even if they go into industry or government since logical organization and presentation of material are important in all areas.

7 Unfortunately, most research shows that there is almost no correlation between effective teaching and effective research (see Section 17.3 for a detailed discussion). Frequently heard comments to the contrary often appear to be based on examples of good researchers who are also good teachers, while ignoring examples of good teachers who do not do research and examples of good researchers who are poor teachers. This should not be interpreted as a statement that engineering professors should not do research. Ideally, they should strive to do both teaching and research well, and they should be trained for both functions.

8 There are a few courses in teaching in engineering colleges (e.g., Wankat and Oreovicz, 1984; Stice, 1991), and at the University of Texas at Austin the teaching course has been offered since 1972 (Stice, 1991). Many, if not most, universities offer teaching workshops either before the semester starts (e.g., Felder et al., 1989) or during the semester (e.g., Wentzel, 1987). Professional societies such as the American Society for Engineering Education (ASEE) also frequently offer effective teaching programs.

There are additional good reasons for learning how to teach. Teaching when you don't know how may be considered unethical! Canon 2 of the Accreditation Board for Engineering and Technology (ABET) states, "Engineers shall perform services only in the areas of their competence" (see Table 12-1). Since teaching is a service, teaching when one is not competent is probably unethical. Also, the ASEE Quality of Engineering Education Project concluded, "All persons preparing to teach engineering (the pretenure years) should be required to include in their preparation studies related to the practice of teaching" (ASEE, 1985, p. 156).

1.2. THE COMPONENTS OF GOOD TEACHING

Exactly what characterizes a good teacher? Many adjectives come to mind when this question is asked: stimulating, clear, well-organized, warm, approachable, prepared, helpful, enthusiastic, fair, and so forth. Lowman (1985) synthesized the research on classroom dynamics, student learning, and teaching to develop a "two-dimensional model" of good teaching. The most important dimension is intellectual excitement which represents the teacher's "obligation to knowledge and society" (Elbow, 1986, p. 142). This dimension includes content and performance. Since most engineering professors think content is the most important, making this dimension the most important agrees with common wisdom in the profession. Included in intellectual excitement are organization and clarity of presentation of up-to-date material. Since a dull performance can decrease the excitement of the most interesting material, this dimension includes performance characteristics. Is the professor energetic and enthusiastic? Does the professor clearly show a love for the material? Does the professor use clear language and clear pronunciation? Does the professor engage the students so that they are immersed in the material?

The second dimension identified by Lowman is interpersonal rapport which is the teacher's "obligation to students" (Elbow, 1986, p. 142). Professors develop rapport with students by showing an interest in them as individuals. In addition to knowing every student's name, does the professor know something about each one? Does he or she encourage them and allow for independent thought even though they may disagree with the professor? Is the professor available for questions both in and out of class? Although engineering professors do not uniformly agree that interpersonal rapport is important, students consistently include this dimension in their ratings of teachers (see Section 16.3.2). Note that at times the content and rapport sides of teaching conflict with each other (Elbow, 1986).

How do these two dimensions interact? The complete model is shown in Table 1-1. Lowman (1985) divides intellectual development into high (extremely clear and exciting), medium (clear and interesting), and low (vague and dull). He divides the interpersonal rapport dimension into high (warm, open, predictable, and highly student-oriented), medium (relatively warm, approachable, democratic, and predictable), and low (cold, distant, highly controlling, unpredictable). To interpersonal rapport we have added a fourth level below low—punishing (attacking, sarcastic, disdainful, controlling, and unpredictable)—since we

TABLE 1-1   TWO-DIMENSIONAL MODEL OF TEACHING (Modified from Lowman, 1985)

Intellectual Excitement	Interpersonal Rapport			
	Punishing	Low	Moderate	High
High	6'. Intellectual Attacker	6. Intellectual Authority	8. Masterful Lecturer	9. Complete Master
Moderate	3'. Adequate Attacker	3. Adequate	5. Competent	7. Masterful Facilitator
Low	1'. Inadequate Attacker	1. Inadequate	2. Marginal	4. "Warm fuzzy"

have observed professors in this category.

The numbering system in Table 1-1 indicates that professors improve their teaching much more quickly by increasing their intellectual excitement than by developing greater rapport with students. For example, a professor who is high in interpersonal rapport and low in intellectual excitement (position 4) will be considered a poorer teacher than a professor who is high in intellectual excitement and low in interpersonal rapport (position 6). Because their strengths are very different, these two teachers will excel in very different types of classes. The professor in position 4 will do best with a small class with a great deal of student participation, whereas the professor in position 6 will do best in large lecture classes. Our impression based on a very unscientific sample is that most engineering professors are in the broad moderate level of intellectual excitement and are at all levels of interpersonal rapport. The difference between these teachers and those at the high level of intellectual excitement is that the latter either consciously or unconsciously pay more attention to the performance aspects of teaching. Fortunately, all engineering professors can improve their teaching in both dimensions, and position 5 (competent) is accessible to all. Although becoming a complete master is a laudable goal to aim for, teachers who have attained this level are rare.

Hanna and McGill (1985) contend that the affective aspects of teaching are more important than method. Affective components which appear to be critical for effective teaching include:

- Valuing learning
- A student-centered orientation
- A belief that students can learn
- A need to help students learn.

These affective components are included in the model in Table 1-1. High intellectual excitement is impossible without valuing the learning of content and a need to present the material in a form which aids learning. High interpersonal rapport requires a student-centered orientation and a belief that students can learn.

A few comments about the punishing level of interpersonal rapport are in order. Since most students will fear such a professor, they will do the course assignments and learn the material if they remain in the course and aren't immobilized by fear. However, even those who do well will dislike the material. In our opinion and in the opinion of the American Association of University Professors (see Table 17-3), this punishing behavior is unprofessional. The only justification for a punishing style is to train students for a punishing environment such as that confronted by boxers, POWs, sports referees, and lawyers. Professors who stop attacking students immediately move into the level of low interpersonal rapport and receive higher student ratings.

1.3. PHILOSOPHICAL APPROACH

Teaching is an important activity of engineering professors, both in regard to content and in relation to students. New professors are usually superbly trained in content, but often have very little idea of how students learn.   This book is based on what may possibly be a revolutionary hypothesis: Young professors will do a better job teaching initially if they receive education and practice in teaching while they are graduate students or when they first start out as assistant professors. They will be more efficient the first few years and will have time for other activities.

The teaching methods covered in this book go beyond the standard lecture format, although it too is covered. Unfortunately, for too many teachers lecturing is often synonymous with teaching. In an attempt to broaden the reader's repertoire of teaching techniques, we include other teaching methods which may be more appropriate for some courses. Because advising and tutoring are closely tied to teaching, we also include these one-to-one activities. And since we believe that learning to become a good problem solver and learning how to learn are two major goals of engineering education, we also cover methods for teaching students to attain these goals.

Engineering professors invariably serve as models of proper behavior. Thus, an engineering professor should be a good engineer both technically and ethically, not using his or her position to persecute or take advantage of students. We agree with Highet (1976, p. 79) that in general students are likely to be immature and that "our chief duty is not to scorn them for this inability to comprehend, but to help them in overcoming their weakness." A well-developed sense of fairness is almost uniformly appreciated by students.

Our position on human potential is that people want to learn. Therefore, we search for ways to stop demotivating students while realizing that a few discipline problems always exist. Teaching is an important activity of engineering professors. Since they must also be involved in varying amounts of research, administration, advising, committee work, consulting, and so forth, we emphasize both effectiveness and efficiency.

1.4. WHAT WORKS: A COMPENDIUM OF LEARNING PRINCIPLES

Throughout this book we will base teaching methods on known learning principles. Many comments on what works in teaching are scattered throughout. In this section we will list many of the methods that are known to work. The ideas in this section are based on Chapters 13 to 15, papers by Chickering and Gamson (1987), Durney (1973), Irey (1981), and Wales (1976), books by Lowman (1985), Elbow (1986), McKeachie (1986), and Peters and Waterman (1982), and the government brochure *What Works* (1986).

1 Guide the learner. Be sure that students know the objectives. Tell them what will be next. Provide organization and structure appropriate for their developmental level.

2 Develop a structured hierarchy of content. Some organization in the material should be clear, but there should be opportunities for the student to do some structuring. Content needs to include concepts, applications and problem solving.

3 Use images and visual learning. Most people prefer visual learning and have better retention when this mode is used. Encourage students to generate their own visual learning aids.

4 Ensure that the student is active. Students must actively grapple with the material. This can be done internally or externally by writing or speaking.

5 Require practice. Learning complex concepts, tasks, or problem solving requires a chance to practice in a nonthreatening environment. Some repetition is required to become both quick and accurate at tasks.

6 Provide feedback. Feedback should be prompt and, if at all possible, positive. Reward works much better than punishment. Students need a second chance to practice after feedback in order to benefit fully from it.

7 Have positive expectations of students. Positive expectations by the professor and respect from the professor are highly motivating. Low expectations and disrespect are demotivating. This is a very important principle, but it cannot be learned as a "method." A master teacher truly believes that her or his students are capable of great things.

8 Provide means for students to be challenged yet successful. Be sure students have the proper background. Provide sufficient time and tasks that everyone can do successfully but be sure that there is a challenge for everyone. Success is very motivating.

9 Individualize the teaching style. Use a variety of teaching styles and learning exercises so that each student can use his or her favorite style and so that each student becomes more proficient at all styles.

10 Make the class more cooperative. Use cooperative group exercises. Stop grading on a curve and either use mastery learning or grade against an absolute standard.

11 Ask thought-provoking questions. Thought-provoking questions do not have to have answers. Posing questions without answers can be particularly motivating for more mature students.

12 Be enthusiastic and demonstrate the joy of learning. Enthusiasm is motivating and will help students enjoy the class.

13 Encourage students to teach other students. Students who tutor others learn more themselves and the students they tutor learn more (*What Works*, 1986). In addition, students who tutor develop a sense of accomplishment and confidence in their ability.

14 Care about what you are doing. The professor who puts teaching "on automatic" cannot do an outstanding job.

15 If possible, separate teaching from evaluation. If a different person does the evaluation, the teacher can become a coach and ally whose goal is to help the student learn.

1.5. CHAPTER COMMENTS

At the end of each chapter we will step aside and look philosophically at the chapter. These "metacomments" allow us to look at teaching from a viewpoint that is "outside" or "above" the teacher. In class we use metadiscussion to discuss what has happened in class. In this chapter we set up a strawman who argued against courses on teaching methods and then knocked him down. The strawman is real in some universities, and we have met him many times while developing the course this book is based on. This book is written in a pragmatic, how-to-do-it style. There are philosophical and spiritual aspects of teaching which are given little attention. A good counterpoint to this book is Palmer's (1983) book on the spiritual aspects of education.

1.6. SUMMARY AND OBJECTIVES

After reading this chapter, you should be able to:

- Discuss the goals of this book.
- Answer the comments of critics.
- Explain the two-dimensional model of teaching.
- Discuss some of the values which underlie your ideals of teaching.
- Explain some applications of learning principles to engineering education.

HOMEWORK

1 Many additional critical comments can be made about the need for a teaching course. Develop both a critical comment and your response to this comment.

2 Good teachers must remain intellectually active. Brainstorm at least a dozen ways a professor can do this during a forty-year career.

3 Discuss the values which influence your teaching.

4 Determine the positions in Table 1-1 of engineering professors you have had as an undergraduate or graduate student. What could these professors have done to improve their teaching? (If this assignment is turned in, do not identify the professor by name.)

REFERENCES

ASEE, "Quality of Engineering Education Project," *Eng. Educ.,* 153 (Dec. 1985).Durney, C. H., "A review: Principles of design and analysis of learning systems," *Eng. Educ.,* 406 (March 1973).

Chickering, A. W. and Gamson, Z. F., "Seven principles for good practice in undergraduate education," *AAHE Bull.,* 3 (March 1987). (AAHE is the American Association for Higher Education.)

Elbow, P., *Embracing Contraries: Explorations in Learning and Teaching,* Oxford University Press, New York, Chapter 7, 1986.

Felder, R. M., Leonard, R., and Porter, R. L., "Oh God, not another teaching workshop," *Eng. Educ.,* 622 (Sept./Oct. 1989).

Hanna, S. J. and McGill, L. T., "A nurturing environment and effective teaching," *Coll. Teach.,* 33(4), 177 (1985).

Highet, G., *The Immortal Profession: The Joys of Teaching and Learning,*Weybright and Talley, New York, 1976.

Hodgkinson, H. L., "All One System. Demographics of Education: Kindergarten Through Graduate School, " The Institute for Leadership, Washington, DC, 1985 (pamphlet).

Irey, R. K., "Four principles of effective teaching," *Eng. Educ.,* 285 (Feb. 1987).

Lowman, J. *Mastering the Techniques of Teaching,* Jossey-Bass, San Francisco, 1985.

McKeachie, W. J., *Teaching Tips,* 8th ed., D.C. Heath, Lexington, MA, 1986.

Palmer, P. J., *To Know as We Are Known: A Spirituality of Education,* Harper Collins, San Francisco, 1983.

Peters, T. J. and Waterman, R. H., Jr., *In Search of Excellence: Lessons from America's Best-Run Companies,* Harper and Row, New York, 1982.

Reynolds, W. A. and Oaxaca, J. (Co-chairs), "Changing America: The new face of science and engineering," Interim Report of the Task Force on Women, Minorities, and the Handicapped in Science and Technology, Washington, DC, 1988.

Stice, J., "The need for a 'how to teach' course for graduate students," *Proceedings ASEE Annual Conference,* ASEE, Washington, DC, 65, 1991.

Van Ness, H. C., "Chemical engineering education: Will we ever get it right?" *Chem. Eng. Prog.,* 85 (1), 18 (Jan. 1989).

Wankat, P. C. and Oreovicz, F. S., "Teaching prospective faculty members about teaching: A graduate engineering course," *Eng. Educ.,* 84 (1984).

Wales, C. E., "Improve your teaching tomorrow with teaching-learning psychology," *Eng. Educ.,* 390 (Feb. 1976).

Wentzel, H. K., "Seminars in college teaching: An approach to faculty development." *Coll. Teach.,* 35, 70 (1987).

What Works, U.S. Department of Education, Washington, DC, 1986. (Copies can be obtained by writing to *What Works,* Pueblo, CO, 81009).

CHAPTER 2

EFFICIENCY

Efficiency has long been a topic of considerable interest in the popular press. If you pick up an airline magazine out of boredom as you circle the airport for the fifth time, chances are there will be an article on some aspect of efficiency. Despite this popular interest, however, most professors and students are inefficient. A little formal study of efficiency and some practice can tremendously improve one's productivity.

Most new professors work long hours and still feel they don't have time to do everything they want or need to do. By being more efficient they could do more research and do a better job of teaching. An efficient teacher can do a good job teaching a course in less time than it takes an inefficient teacher to do a mediocre job. It is also important to train both graduate and undergraduate students to become professionals. Engineers are more effective if they are trained to be efficient. We contend that part of this training should be done in school (it does not hurt to be an efficient student either).

Being efficient requires both an attitude and a bag of tricks. We have placed this chapter near the beginning so that the importance of efficiency can be emphasized throughout the book. The bag of tricks will be discussed in this chapter and to a lesser extent throughout the book. This chapter is a considerable extension and revision of Wankat (1987), and it draws upon the books by Lakein (1973) and Covey (1989) for many of the basic ideas.

2.1. GOAL SETTING

People need a reward for being efficient. What will you gain if you get the task done well in less time? To achieve what you want, you first need to set goals. If you are serious about developing a more efficient and productive work style, you need to set both short- and long-term goals. Do this for work and leisure. To illustrate, a young professor's lifetime goals may include the following:

Be promoted to associate professor and then to professor
Become a recognized technical expert
Be recognized as an outstanding teacher
Provide for children's education
Spend a sabbatical in Europe
Remain in good health
Develop a happy marriage

This is a reasonable but certainly not all-inclusive list. Your goals may be different, of course, because only you can develop that list.

A lifetime is, one would hope, a long time. Action plans are easier to develop for shorter-term goals, so a two- or three-year list of goals such as the following may be useful.

Remain in good health
Publish five papers in refereed journals
Be promoted to associate professor
Take a Caribbean cruise

Even shorter term lists such as semester lists are useful. Achieving just one or two major goals in a semester requires an unusual level of persistent effort. Lakein (1973) and Covey (1989) have much more information and examples on setting goals. In order for this chapter to be useful you need to write down your own goals (and later activities and priorities). Either start now or do the homework when you are finished with the chapter.

Once various goals have been listed, it is time to set priorities. This involves juggling the order of the goals until you find an order which satisfies you now. Don't try to set priorities for all time. Goals are made to be changed. A reasonable choice for a number-one priority is to maintain good health since it makes achieving the other goals much easier.

Lists of goals have the advantage of keeping you focused on the big picture. However, they don't tell what you need to *do*. For this you need a list of *activities* which will help you achieve your goals. For example, the following list will probably help someone achieve the goal of good health:

Stop smoking
Lose ten pounds
Jog or swim three times per week
Control stress and learn relaxation techniques (see Section 2.7.)
Have a physical examination

Activity lists can be developed for each of the goals. In some cases a certain amount of ingenuity may be required to develop a list of appropriate activities. When the desired goal requires a decision by others, such as being promoted, it is helpful to determine what the requirements are for achieving this goal. Unfortunately, these requirements are often moving targets, and it is impossible to get a firm commitment on what is required. For instance, the criteria for promotion usually do not list the number of papers required. However, by asking several full professors you should be able to get an approximate idea of the number and type of publications required. This gives you information for your activity lists which can be used in planning the right activity for reaching your goal.

2.2. PRIORITIES AND TO-DO LISTS

Once you have worked out goals and activities, you need to set priorities for the activities. Not everything can be done at once. The professor desiring promotion may give that goal a higher priority than taking a long vacation. The long vacation can be seen as a reward for accomplishing the first goal. Professors usually must work on several goals at once. If research is a major priority, the most hours may be put into this activity, but other activities also must be worked on. Maintaining good health requires a steady commitment. At the same time, courses must be well taught. Committee meetings must be attended, and so forth.

Meeting goals is a day-by-day commitment. An ABC system can be used to set priorities (Lakein, 1973). List the important items to do in the near future as A's. Include work items which have to be done such as writing a series of lectures or a proposal. Also include activities which will help you achieve your lifetime goals and which you chose to work on this week. Note that writing a proposal eventually helps you achieve the goal of being recognized as a technical expert. It is also important to include on the A list large, long-term projects such as writing a book. A mix of things that you have to do and things that you want to do makes work more pleasurable. The A jobs should be worked on during periods of peak efficiency. Putting an item on the A list does not mean that you will finish it today or this week or even this year. Instead, think of it as a commitment to spend a minimum of five minutes on the activity. The purpose of this is to break down overwhelming tasks into little pieces to prevent procrastination. The five minutes may grow into several hours of effort once you get started.

The A items can be listed in order of priority, A1, A2, and so forth. Lakein (1973) suggests this ordering, but we've never found it to be necessary. B items are either less important or less urgent. If there is time, you can work on them this week. If not, the B's and perhaps some of the A's will wait for next week. C items are even less important and are held in reserve. Sometimes these items take care of themselves and there is no need to work on them. Priorities change. A paper due August 15th may be a C in June, a B in July, an A in August, and an A1 on August 14th. Personally, I prefer to make the rough draft an A in June, the final draft an A in July, and finish everything two weeks ahead of time.

It is useful to realize that importance and urgency are not necessarily equivalent. Keeping up with the literature in your speciality is important, but it is rarely urgent. Priorities help you to be sure that these important but not urgent things are done. There are urgent but less important chores such as committee work, writing thank-you notes, and preparing expense reports which must be done. Do these all at one time when your energy is running low and you need a break from important activities. In setting up priorities it is useful to think about *critical paths* for large projects. Think about what needs to be done in what sequence so that the whole project can be completed quickly. This is illustrated in Fig. 2-1 for an experimental research project. It is important to do the preliminary design quickly so that equipment can be ordered. New graduate students often do not realize that it may take from one month to more than a year for equipment to arrive. If ordered early, the equipment may be available when the experimenter is ready for it.

The tools for ensuring that high-priority items are worked on are *to-do lists* and desk *calendars*. A to-do list delineates the activities that you want to work on within a given time period. Good choices are daily, weekly, and semester to-do lists. A semester to-do list, which is the least detailed, includes only major projects such as papers, proposals, and books. This list is glanced at when weekly lists are prepared. A weekly to-do list includes the activities you want to do that week. Many of the activities may be assigned duties. These assigned duties are indirectly related to your lifetime goals since doing them well will help you keep your job and perhaps be promoted. Include some discretionary activities related to your high-priority goals. Also include nonwork activities which are important to reaching your goals, such as swimming three times a week.

Begin the week by listing the highest priority activities. Put these on daily to-do lists on a desk calendar or appointment book. If you don't get to an activity on Monday, work on it on Tuesday. On Friday, check to see what A's haven't been worked on. Either work on them then, or move them to next week's list. You may find that you no longer want to bother listing B or C items since you'll likely always have more A items than you can finish. You also may want to omit routine meetings and class meetings. Routine meetings can be put on a desk calendar or appointment book and taken care of as they occur. Phone calls and letters can also be recorded on the desk calendar. One suggestion is to arrange your schedule so that you have no meetings on Tuesday mornings and Friday afternoons. This gives you a chance to work on items on the to-do list early in the week, and a chance to clean up at the end of the week.

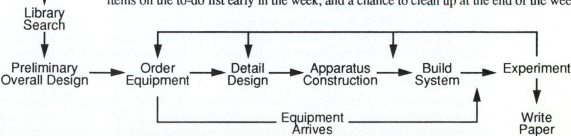

FIGURE 2-1   CRITICAL PATH FOR EXPERIMENTAL RESEARCH PROJECT

One problem with priorities and to-do lists is that you may become too work-oriented and forget to stop and "smell the roses." If this continues, you will start to burn out. Lakein (1973) suggests writing "smell the roses" on the to-do list. Loosening up on the rigidity of the list may also work. Consider most items on the list as a guide and don't worry if you don't work on a particular task. Try to be productive without being rigid about following a schedule. When you become saturated with one project, switch to something else. This is often a good time to initiate people contact or to do nonurgent but important chores.

2.3. WORK HABITS

Once we have set goals and developed activity lists to help us reach those goals, we are ready to consider the details of how we do our work. These work habits have a major effect on how efficiently we satisfy our goals and thus are the subject of many books on time management and efficiency (e.g., Covey, 1989; Lakein,1973; Mackenzie, 1972).

2.3.1. Interactions with People

Visiting.   Since much of a professor's time is spent interacting with various people, your work habits involving people are important. You need to determine when and where you work most efficiently by yourself and with others. Some professors prefer to have blocks of time in the morning to work alone, while others prefer the afternoon. For some an hour at a time is sufficient, while for others much longer periods are desirable. Some professors find interruptions very disturbing, while others enjoy them. When you work with others, do you prefer a formal schedule or an informal drop-in policy? These individual preferences are something that only you can determine. A useful way of looking at these individual preferences is with the Myers-Briggs Type Indicator (MBTI) which is discussed in Chapter 13.

Once you have discovered the most efficient way to work, arrange your schedule and develop methods to control interruptions and visitors. Listed office hours are very useful. If a student comes in at another time when you are busy, say, "I only have a couple of minutes now, but I'd be happy to spend more time with you during my office hours." This approach is most acceptable to the student if you have office hours four or five days a week and you have the reputation of being in your office for your office hours. A second method to control interruptions is to say *no*. It is easier to say no if you have a good reason such as preparing a class in one hour (share this reason with the student), and if you can offer the student an alternate time. If access to your office is controlled by a secretary, he or she can say no to interrupters.

Another method for controlling interruptions is to hide. A second office or an office at home can be a good place to do work which requires solitude. For some reason, most students do not become upset if they can't find a professor, although they may become very upset if they find the professor and he or she does not have time to talk. Controlling the length of visits is also important. Students and colleagues often want to chat. They may not be busy and may not realize that you are. When the visit has lasted long enough, stop it. Stand up. Say, "It's been nice talking to you, but I have to get back to work." Escort your visitor to the door. This can be done politely but firmly.

Secretaries. Unless you have had industrial experience, you probably have never worked extensively with a secretary and have never been in charge of a teaching assistant (TA). Thus working with these people is your first chance to be a manager. The situation is further complicated since you are undoubtedly not the only boss and are probably one of the less important bosses from the viewpoint of your secretary and your TA. How can you best use their capabilities to help both of you do your jobs better?

Peters and Waterman (1982), in their best selling book *In Search of Excellence*, note that outstanding companies obtain productivity through people. A productive professor treats secretaries and TAs with respect. Realize that they have other things to do besides your jobs. Plan ahead and help them plan ahead. Develop a "win/win" atmosphere where both you and the secretary or TA can work efficiently (Covey, 1989). Give your secretary class assignments a day or two before they have to be handed out, not fifteen minutes before class starts. Tell your secretary clearly when they are due. Build in sufficient time so that you can proofread the papers before they are copied. Ask your secretary to proofread the material before it is given to you. Work with your secretary so that he or she understands what you want. For instance, Greek letters may be a mystery to your typist. Explain what they are and point them out on a template.

Try to make your secretary a partner with you even though he or she also works for five other professors. Ask if one time is better for getting a project done than another. Give a warning when there is a big project such as a proposal coming up. If something is not needed quickly, tell your secretary when it is due. (It hardly seems fair to let something sit on your desks for months and then demand that the secretary finish it immediately.) If you consistently give materials on time, then when there really is a big rush, your secretary will reward your fairness with an all-out effort. When someone has really gone out of the way to help you with a project, reward him or her appropriately. Praise never goes out of style. Finally, remember that "please" and "thank-you" are magic words.

Some universities do not provide secretarial assistance to professors because of budgetary constraints. This is a very short-sighted view which squanders the much more valuable professorial resource.

Teaching Assistants. Teaching assistants can be extremely helpful to professors, particularly in large classes. However, new teaching assistants often have no experience in grading and they need to be trained. Your goal is to make the teaching assistant a partner in teaching the course. Discuss the following with the TA before the semester starts.

1 Your expectations. TAs have a paid job and should be expected to earn their money. Usually their duties start before the semester starts and continue until grades are due. The TA may not realize that he or she has contracted for this time.

2 Attendance and note taking at your lectures. Otherwise, the TA will be very rusty in grading and helping students.

3 Proctoring tests and recording grades.

4 Office hours. Help the TA set required office hours at times that are convenient to both the students and the TA. Expect the TA to be available during office hours but protect him or her from excessive demands from students at other times.

5 Grading. Explain in detail how you want grading done. Remember this is probably a learning experience for the TA also. For the first few assignments grade a few problems to serve as examples. Check over the TA's grading and give feedback so that he or she can improve as a grader. Expect a reasonable turnaround on grading, but tell the TA in advance when a heavy grading assignment will be coming. If students ask for regrades, work with the grader. Listen to the TA's reasons for assigning grades. Try to balance consistency in grading with fairness.

6 Student interaction. If laboratory or recitation sections are involved, encourage the TA to prepare ahead of time and to learn the names of students.

7 Efficiency. Arrange the TAs workload so that it can be done in the amount of time the person is being paid for.

8 Communication. Students who cannot communicate in English should not be used in positions where they will have extensive contact with students.

9 Personal behavior. For foreign student TAs you may need to explain clearly that U.S. standards of behavior towards women are different from those in many other countries. Explain these standards and note that they will be enforced.

10 Training program. If one is available, encourage or insist that your TA enrolls. If one is not available, consider starting one (Righter, 1987).

Other Support Personnel.   There are always other personnel in the department who can be helpful to you or who can cause you problems. They include janitors, shop personnel, laboratory instructors, instrumentation specialists, storeroom clerks, business office personnel, computer systems people, and so forth. If you treat them and their work with respect, then they will be helpful. In some departments they have significant student contact, and they may know the students better than do most of the professors. If this is the case, they can be very helpful if you have any problems with particular students.

Whiting (1987) makes the point that a professor must be honorable and honest in all dealings with secretaries, TAs, and other support staff. Thus, do not ask them to do personal favors or anything illegal or unethical. Respect their privacy and what little personal space they have. Ask permission before you borrow any equipment or use any of their equipment such as personal computers. Finally, be sure that your TAs and research assistants also treat the support staff appropriately.

2.3.2. Using a Computer

A computer can be an excellent time saver if it is used for activities that require a great deal of time. Professors often spend a large fraction of their time writing. Composing a draft on a word processor or typewriter is much faster than writing it by hand if you can touch-type. Prepare the first draft of all manuscripts on a computer. Make a hard copy and write corrections and additions on it. Then ask your secretary to make the corrections. This procedure will take significantly less time than hand writing the first draft, and it results in a better final manuscript since revisions are easier to do with a word processor. Your secretary will also find this procedure faster and less work than typing from a handwritten manuscript.

Obviously, efficient use of a word processor requires that you touch-type. The advantages of typing the rough draft of manuscripts are so great that it will pay you to spend the time needed to learn to touch-type and to learn to use a word processor. In addition, all students interested in engineering should be strongly encouraged to learn to touch-type (now called *key boarding*). Even if you know how to touch-type and how to use a word processor, composing on a computer will seem awkward at first. Use an extensive outline as a guide while composing on a computer. Most people can write both quicker and better if they write the first draft as quickly as possible and then spend a significant amount of time revising (Elbow, 1986).

Computer graphics used to be a time sink since the programs were difficult to use, and unless you used a program often you had to relearn it every time. If this is the only software available to you, it would probably be more efficient to have someone else prepare figures. Computer graphics on a user-friendly computer such as an Apple Macintosh is much easier to learn and to remember. In this form computer graphics has become a time saver. However, if someone else is available to do the graphics it would probably be more efficient to give them a rough sketch to be drawn on a computer.

Spreadsheets are starting to be used to a significant extent in engineering education. They are easier to use than programming from scratch and thus tend to be more efficient. Regular use of a spreadsheet will enable the professor to become quite proficient with it, and the spreadsheet will be a time saver. If the spreadsheet is used only on rare occasions, then it is likely to be a time sink instead of a time saver. Students should also be taught to use spreadsheets (see Chapter 8).

In many areas a computer can be a time waster and not a time saver. Programming is a very time-consuming process. Writing programs for classes or developing computer-aided instruction is unlikely to save time because programs have to be polished before being given to students. The use of unpolished programs in a class often results in breakdowns (always at inconvenient times) which the professor must try to fix. It is certainly more efficient to use programs that someone else has written and debugged. If you *must* write your own programs, write fewer programs and spend more time on them so that they work even when abused by students. Setting up files on a computer is a second area where the computer can become a time waster. The problem with files is that they can become an end unto themselves, and they are seldom used for a productive purpose. The files, a C-priority, become more important than the A-priority jobs which the files should support. A third trap which is common for students but not professors involves computer games. Games are fun for relaxation, but don't fool yourself by thinking they are educational. There are many other examples of nonproductive uses of a computer which any computer hacker can list.

2.3.3. Miscellaneous Efficiency Methods

Covey (1989), Lakein (1973), Mackenzie (1972) and Roberts (1989) suggest a variety of methods for improving the use of time. One of the most important is to avoid perfectionism. Manuscripts can be revised forever, and the reader will never think they are perfect. At some point you have to let go and put out a less than perfect, but not sloppy, manuscript. This same reasoning is applicable to other work such as lectures.

A second very important principle is to reward yourself and take breaks. Most people become very inefficient if they try to work all the time. You might recommend to your graduate students that they take at least one day a week off and do no work on that day. This will pay off in terms of long-term efficiency, and overall work production will actually increase despite working fewer hours. Most people also need vacations (even assistant professors). Over a five- or six-year period an assistant professor will probably enjoy life more and get more done if he or she takes at least one week of vacation every year instead of working all the time.

One important efficiency method is to use the same work several times: a process called *piggy backing*. The most obvious application of this is teaching the same course several times. Then the work spent in setting up the course is reused when you teach it the second and third times. Another related application of piggy backing is teaching courses in your research area. Then time spent on research will help you present a more up-to-date course, and time spent on the course will help you better understand your research area. Another example is to prepare a literature review. This work can be published, serve as the literature review of a proposal, be presented as a paper, or serve as the basis for several lectures.

Change your work environment or your task when you get bogged down. Carrying work to the library, college union, or local hangout can provide just the change you need. Switching tasks can also provide a needed break. If proofreading has you down, try reading a technical journal for half an hour.

Another suggestion is to use odd moments to do useful work. Can you do useful work while you commute to work? (Note that relaxation may be the most useful thing to do.) Plan work for trips (see Section 2.4). Take a book or papers to grade to the doctor's office. Figure out what works for you for those ten- or fifteen-minute periods which are not long enough for a "serious" project.

Mail can be handled more efficiently. The general rule is to minimize the number of times you handle it. There is no law which says that you must open junk mail. If you *do* open a piece of mail, try to complete your response immediately. If this is not possible, at least be sure that you do something to move it forward each time you pick it up. Very often a phone call or a Fax will be the most efficient way to take care of mail. You can help your correspondents be more efficient by putting your telephone number, your Fax number, and your computer address on your correspondence.

Carry a small notebook or pocket calendar at all times. Then you can write down appointments and transfer them to your desk calendar later. This helps you to avoid missing meetings. The notebook can also be used to jot down ideas, record references, list names of people you meet, and so forth.

2.4. TRAVEL

Travel can consume so much time that a separate section on travel efficiency seems to be called for. It can be exhilarating and broadening, but also exhausting. Many of the effects of travel are shown in Fig. 2-2. The interest and energy generated is very high when you seldom travel (say, once or twice a semester). As you travel more often, the interest in each trip decreases. The first trip to Europe is very exciting; the fifth trip in the same year is a lot less so. Every trip involves a certain amount of hassle in developing plans, buying tickets, arranging for classes while you are gone, and so forth. In addition, when you return you have to catch up on the work you missed while you were gone. These hassles and the work you have to make up lead to a tiredness factor. Cumulatively, tiredness increases as you make more and more trips. The combination of interest and energy generated by the trip and energy drained by the trip is the efficiency curve shown in Figure 2-2. This curve goes through a maximum at a certain number of trips per semester. An additional factor is the effect of your travels on your spouse or significant other. (Even some pets don't like to be left alone.) However, a spouse who travels with you may be more positive about traveling.

There are no numbers on Fig. 2-2 since the values depend upon individual circumstances. If you're not feeling well, one trip may be too many. If your home life is unhappy, getting away may energize you, and the more trips a semester the better. Also, extroverts tend to like traveling more than introverts do, probably because the hassles are not as draining for them. The point of Fig. 2-2 is that there is probably an optimum amount of travel for you.

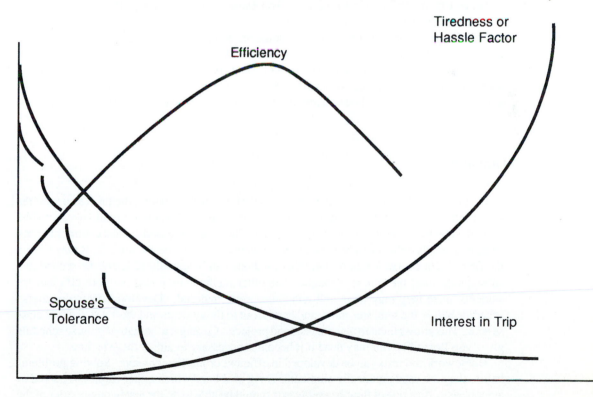

FIGURE 2-2   EFFECTS OF TRAVEL

From the point of view of your career and teaching, travel can be either over- or underdone. Not traveling may lead to stagnation, parochialism, and a lack of name recognition for your research. There are several dangers in traveling too much. Certain responsibilities such as office hours, committee meetings, and academic advising really cannot be made up. Professors who are gone too much risk the danger that their classroom effectiveness may decline (see Section 2.5). The important question to ask is, does this travel help me reach my long-term goals? Sometimes travel may help you reach some goals, such as seeing the world, but hinder your reaching other goals such as writing a book. It will probably take one day to catch up for each day you are gone. If you are gone a week, it will take a week to catch up, and that will be two weeks where you do only routine and urgent tasks and don't get a chance to work on important goals. If you decide that you are traveling too much, then learn how to say no to the less important trips. It helps to develop a standard letter for declining invitations.

Once you know that you are going to travel there are some tricks to increasing your efficiency. First, a good travel agent is very important. Not all travel agencies and not all agents within a given agency are equal. Shop around until you find one who will work with you, and then stay with her or him. If you work with a large agency, be sure to get the name of the agent so that you can always contact the same person. Currently, planning ahead, getting your tickets early, and being flexible as to the dates you travel can save money. Registration fees at conferences are lower if you register early.

Use the time spent on airplanes to get some work done. A long flight may represent the longest period of uninterrupted time that you'll have in months. Bring a combination of writing projects and reading, such as a book or some articles to review. If possible, also bring some light technical reading. When the flight is at night after a busy day, you may decide that a review of the day and relaxation are more important goals than doing more work.

2.5. TEACHING EFFICIENCY

Courses can be organized so that they are efficient for the student, the professor, and the TA. First, develop the goals and objectives for the course. Coverage should be reasonable. Then decide upon the basic course organization. The lecture method is most commonly used since it is widely believed to be the most efficient use of a professor's time. This may be true the first time a course is taught, but other methods can be equally efficient the second and subsequent times the course is taught. The other methods may also be more efficient for students since they may learn and remember more material. Develop a tentative course schedule before the semester starts, and hand it out to the students and the TA at the first class session. This allows them to plan for tests and projects. Calling it a "tentative" course schedule gives you flexibility you may need if it becomes necessary to adjust the schedule.

Homework and tests can be developed in efficient or inefficient ways. Solving problems before they are used practically eliminates using problems which either cannot be solved or are too easy. As a rule of thumb, a professor should be able to do the test in one-quarter of the time the students will take. Occasionally using a homework problem or lecture example on a test not only saves time but also emphasizes the importance of doing homework and paying attention during the lecture (Christenson, 1991). Ask TAs to solve some of the homework problems, but check their solutions. On tests give the TA a solution plus a grading scheme. Requiring written requests for regrades drastically reduces the number of arguments you have to confront.

Preparing for a lecture immediately before the period it will be given ensures that you are fresh. When presenting a lecture given previously, allow yourself one-half to three-quarters of an hour to revise and prepare for the lecture. For totally new lectures or major revisions plan

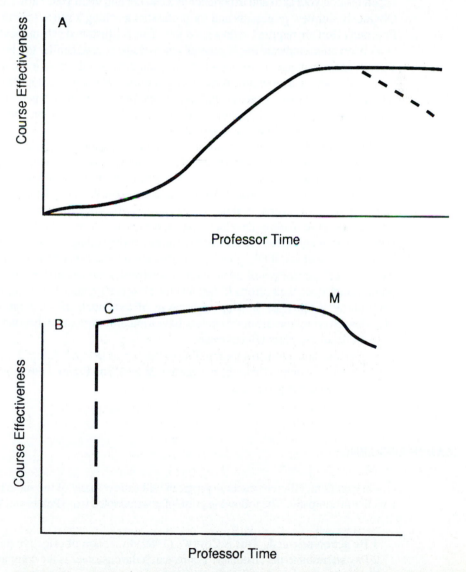

FIGURE 2-3   EFFECTS OF PROFESSOR'S TIME AND EFFORT ON COURSE EFFECTIVENESS.
A. New professors or new courses. B. Experienced professor with established material.

on preparing a fifty-minute lecture in two hours or less. This prevents Parkinson's law (work expands to fill the time available) from controlling your time. Of course, if you don't understand the material, much more time may be required. Some time can be saved in lecture preparation by using examples from other textbooks (Christenson, 1991). This is preferable to just repeating an example from the assigned textbook. Most new faculty members drastically overprepare and spend much more time than we have suggested here (Boice, 1991).

How much time needs to be spent on a course before it deteriorates? The answer to this depends upon your skill and experience as a teacher and upon your knowledge of the content. Obviously, for new professors and for professors teaching a subject for the first time, more time and effort are required to do a good job. This is illustrated schematically in Figure 2-3a. Our observations indicate this is generally an S-shaped curve similar to a breakthrough curve. Effectiveness increases rather slowly at first and then speeds up as more time is put into the course. As more and more time is spent on the course, effectiveness approaches an asymptotic limit and may actually decrease slightly. As the professor gains experience in teaching and becomes more familiar with the material, the curve sharpens and moves upward and to the left (i.e., to higher effectiveness with less effort).

The hypothetical curve for an experienced professor is shown in Figure 2-3b. Our experience is that there is a very broad range of professorial effort where course effectiveness is quite satisfactory. However, at critical point C there is a discontinuous drop in course effectiveness and the course drops below acceptable levels. This drop occurs because teaching, unlike research, is always a "what have you done for me lately" activity. All the rapport and good feeling developed one semester has to be rebuilt the next semester. A professor with a good reputation will have an easier time doing this than a professor with a bad reputation. However, if the "good" professor does not put in enough time or is gone too often, the course effectiveness will crash. Experienced professors can hover in the flat plateau above point C and adjust their efforts if they feel the class is slipping. This is somewhat dangerous, particularly if the class is slipping because of too much travel. Note in Fig. 2-3b that experienced professors are more likely to have a maximum, point M, beyond which extra effort actually decreases class effectiveness.

In general, students also appear to follow the curves in Fig.2-3. Fig. 2-3a refers to students who are not experienced learners in a particular area, and Figure 2-3b refers to students who are experienced in a given area.

2.6. RESEARCH EFFICIENCY

An excellent, efficient research program will follow many of the same basic principles as a well-run company. The following principles are adapted from Peters and Waterman (1982).

1 Be action-oriented. This is Covey's (1989) first habit of effective people.

2 Pay attention to the customer. For research the customer is the company, foundation, or government agency supporting the research.

3 Within broad guidelines, give graduate students control and responsibility. Do not spell out the nitty-gritty details.

4 Show respect for each student. One way to do this is to listen more and talk less.

5 Be hands on and driven by values. Visit the graduate student's laboratory or office. Continually set the basic value of the research group (e.g., "innovation" or "careful experimentation").

6 Stick to the "knitting." Stay fairly close to your area of expertise but don't continually repeat the same research. Before starting a new project ask, "Do I have the skills, time and energy to do a good job?"

7 Develop an intense atmosphere with the expectation of regular contributions from every group member.

Supervision of graduate students should aim for a happy medium between too little and too much. Regularly scheduled meetings can prevent excessive procrastination. Occasional visits to the laboratory give the professor a chance to provide intermittent reinforcement. Students work hardest when they feel commitment to a project. This can be attained by having the student develop and work on his or her own ideas. Another method is to ask the student if he or she wants to present a paper at a meeting. A student who makes the commitment to do this will work very hard to complete the paper. Since research efficiency is usually measured on the basis of the number and quality of the papers published, it is important to complete the work and publish the papers.

It is easiest to get results and write publications when you work on new ideas instead of following the well-beaten research track. Thus, time spent on generating new research ideas usually pays off. Many articles and books have been written on creativity and problem solving (Adams, 1978; de Bono, 1973; Wankat, 1982) (see Chapter 5 also). Application of these creativity methods can lead to more efficient research.

A useful method to help you determine semi-quantitatively if a particular project is worth doing is a *cost-benefit analysis*. Cost-benefit analyses can be done for projects other than research, but they are easy to illustrate for projects such as proposals where monetary value is involved. The method is easy to illustrate with an example comparing two proposals. The benefit-to-cost ratio in dollars per hour for writing a proposal can be estimated from

$$(\text{Benefit/cost})(\$/\text{hour}) = \text{(dollars received) (funding probability) / (hours of writing)} \qquad (1)$$

where the hours required to write the proposal is approximately

$$\text{Hours of writing} = k \times \text{number of pages} \qquad (2)$$

The value of the proportionality factor k depends on your speed. The probability of funding is harder to estimate, particularly initially when you have no experience. Some idea of this value can be determined by talking to experienced professors or by talking to the agency. It is desirable to maximize the benefit-to-cost ratio in Equation (1).

Consider two sources of funds: one offering a small amount of money but having a high probability of success, and another offering significantly more money but having a lower probability of success. The following approximate comparison can be done.

Source A $10,000, requiring a ten-page proposal, and having an 80 percent chance of funding:

Benefit/cost = $\dfrac{\$10{,}000}{10 \text{ k}} \times 0.8 = \dfrac{800}{k}$

Source B $150,000 (for 3 years), requiring a twenty-page proposal, and having a twenty-five percent chance of funding

Benefit/cost ($/hour) = $\dfrac{\$150{,}000}{20 \text{ k}} \times 0.25 = \dfrac{1875}{k}$

On the basis of the cost-benefit ratio alone, source B looks more advantageous. However, there may be other reasons for trying source A first:

1 Need to quickly show success in obtaining funding.
2 Grant is small but prestigious.
3 Grant is for a proof of concept and could easily lead to much more money later.

It may be possible to send very similar proposals to both organizations, but this is ethical only if you inform the agencies of your intention.

A final comment on writing papers. A large fraction of the citations in the literature have some mistake. Therefore, *always check your references.*

2.7. HANDLING STRESS

Professors and students often feel a great deal of stress. Modest stress may increase your efficiency and not be harmful to your health. But after some point stress can decrease your efficiency and become harmful. When this occurs is obviously an individual matter. Some people can thrive in an environment which is very stressful for others. An extreme example of this is the response of soldiers to combat. Some soldiers find combat exhilarating, others find it stressful, and some find it so stressful that they break down. In this section three approaches to handling stress are discussed: change of environment, change of perception, and relaxation methods.

Changing the environment may not be easy, but it is a very effective way to reduce stress. Sometimes all that is needed to change the environment is the realization that there are alternatives. For example, some professors find lecturing to be very stressful. These professors can use other teaching methods once they realize there are other methods. A professor who finds the noise from a student lounge to be annoying can ask to be assigned another office. A professor may find that part of his or her lifestyle is increasing stress, and this stress can be reduced by changing lifestyles. Even excessive coffee drinking may increase stress. People with certain medical conditions may find that some weather patterns cause them physical stress. Alleviation of this problem may require moving to another university in another section of the country. Some people find all aspects of a professor's life stressful. Their only solution may be to find a job in industry or in a government laboratory. Often a stressful part of the environment can be changed only by a major move, and other aspects of the position make such a move undesirable. In cases such as this it is important to learn to live with the stress.

Another way to deal with stress is to change your perception. You do not change the actual incidents but instead change how you feel and react to them. Everyone has a surprisingly large degree of control over how they feel and react to situations and conditions. Some professors feel that they have to be perfect and thus become very upset if a class does not go well or a research paper is harshly criticized. Being upset over these "failures" is not a problem. The problem lies in being so upset that the person is unable to function. Individuals with a need to be perfect will be happier and more efficient if they learn to accept some imperfection in their lives (see Appendix 2A).

A similar problem arises with professors who feel responsible for the actions of others. For example, most professors do not enjoy flunking students, but some professors find doing so to be extremely stressful. They feel that the F is their responsibility instead of the student's. It is much less stressful and fairer to assign this responsibility to the student where it belongs. Alleviating the problem of assigning yourself too much responsibility is possible by the same methods which work for overperfection.

Another related problem is the *catastrophe syndrome*, that is, believing that a catastrophe will occur whether something happens or does not happen. The something can be rejection of a paper, low teaching ratings, denial of tenure, or whatever the professor wants to name. Admittedly, none of these are pleasant, but they are catastrophes only if they are perceived that way.

There are many psychological methods which can be used to overcome perception problems which increase stress. Many of these methods require the help of a counselor or psychologist. There are some methods such as rational emotive therapy (RET) which can be learned and applied to oneself (Ellis and Harper, 1961, 1975; Ellis, 1973, 1988). Essentially, RET postulates that we think irrational, unhealthy thoughts and it is these thoughts that make us feel bad. The solution proposed by RET is to rationally attack the irrational thoughts and change our thinking patterns. The RET approach is briefly outlined in Appendix 2A.

The perception of stress can also be reduced by desensitization procedures (Humphrey, 1988). Desensitization involves repeated exposure to the stress-causing stimulus, but in a relatively supporting and nonthreatening environment. In a clinical setting the exposure is usually obtained by imagining the stress-causing event. In classical applications of desensitization the stimulus is first present at a very low level, and then gradually the level is increased. This may sound complicated, but it is not uncommon for professors or department heads to apply a similar procedure. For example, a new professor may first teach a graduate class with ten students, then an elective class with thirty students, and finally a required sophomore elective with 150 students. Unless he or she suffers from a more deeply rooted problem, this individual can become desensitized to the stress of presenting a lecture to a large audience. A professor who gives many quizzes in class is in effect desensitizing students who have problems with test anxiety. This method is most effective if the first quizzes are worth a smaller proportion of the course grade than later quizzes, or if a practice test is given.

Relaxation techniques are useful for reducing excessive stress while it is occurring (Humphrey, 1988; Jacobson, 1962; Whitman et al., 1986). There are many methods which are useful in helping one to relax. Physical activity such as jogging, tennis, swimming, or walking is a good way to get away from the pressures of being a professor or a student. Regular weekend activities, particularly those which get you outside and involve some physical activity, are useful to keep stress from building up. It is important to *get away* and not carry work with you. Professors at least have the advantage that they do not regularly carry paging devices with them. When professors start to carry beepers, stress levels and burnout will increase [see Baldwin (1985) for a discussion of the effects of modern technology on stress].

There are other relaxation methods which are useful on a daily basis. Although less popular now, transcendental meditation (TM) or repeating a mantra works for many people (Humphrey, 1988; Benson, 1974). A westernized version of TM is given by Benson (1974). Various breathing exercises are also useful to help someone who is very stressed relax. These can be as simple as taking a deep breath, holding it for ten seconds, and then slowly letting it out. This simple exercise is useful to hold in reserve in case a student becomes extremely anxious during an examination. Various stretching exercises and methods to relax one set of muscles at a time are also useful and easy to learn. Humphrey (1988) presents a variety of simple exercises which can be used to reduce stress.

Excessive stress can also be very detrimental to students. It is helpful to be able to recognize this and help the student to cope with the stress. The procedures for doing this are similiar to those for coping with your own stress. These procedures are discussed in detail by Whitman et al. (1986).

2.8. LIMITATIONS

Efforts at efficiency can be overdone, and many things cannot be done extremely efficiently. Most activities which require personal contact with other people have some built-in inefficiencies. Examples include:

Starting a class period
Tutoring
Advising students
Mentoring graduate students
Building consensus (e.g., within a department for a curriculum revision)
Marriage
Raising children

If you try to do these activities in a very efficient manner, then others may feel rushed and devalued. The net result is a rapid transaction which may minimize your time, but it is not efficient since what needs to get done doesn't get done. A classic horror story which may be true involves a professor who set a three-minute egg timer whenever a student came in to ask a question. After a short period most students stopped coming in, and the professor saved himself time, but he did not help students learn. Professors can limit interruptions by scheduling these personal contacts at specified times of the day. This will help overall efficiency even though the individual interactions are inefficient.

Innovation and creativity in research and teaching tend to be messy and not particularly efficient processes. (It is hard to sit down and say, "In the next ten minutes I will have a brilliant idea.") The paradox here is that being innovative and creative can drastically increase your overall efficiency even though the processes themselves are inefficient. Once a professor has a great idea for research, actually conducting the research may be relatively quick and easy. In addition, the research may have considerable impact. To a lesser extent, the same is true of creative ideas in teaching. Students enjoy a bit of change and creativity in their classes.

The planning of an entire career does not appear to be an efficient process despite many books and courses on career planning. Many biographies and autobiographies tell of famous people who go through a period of wandering about before they seize upon their life's work. There may be false starts and failures or successes until they settle down into their great work.

It is useful to separate tactics from strategy. Efficiency is almost always a good idea in day-by-day tactical concerns. An example is preparing for a class. If you want to break new ground or be able to respond to new areas such as developing a new research program or course on superconductivity, it is probably not possible to have an extremely efficient long-range or strategic plan.

Relaxation is necessary to be efficient over long periods; however, relaxation itself almost appears to be the opposite of efficiency. As noted in Section 2.7, we can learn to relax better or more efficiently. During the period of relaxation, it appears that nothing useful is occurring, yet useful things must be occurring. The paradox that we must learn to live with is that only by allowing for inefficiencies can we truly be efficient as professors and as human beings.

2.9. CHAPTER COMMENTS

One of the common problems in designing a course or a textbook is that there is no order that really works. There is always some part of the subject which should be discussed before covering the current topic, but not everything can be last. In addition, for motivational purposes it is useful to present a practical part of the course early so that students know why they are studying the theory. Not all aspects of this chapter will be completely clear until other chapters have been covered, and some won't be clear until after you have had experience as a professor. We decided to put this chapter early to help professors think about being efficient when designing courses. In addition, putting an important chapter early in the course ensures that sufficient time will be devoted to it. This illustrates a second problem in course design: The most interesting and most useful material such as efficiency and creative design is often left until last so that all prerequisite material can be covered first. When this is done, the interesting material is crowded into the end of the semester when there is never enough time and everyone is tired.

Teaching efficiency in class is a challenge. The concepts are simple and often just common sense. The hard part is applying the principles. Perhaps the best approach would be to not lecture but require students to apply one or two principles to their lives. Then three or four weeks later require an informal oral report on the results. We have found that groups of experienced professors are much more attentive and receptive to a lecture on efficiency than graduate students.

2.10. SUMMARY AND OBJECTIVES

After reading this chapter, you should be able to:

- Set goals and develop activities to meet those goals.
- Prioritize the activities and use to-do lists.
- Improve your work habits with respect to people interactions, computer use, and other activities.
- Analyze your travel patterns and improve your time use during travel.
- Explain how time spent preparing to teach affects course effectiveness, and use methods to improve your teaching efficiency.
- Improve your research efficiency and apply approximate cost-benefit analyses.
- Use methods to control stress.
- Discuss situations when a strict application of efficiency principles may not be the most efficient in a global sense.

HOMEWORK

1 Develop your lifetime goals as of now.

2 Develop your goals for the next three or five years, whichever time frame appears more appropriate.

3 Develop activities which will help you attain your lifetime goals.

4 Develop your activities to help you attain your goals for Problem 2.

5 Assume one of your goals is to become a good teacher
  a Define the term "good teacher."
  b Develop your activities to reach this goal.
  c How will you know when you have reached this goal?

6 Develop a semester to-do list. Be sure to include some of your activities to reach your goals on this list.

7 Analyze your use of computers. What could you do more efficiently? Develop a plan to increase your computer efficiency.

8 Explain why a professor's effectiveness in teaching a class or a student's effectiveness in taking a class will crash if some minimum amount of time is not put into the course.

9 Learn one relaxation method and practice it every week.

REFERENCES

Adams, J. L., *Conceptual Blockbusting,* Norton, New York, 1978.

Baldwin, B. A., *It's All for Your Head: Lifestyle Management Strategies for Busy People*, Direction Dynamics, Wilmington, NC, 1985.

Benson, H., *The Relaxation Response,* William Morrow, New York, 1974.

Boice, R., "New faculty as teachers," *J. Higher Ed.,* 62, 150 (March/April 1991).

Christenson, R., "Time management for the new engineering educator," *Proceedings ASEE Annual Conference,* ASEE Washington, DC, 216, 1991.

Covey, S. R., *The 7 Habits of Highly Effective People,* Simon and Schuster, New York, 1989.

de Bono, A., *Lateral Thinking: Creativity Step by Step,* Harper and Row, New York, 1973.

Elbow, P., *Embracing Contraries. Explorations in Learning and Teaching,* Oxford University Press, New York, 1986.

Ellis, A., *Humanistic Psychotherapy,* McGraw-Hill, New York, 1973.

Ellis, A., *How to Stubbornly Refuse to Make Yourself Miserable about Anything!*, Carol Communications, New York, 1988.

Ellis, A. and Harper, R.A., *A Guide to Rational Living,* Prentice-Hall, Englewood Cliffs, NJ, 1961.

Ellis, A. and Harper, R.A., *A New Guide to Rational Living*, Wilshire Book Company, North Hollywood, CA, 1975. (This and the original edition are classics in the self-help literature.)

Humphrey, J. H., *Teaching Children to Relax,* Charles C. Thomas, Springfield, IL, 1988. (There are many useful methods for adults in this excellent book.)

Jacobson, E., *You Must Relax,* 4th ed., McGraw-Hill, New York, 1962.

Lakein, A., *How to Get Control of Your Time and Your Life,* Signet Books, New York, 1973.

Mackenzie, R. A., *The Time Trap: How to Get More Done in Less Time*, McGraw-Hill, New York, 1972.

Peters, T. J. and Waterman, R. H., Jr., *In Search of Excellence: Lessons from America's Best-Run Companies,* Harper and Row, New York, 1982.

Righter, R., "Training for Teaching Assistants," *Eng. Ed., 77* (10), 135 (Nov. 1987).

Roberts, M., "8 ways to rethink your workstyle," *Psychol. Today,* 42 (March 1989).

Wankat, P.C., "How to overcome energy barriers in problem-solving," *Chem. Eng.*, 89 (21), 121 (1982).

Wankat, P. C., "Efficiency for Professors," *Proceedings ASEE/IEEE Annual Frontiers in Education Conference*, IEEE, Washington, DC, 207, 1987.

Whiting, P. H., "Getting Started: A Potpourri of Suggestions for the New Engineering Educator," *Proceedings ASEE Annual Conference*, ASEE, Washington, DC, 874, 1987.

Whitman, N. A., Spendlove, D. C., and Clark, C. H., *Increasing Students' Learning: A Faculty Guide to Reducing Stress Among Students*, Association for the Study of Higher Education, Washington, DC, 1986.

APPENDIX 2A. THE RATIONAL-EMOTIVE THERAPY APPROACH

The rational-emotive therapy (RET) approach will be briefly outlined here. RET postulates an ABC method of viewing human reactions. The *activating experience* A is the outside stimulus that the person reacts to (e.g., bad reviews of a research paper or the acceptance of a paper). Step B consists of the *internal beliefs* which lead the person to interpret what has happened. These beliefs can be rational or irrational. For example, a rational belief is that a rejection is unfortunate since more work will be required, but that the rejection is not a catastrophe. An example of an irrational belief is that a rejection is a catastrophe which must not happen. Eventually, the person experiences an *emotional consequence* C which he or she thinks is caused directly by activating experience A. An example of an emotional consequence C is anger and depression. Thinking that A caused C is irrational. This must be irrational since another person will react to the same A in a very different way and experience a completely different C. The emotional consequence C is caused by the beliefs B which the person has. If these beliefs are rational, then C will be reasonable (e.g., if the belief is that bad reviews are unfortunate since they will require additional work, then C will be a mild displeasure). If the beliefs are irrational, then C can be an extreme reaction.

Most people have irrational beliefs about something. Examples of irrational beliefs which people subconsciously carry around in their heads are:

1 It is horrible to be rejected.
2 I have no control over my feelings.
3 I must be liked by everyone.
4 All my lectures must be perfect.
5 It is catastropic if something I do is not perfect or is criticized.
6 All snakes are dangerous and bad.

The amount of disruption these beliefs cause in the person's life depends upon the belief and how strong it is.

The RET approach is a method for combatting the dysfunctional aspects of emotional consequences caused by irrational beliefs. The method involves rigorously analyzing your thoughts to determine the irrational beliefs and to replace them with rational beliefs. For example, suppose you just received a letter from a funding agency rejecting a proposal which you thought was very good. Your first reaction is to become angry and you know that you will be depressed and angry for several days. With the RET approach you first stop and listen to what you are saying to yourself.

Ask, "Why am I angry?" Then listen to your own response which might be, "Because I was turned down." Now the RET approach pushes deeper. Ask yourself, "But why does being turned down make you angry?" Here the response might be, "Because I'm not supposed to be turned down." A further push could be, "Why aren't you supposed to be turned down?" "Because I should be perfect." "And what else?" "Well, everyone should always approve of my work."

The beliefs that one is supposed to be perfect and have everyone approve of one's work are clearly irrational. These and probably other irrational beliefs are causing the anger and depression. RET postulates that the appropriate place to intervene is in the irrational belief system. Continuing our example, you could respond to yourself, "Perfect! No one is perfect. That just is not rational. It's also not rational to expect everyone to like your work even if that work is very good." Now you need to substitute a rational belief for the irrational one. For example, "A rational approach is that it is nice and certainly preferable if your work is good and is approved by everyone. However, it is not a catastrophe if you occassionally do some work which is not up to your high standards or if someone does not like your work. It is unfortunate that the proposal was not funded since you will have additional work to do to resubmit it, but it is not worth becoming angry and depressed over for days."

This approach may sound simplistic or too good to be true. The method does work but requires considerable work and practice. The irrational beliefs have been there for a long time and are usually difficult to eradicate. However, if these beliefs are attacked logically every time they appear, they become weaker. In addition, as one practices RET on oneself, one becomes much more adept at spotting irrational beliefs and at fighting them. Readers interested in this method should read one of the books by Ellis (Ellis and Harper, 1961, 1975; Ellis, 1973, 1988).

DESIGNING YOUR FIRST CLASS

You've started your first position as an assistant professor and have been assigned your first class with real students.

- What do you do?
- What teaching method do you use?
- What level do you aim for?
- How do you structure the class?
- How do you pick a textbook?
- What do you ask on tests?
- How do you know how much you can cover in a semester?
- How many tests and how much homework do you require?
- How do you grade?
- How do you behave toward the students?
- How much time will this take you?
- Why didn't someone tell you how to do this?

This chapter provides an overview of what a professor does in designing and teaching a course, and it raises a number of questions about the process. The remainder of this book will be spent finding some answers to these questions.

3.1. TYPES OF COURSES

Engineering professors teach a variety of courses. Since course design is often different for different types of courses, it is useful to categorize them.

Required undergraduate courses which are prerequisites for other required courses tend to have the most structured content. It's likely that a curriculum committee will select the content and even the textbook. Professors who teach succeeding courses care about how well the introductory or prerequisite course is taught and the extent of the coverage. If the XYZ transform is not taught and they expect it to be taught, they will let you know about the omission. So it's a good idea to ask other professors what they expect from a course. In teaching these courses you'll likely have less freedom in coverage. Balancing this, particularly for new professors, it is highly likely that past syllabi, homework, tests, and a recommended textbook will be available. Past instructors will be available for some assistance if they are asked, but you'll probably have to ask since few faculty will volunteer teaching help unless asked. Often, these classes may be rather large, and student abilities will vary widely.

Required undergraduate courses which are not prerequisites for other courses are similar but have a few differences. The course content is probably a bit less rigid, and other professors have less of a vested interest in what is taught. These are often senior courses, which means that very weak students will not have made it this far. However, graduating seniors are notoriously difficult to motivate. Past syllabi and a textbook are probably available, but there will be less pressure to follow them closely.

Required or core graduate-level courses have all graduate students in them, and class size varies from small to medium. A syllabus probably exists, but the professor usually has some freedom in rearranging it. Invariably, the amount of material to be covered is staggering, and textbooks may or may not be available. The research professors in the department are often very interested in the content and how well the students learn the material. These courses often give the professor a good opportunity to get to know and impress new graduate students before they pick research topics. Thus, in some departments these courses may be considered "plums."

Undergraduate electives and dual-level undergraduate-graduate electives are usually regularly offered courses with a sample syllabus, textbook, and tests available. Professors who have taught the course in the past are probably available for advice. Since electives are not prerequisites for other courses, the professor can usually change the syllabus and textbook, and these courses have a comfortable combination of guidance and freedom. Class size is usually small to medium, and students tend to be interested since they have selected the course. Overall, these courses offer a new assistant professor a good beginning to an academic career.

Graduate-level electives and seminars are the most open in coverage of content. These courses may be very specialized, and other professors in the department often pay little attention to what is covered as long as their graduate students don't complain too loudly. The freedom involved in selecting course content is very liberating but also somewhat daunting since well-developed syllabi, homework and test examples, and a textbook probably are not available. The classes are usually small, and the students are likely to be both intelligent and interested. The teaching of graduate electives in one's research area is an effective way to integrate teaching and research. However, faculty often compete to offer these courses, and new professors may not be given the opportunity immediately.

Design classes, particularly capstone design classes for undergraduates, tend to be somewhat different from other classes. Economics may be a required part of the course. The course may be taught with case studies and often is loosely structured. The workload is often high because of grading demands and the need to develop new case studies. Design classes are also sometimes associated with laboratories, which can further increase the large workload. Professors with industrial experience are often assigned to these classes. Chapter 9 deals with design classes in more detail.

The laboratory course usually differs markedly from all the other types of courses which are often referred to as "lecture courses" (see Chapter 9). It may be attached to another course and may or may not be administered separately. It also tends to be tightly structured since the experiments or projects are limited by the available equipment. Unfortunately, the equipment is often old and may not work well. Experimental write-ups are available, but always need modification. For safety reasons, the section size is usually controlled. And, for various reasons, new professors commonly find themselves assigned to a laboratory course. It is easier to step into a lab course without much preparation, and since teaching lab courses tends to be an unpopular assignment, the department head may also staff the lab with new professors since they are more likely to accept the assignment gracefully. Although the course is a required one, the material covered is usually not a critical prerequisite for follow-up courses. Teaching involves a great deal of informal contact with students and extensive grading of laboratory reports; little if any lecturing is done. Some schools have adapted the lab course and added a communication component by adding a credit hour and a lecture on writing and speaking. The often extensive report writing in the course makes it a natural place for teaching such skills, one advantage being the focus of the report on technical material which adds relevance to the student's job.

3.2. BEFORE THE COURSE STARTS

Several tasks normally need to be completed efore the course starts (exceptions occur when students are heavily involved in planning the course). Some may be done for the professor if the course is well-established, but with new courses all these tasks need to be at least partially completed before classes start.

3.2.1. Knowing the audience

Clearly, it is helpful to understand your students. Are they sophomores, juniors, or graduate students? What prerequisite courses have they had? What are they capable of doing? Are they mostly full-time or part-time students? Are they majoring in your field? Have they made a definite commitment to be an X-type engineer, or are they still searching for a major? How mature are they? In general, is it likely to be a good or poor class? The more you know about the students, the better you will be able to plan the course and select the appropriate level for the material. Student characteristics are discussed in detail in Chapters 13 and 14.

3.2.2. Choosing Course Goals and Objectives

What should the students know and be able to do at the end of the semester? This question includes both coverage of the content and the ability *to do something* with the content. Goals are relatively broad, while objectives tend to be quite specific. For example, your goal may be that students understand the control of systems, whereas an objective may be that they know how to use the Laplace transform in the analysis of linear control problems. The goals and objectives must satisfy what is expected for any subsequent courses. The development of goals and objectives for a course is important since it controls the coverage and, to a lesser extent, the teaching method. The most important part of a class is the content covered because it makes no sense to do a wonderful job teaching unimportant material. A part of the goals and objectives for the course is the choice of the level at which to present the material. New professors, particularly those fresh out of graduate school, are notorious for setting the level too high and being too theoretical. The choice of an appropriate level for a class is actually a complicated question (see Chapters 14 and 15). Appropriate goals and objectives for a course may be somewhat subtle; for example, if one departmental goal is to produce graduates who are good communicators, then writing and speaking have to be incorporated into the engineering courses in the department. Goals and objectives are discussed fully in Chapter 4. Existing courses may not have explicitly stated goals and objectives.

3.2.3. Picking a Teaching Method

Once you know what you want to accomplish, you can choose a teaching method that is congruent with your style and with the students' learning styles. The teaching method should also satisfy certain learning principles so that the method will be effective. Learning styles and learning principles are the subjects of Chapters 13 and 15. Lecturing and various modifications of lecturing are by far the most common teaching methods and in most universities will be acceptable to the other professors in the department. Lecturing is also one of the easiest methods to use the first time you teach a course partially because everyone is familiar with the method. Unfortunately, lecturing is not the best method for some of the goals of engineering education. If one goal of the course is to have students become proficient in working in engineering teams, then lectures need to be supplemented with group work. No matter which teaching method is chosen, you need to check the classroom ahead of time to be sure that appropriate equipment such as a blackboard or an overhead projector will be available. Teaching methods are discussed in Chapters 6 through 10.

3.2.4. Choosing a textbook

The quality of the textbook will have a major effect on the quality of the course and on what can be conveniently covered. Unfortunately, the textbook may have to be selected months ahead of time because of bookstore requirements. For new professors who arrive a week before the semester starts, the book has probably been chosen by someone else. For required undergraduate courses a committee may select the book. If you do not like the textbook, you can work at changing it for subsequent semesters, but do *not* tell the students that it is a poor book. Textbook selection is discussed in Chapter 4.

3.2.5. Preparing a Tentative Course Outline

An outline of the entire course in advance is helpful but not essential. If time is short, outline at least the first month so that you and the students know where you are going. The course outline should list topics for each day. This requires that you estimate the rate at which you can cover topics. New faculty find this to be very difficult. It is useful to build in one or more open periods, or periods which can be skipped before major tests. Tests, quizzes, and student presentation days should be noted. This requires that you decide on the number of tests and quizzes initially. Every school has breaks, student trips to conventions, major extracurricular activities such as homecoming, and student professional activities such as the Engineering-in-Training examination. Do you want to adjust your schedule for these events? Also, now is the time to look at your calendar and adjust the class schedule if you will be out of town for one or two classes. In a required undergraduate course the existing course outline will probably be adapted.

3.2.6. Deciding on a Grading Scheme

You can be sure that on the first day of class the students will ask how the course will be graded. Yes, our schools put too much emphasis on grading. Yes, it would be better if the students were there mainly to learn. No, you cannot ignore the students' desire to know how they will be evaluated. They want to know how much quizzes, tests, a final, homework, computer problems, and projects will count. Will there be extra credit? Will you follow a 90-80-70-60 scale or will you use a curve? Are you an easy or a tough grader? Are you fair? And so on. In our experience most students are satisfied if given the major outline of the grading method. Grading is discussed in Chapter 11.

3.2.7. Arrange To Have Appropriate Material Available

Appropriate material includes the textbook and supplementary books, handouts, and solutions to homework and tests. Copies of these materials should be available to students in a library or learning resource center. Be sure the bookstore has enough copies of the textbook.

3.2.8. Preparing for the First Class

Preparation for the first class is discussed in detail in section 3.3.

3.2.9. Developing Your Attitude or Personal Interaction Style

What is your attitude toward teaching and toward students? It helps to be enthusiastic and to believe that teaching is an important and even noble activity. How much personal warmth and caring will you show to the students? What, if any, is your responsibility for helping students grow? Is it important to you to be loved, or is respect sufficient? Do you believe all students can learn, or is removing students who cannot learn part of your job? Great teachers are marked by a deep commitment to their students. Your attitudes and style will have a major effect on your rapport with students. Once this and the other tasks are taken care of, you are ready to start. It is usually desirable to have these chores done before class starts, but fortunately some of them can be partially delayed until after the semester starts.

3.3. THE FIRST CLASS

It is traditional to start the first class with "housekeeping chores." There are other ways to start a class and these will be discussed at the end of this section. Housekeeping chores are routine and nondemanding. They allow students to get settled in, but this is not an exciting way to start a class. Use the blackboard to write down information or pass out a paper with the information on it or do both. Then latecomers (and there are always latecomers on the first day of class) can get the information without interrupting the class. Give the students all the information about the course structure that you can. The following items probably should be included:

1 Course name and number. You will be surprised by the number of students who come to the wrong classroom. Also list the hours and the class location, particularly if two locations are used.
2 Professor's name, office location, office phone number, and office hours. The way you present your name is important. If you write it as Professor Jones, the students will call

you Professor Jones. If you write Carol or John, they will call you Carol or John. You need to select your office hours before the class starts. Try to choose office hours which will be available to most students. If you welcome phone calls at home, give your home phone number. If you don't want to be called at home, don't give your home phone number. If computer mail is to be used, give your e-mail address.

3 TA names, office hours, and office location. Introduce the TAs if they are present.

4 Prerequisites. Discuss how important these prerequisites are. Will you accept an F or a D or an incomplete in a prerequisite course? (Check your school's policy on this ahead of time.)

5 Textbook. Discuss any other supplementary material that the students should buy or that is available in the library. Pass out a reading list if you use one.

6 Tentative course outline. Discuss the course outline and note the dates of tests and due dates of major projects. The earlier you give this information to the students, the fewer problems you will have with conflicts.

7 Teaching method and expectations of the students. If 99 percent of the students' courses have been lectures and you will lecture, this can be very brief. If your course will be different, added discussion will be valuable.

8 Grading scheme. If you don't discuss the grading scheme, the students will ask about it. Be prepared for a question on extra credit.

9 Seating arrangements and names. Begin to learn names. If there is to be a seating chart, describe how it will be set up. Start learning student names unless there will be a large turnover the first week. Note that seating is not discussed at the beginning of the period since not everyone will be on time. If it is important to you that the students know that you care about them as individuals, then you *must* learn their names fairly quickly. Some teachers memorize the seating chart, others use photographs of students, some call the roll the first few weeks, and some ask questions using the class list. If discussion or group work will be important, you may want to use some method which introduces students to each other. Various types of "name games" can be used to do this. Students can introduce themselves or others. Many students will appreciate a copy of the class roster.

10 General discipline policies. Always enter the class with a positive attitude toward your students. However, since it's likely that one or two of them may pose problems, briefly discuss the rules the class must live by. What is the policy about cheating, being late, being absent, and reading a newspaper or sleeping in class? What will your policy on makeups be? Be sure that your policies do not conflict with university and school policies.

11 Student questions about course structure and policies. If there are no immediate questions, erase the board to give the students time to formulate questions.

The housekeeping chores are complete at this point. Some professors dismiss the students at this point, but this may be a mistake. Instead, start in on the course. Send the message that you mean business. The students will not be ready to start business, but they are never ready until you get them started. What should you do with the remaining fifteen to thirty minutes? Use it for lecture, discussion, or whatever teaching method you intend to use in class. What content should be covered? One possibility is to give an overview of the entire course and to explain the importance of the material. A second possibility is to review a previous course which is an important prerequisite for the course. A third possibility is to start the first lesson. Regardless of the content, it should be presented with enthusiasm and a sense of excitement so that the students will know that you consider the material to be important. Leave enough time for a short summary.

Finally, give the first homework assignment. Homework on the first day! Yes, at the very least the students should start reading. You know that they will not be very busy with homework the first week, and you want them to take your course at least as seriously as the competition. Pass out a sheet with the homework assignment on it. There will be fewer misunderstandings about what is due when. This completes the first class. Tell the students you will see them on Wednesday (or whatever) and signal them that class has ended. A clear signal that the class is over, such as picking up your books or saying goodbye, will be useful throughout the semester.

If you don't like housekeeping, there are other ways to start the first class. The first class period can be used to develop a course outline with the students' input. This may be appropriate in some elective courses. A test on prerequisite material can be given, but this will be unpopular and probably won't be extremely valid since no one has reviewed for it. The students can be given a problem which they will be able to solve at the end of the course, and they can be asked to work on it in teams. This works if the importance of the problem is clear. The students can be required to write and turn in a paragraph on why they are taking the course. This is usually meaningful only in electives. It does give you a writing sample and makes the students think. Your creativity can guide you to other possibilities (see problem 6).

Your attitude toward teaching is very important. If you are enthusiastic and look forward to the class, then the students will tend to do the same. If you have the attitude that it is your job to help student's learn and earn a good grade, then you've taken a big step toward building rapport.

3.4. THE SECOND CLASS

The second class is surprisingly important since many students consider it the first "real" class of the semester. The first and second classes set the tone for the remainder of the semester. Thus, it is very important for you to be well prepared for it. "Winging it" is always a mistake, but can be a disaster if done while you are still setting the course tone and student expectations. Enter the class with a sense of excitement and be enthusiastic. It should be obvious that you should avoid scheduling trips the first week of the term so that you can meet with your classes.

Start the class slowly but on time. Classes should always be started slowly so that students can switch gears and start thinking about this class. For starting this and other classes you might:

• Collect homework.
• Practice the names of students.
• Review the previous class.

• Add a bit of humor if you can do so naturally.

• Show a cartoon related to the day's subject on the overhead projector. A little entertainment before the class gets started will not detract from the seriousness of your message.

• Make announcements from student organizations.

• Answer student questions from previous classes, reading, or homework.

• Mention a current event which relates to the class. Examples are a strike at a plant, the sale of sensitive computer parts to unfriendly countries, a new automotive design, an explosion and fire at a chemical plant, and a nuclear protest. Be sure to explicitly relate the event to the class.

A slow start is important, but these activities should last only a few minutes. Don't allow the students to lead you off on extensive tangents. During the remainder of the class cover the content listed in the course outline using the teaching method of your choice. It is very important in this class section to work hard at getting the students to be active since you are setting the tone for the rest of the semester (see Chapter 15 for a discussion of why students should be active). Toward the end of the period set aside time for student questions. Then summarize what has been covered in class. Pass out the homework and reading assignments or remind the class of the assignments. Remind students of your office hours and invite them to stop in and see you. Ask anyone who missed the first class period to see you after class. Dismiss the class slightly before or at the bell.

In general, it is useful for you to leave the classroom very slowly. This gives students time to ask you questions. Many of these questions will be questions that could have been asked during class. Answer them this time, but encourage students to ask similar questions in class next time. Most classes need a good deal of encouragement to ask questions. Some student questions pertain only to a particular student and should be handled privately.

3.5. THE REST OF THE SEMESTER

With the semester under way, classes develop a routine which is punctuated by tests and large projects. You prepare for class, develop homework assignments and tests, present lectures or use another teaching method, grade or arrange for grading of homework and tests, have office hours, and deal with any problems which may arise. Then at the end of the semester you assign course grades, post them on the wall, and run off to a meeting or vacation. This appears straightforward, but conceals many issues.

If you are lecturing, you will need to prepare each lecture before class. The way you go about this often requires a little experimentation before you obtain a feel for how to proceed. Do you need to write everything out, or are just a few notes sufficient? Can you accurately reproduce equations without notes? Are your presentations clearer when you use an overhead projector or a blackboard? How much material can you comfortably cover in a class period? What is a good balance between theory and examples? Students always want more examples. How closely should the lectures follow the textbook? Students will complain if you follow the textbook too closely or if the lecture has little connection to the textbook. What material is important and should be emphasized? Every textbook (including this one) has both trivial material and material which is becoming obsolete. It is your job to weed this material out.

To keep students actively involved with the material, have them take notes, ask and answer questions, discuss the material, work in groups, write short summaries of the lecture, solve problems at the board or at their desks, hunt for "mistakes" made by the professor, and so forth. Active learners learn better. Encourage questions by allowing time for them, acknowledging the student by name, repeating the question so that everyone can hear it, and then answering it as appropriate. If the question will be covered later in the lecture, you can tell the student to wait a few minutes.

How should homework assignments be distributed throughout the semester? How long should problems be and how many problems should there be in each assignment? Should homework problems be done solo or should you encourage group effort? Do all problems have to be turned in and graded? Should a particular format be required? How do you or a TA grade a large number of homework problems? There is no one set of correct answers for these questions; however, if you want your students to spread their efforts throughout the semester, you must spread homework and quizzes throughout the semester. Students generally consider quizzes and tests the most important part of a course. Many questions can be asked about the form and method of testing, but the most important questions are on content. What material do you test on? There should be a correspondence between course objectives and the tests. Testing for memory is easier than testing for problem-solving skills but probably is much less important. If you want students to be able to solve problems, testing must include problem solving.

The methods used in testing must also be examined. How many quizzes and tests should you give? How much should they be worth? How many problems should be on each quiz or test? Is it acceptable to use multiple-choice questions? Students appreciate help sessions before tests. Should you have them? If so, who should lead them, and when and where? Do you want to give partial credit? If so, how do you decide how much credit to give? Tests should be graded rapidly and as fairly as possible. In an ideal class graded tests are returned during the next class period, and the professor goes over the correct solutions then. Students do prepare for tests, and students do become anxious before and during exams. Unfortunately, many students stop work on an area once the test is over. How do you get students to learn from their mistakes on tests? You will get requests for regrades, so it would be wise to develop a regrade policy ahead of time. Do you want to give a final? Finals provoke a great deal of anxiety, but they also force students to review the entire semester and to some extent integrate the material they have learned.

How do you establish and maintain rapport with students? Studies of effective teachers show that the best teachers have a good rapport with their students even in large classes (e.g., Lowman, 1985). Students prefer professors who are enthusiastic, are accessible and care about them as individuals. Perhaps most importantly, students want an instructor who is fair. Your challenge is to establish rapport while maintaining some professional distance so that evaluations of the students are fair. The goal is to develop a cooperative atmosphere where you and the students work together to maximize learning.

Office hours give students the chance to clarify questions and to get additional help when they need it. It is helpful if both the TA and the professor have office hours. Some students want to talk to the professor solely, while others are scared to death of the professor. Office hours are useful since they give you feedback on what the students are not understanding and on what problems they cannot do. Keep your office hours. A problem is how to get students to come in for office hours. Continual encouragement and an open and friendly demeanor help. Since students forget, you need to remind them of your office hours and of the TA's office hours. Students will start to use office hours just before the first test or big project. What is the best way to heituation) and are emotionally drained by the experience. For most new pr fessors, learning to teach is on-the-job training (OJT) which is strongly motivating but s not the best way to learn. New faculty often overprepare and spend sixteen to twenty hours per eek preparing for one course, and as much

Suppose that as the semester continues you slowly fall behind and material that should have been covered on Monday isn't covered until Friday. It is important to know why this happens so that the next time it won't. Write yourself notes on a copy of the course outline. For example, if you do not plan to spend a large part of the period after the test going over the test, you may lose half a period. The note to yourself will remind you to allow time for this. Now you know what to do the next time, but what do you do now to cover all the material? If you haven't built an extra period into your course outline, there may be little that you can do other than skip some material. Look at the rest of the semester and decide what to delete. What if you get to the end of the syllabus and the semester is not over? Don't worry; this very seldom happens.

Throughout the semester you may have to deal with discipline problems. Student problems can range from the mildly annoying to the downright dangerous. The most common problems involve chronically late or absent students and passively disruptive students. If lateness bothers you, talk to chronically late students privately. They may have a legitimate reason for being late. However, in all cases start the class on time and do not backtrack for late comers. There is a reasonably strong positive correlation between attendance in class and grades. Chronically absent students will pull down the class average. One possible solution to this problem is to compute the class average and the grade distribution excluding chronically absent students. Then give these students the grade they have earned based on this grade distribution. Then the course grades will not be affected by those who are chronically absent. These students are also most likely to turn in homework and projects late and to be late for or miss a test. Have a policy ready in advance so that you can say, "My policy is to ...." Passively disruptive students include those who talk, sleep, read a newspaper, or wear headphones in class. Remember that the lack of a policy or ignoring these disruptions is also a policy. These problems are discussed in Chapter 12.

As the semester nears the end, you will want to know how well you have done from the students' viewpoint. Ask. Many universities have very elaborate arrangements for evaluating teaching, and some department heads require faculty to have their courses evaluated. If a mechanism does not exist, you can still ask the students for comments on the strengths and weaknesses of the course. The many factors which affect students' evaluation of teaching are discussed in Chapter 16.

At the end of the semester the professor has the privilege and responsibility of assigning grades. How do you do this? If you have given the students a detailed breakdown of grades, you will need to follow that procedure; however, few students will complain if the scale is made easier. If you have been vague about the assignment of grades, make the decision now. Several schemes for assigning grades are discussed in Chapter 11.

Throughout the first year new professors will have many questions about teaching. Talk to other professors—many of them. Since there is no single method of good teaching, you will get varied and occasionally contradictory responses. Sort through these responses and adopt those that fit you. Once you find a kindred spirit, talk to her or him about teaching. If you feel comfortable taking the risk, invite another professor into your classroom. Exploring teaching issues with other professors will help you to learn to teach better and more efficiently, and it will help you maintain your sanity during a very busy first year.

This outline of what a professor does to design and teach a course shows that new professors are very busy their first few semesters. These are also the semesters when you want to start your research program. What really needs to be done? How do you get everything done? The first question is discussed in Chapter 17, while the second is essentially the topic of Chapter 2.

3.6. THE NEW FACULTY MEMBER EXPERIENCE

Most new faculty members feel unprepared to teach (usually a realistic appraisal of the situation) and are emotionally drained by the experience. For most new professors, learning to teach is on-the-job training (OJT) which is strongly motivating but is not the best way to learn. New faculty often overprepare and spend sixteen to twenty hours per week preparing for one course, and as much as thirty-five hours per week (Turner and Boice, 1989). This is much more time than experienced faculty spend; it can be reduced by learning how to teach before teaching the first class. Invariably, new faculty members would like more advice about teaching and handling problem students (Boice, 1991).

Turner and Boice (1989) report on three major problems for new faculty. In order of importance they are:

1 "Adapting to the appropriate pace and level of difficulty for the students."
New professors have forgotten what it is like to learn the material for the first time, and they invariably go too fast and are too theoretical. The lower the level of the students, the more likely this is to occur.

2 "Feeling professionally overspecialized, while not having a well-rounded knowledge of their own discipline."
New faculty members often teach undergraduate courses which have little in common with their Ph.D. research. Although they studied the material as an undergraduate, they are rusty. In addition, a typical new engineering professor has had little or no industrial experience and is not sure what the students will use in an industrial career.

3 "Having trouble establishing an appropriate professional demeanor in their relationships with students."
New professors are asked to make the transition from student to faculty essentially overnight and often are not much older than their students. It is difficult to determine how to

act to help students learn and yet maintain enough distance to be objective in grading. New faculty members must develop a professional demeanor which will allow them to effectively teach and grade both students they like and those they dislike. It takes time to learn the proper distance between oneself and students.

An additional problem of many new faculty members is a fear of not knowing all the answers. It is OK to tell students that you don't know the answer to a question but that you will find out. It is also OK, and actually helps to build rapport, to admit mistakes to the class.

They also are likely to feel that there are not enough hours to get everything done (Boice, 1991). In addition to teaching they have to start a research program, write proposals, finish papers from their thesis, learn the rules of a new university, and adjust to a new city. There is also a psychological adjustment in becoming a professor instead of being a student. It is very useful to have a mentor who knows many of the unwritten rules. Mentoring works best when the procedure is formalized (Sands et al., 1991). Some universities use team teaching of courses to help new faculty. Formal development programs have also been shown to work if new professors will use them (Boice, 1991).

3.7. CHAPTER COMMENTS

The most common method of designing a course and a curriculum in engineering education is to put all the fundamentals first. Once these have been covered, the course or curriculum can proceed to the practical and interesting real-life problems. This approach appears rational but ignores motivation. Most people learn best when they know why they need to learn something. Thus, considering some practical, real-life problems early can help such students significantly. This is one reason why cooperative education works well. This chapter follows this analysis and discusses a real problem before the reader has all the information required to solve it completely.

Although this chapter purposely raises more questions than it answers, clearly spelling out what needs to be done for a class should be helpful to the new professor, and we wanted to provide a chapter that would be immediately useful. One potential problem with enumerating tasks is that they assume a greater importance than attitudes. It is often the teacher's attitude of excitement, enthusiasm, and caring for the student which catch hold of students and fire them up for future work in the discipline. You may be able to do an adequate job as a teacher by just going through the motions, but for excellence you must do more.

Since many professors never ask themselves many of the questions asked in this chapter, one can obviously teach without understanding the process. Instead, the professor mimics former professors. Although strongly discouraged in research, mimicry or plagiarism is encouraged in teaching. Perhaps this observation helps explain why many schools value research more than teaching in their promotion policies.

All books are somewhat idiosyncratic, and this one is no exception. If you want to read alternate approaches to preparing for your first course, try McKeachie (1986) or Dekker and Froyd (1986).

3.8. SUMMARY AND OBJECTIVES

After reading this chapter, you should be able to:

- List the salient features of different types of engineering courses.
- Enumerate the activities which need to be completed before starting a course.
- Discuss how a course is started.
- Explain the importance of the second class period and discuss appropriate activities.
- List the other important activities which occur during the semester. Explain the importance of each of these activities.
- For the preceding items discuss some of the important questions which the professor should consider when designing a course.
- Discuss some of the psychological aspects of becoming a new professor.

HOMEWORK

1 Look through the undergraduate and graduate courses offered in your department. Classify each course. Is the classification system adequate or are there some courses which really do not fit? (Hint: There probably are). If there are, develop new classification categories for the courses which do not fit.

2 What are some sources of information to help you estimate how much material can be covered in one semester?

3 Discuss additional reasons why it is a good idea to learn the names of students.

4 Class size is an important consideration in how you teach a course. List some of the things which are affected by class size.

5 Should you use a seating chart? It seems like a high school practice, but it is almost necessary for large classes if you want to take attendance or learn names. Discuss this issue. Think of alternatives.

6 Brainstorm alternative ways to start the first class. List at least five additional ways.

7 How do you decide what to cover in a course which has never been taught at your school? Brainstorm at least five methods for developing ideas.

8 Do you think it is a bad idea to tell the class that the textbook is a poor book? Explain your answer.

9 What are some of the dangers in teaching a student whom you instinctively like or dislike?

REFERENCES

Boice, R., "New faculty as teachers," *J. Higher Educ.*, 62, 150 (March/April 1991).

Dekker, D. L. and J. E. Froyd, "A Problem-Solving Framework for Course Development," *Proceedings IEEE/ASEE Frontiers in Education Conference,* ASEE, New York, 63, 1986.

Lowman, J., *Mastering the Techniques of Teaching,* Jossey-Bass, San Francisco, 1985.

McKeachie, W. J., *Teaching Tips. A Guidebook for the Beginning College Teacher*, 8th ed., D.C. Heath, Lexington, MA, 1986. (The classic book on teaching for new professors.)

Sands, R. G., Parson, L. A. and Crane, J., "Faculty Mentoring Faculty in a Public University," *J. Higher Educ.,* 62, 174 (March/April 1991).

Turner, J. L. and R. Boice, "Experiences of New Faculty," *J. Staff Program Organ. Develop.,* 51(Summer 1989).

CHAPTER 4

COURSES:
OBJECTIVES AND TEXTBOOKS

What content should be covered in a course? This is obviously a critical question for required courses which are prerequisites for other courses. The answer also depends upon departmental values. In this chapter we will discuss setting goals and objectives for a course, taxonomies of knowledge, the interaction between teaching styles and objectives, development of the content of a course, and finally textbooks.

4.1. COURSE GOALS AND OBJECTIVES

Goals are the broad final result that one hopes to attain during a course. Usually they are stated in broad, general terms. For example, in a thermodynamics course one's goals might be that students should be able to:

- Solve problems using the first law.
- Solve problems requiring use of the second law.
- Understand the limitations of thermodynamics.
- Appreciate the power and beauty of thermodynamics.

Note that content comes first. Engineering education is centered on content, and goals and objectives should focus on content (Plants, 1972). General goals such as these are nonspecific and thus are often fairly easy to agree upon. However, goals are not specific enough to be useful in an operational sense except as an overall guide.

Goals provide an overall guide for what a course is supposed to do. They are helpful to the department in designing the curriculum, to the professor in delineating the boundaries of the class, and to students (particularly intuitive and global learners) in seeing where the class is going. For example, if the department can agree that classical thermodynamics is the goal of the course, then the professor knows that he or she is not expected to cover statistical thermodynamics or irreversible thermodynamics. Professors teaching follow-up courses will know that students who have taken the classical thermodynamics course will not have a background in these subjects. Clearly, this also implies a certain amount of communication and collegiality which does not exist in all departments. Students can look at the curriculum and tell where the course fits. Of course, many students will not bother to do this, but the goals are helpful for those students who do.

Goals can be considered to be global objectives. More specific objectives are useful to guide both the professor and the student in exactly what students will learn, feel, and be able to do after each section of the course is completed. One such popular type of objective is the behavioral objective (Hanna and Cashin, 1987; Mager, 1962; Kibler et al., 1970; Stice, 1976). A behavioral objective states explicitly:

1 What the student is to do (i.e., the behavior).
2 The conditions under which the behavior is to be displayed.
3 The level of achievement expected.

Writing a few behavioral objectives for a class forces the professor to think about observable behavior, conditions, and level of performance. However, we do not know any engineering professors who write out complete behavioral objectives for all their classes. Here is an example of a possible, though cumbersome, behavioral objective for a thermodynamics course:

The student will be able to write down on a piece of paper the analysis to determine the new Rankine cycle performance when the maximum cycle temperature and pressure are changed. This will be done in a timed fifteen-minute in-class quiz, and the student is expected to obtain the correct answer within 1 percent.

Professors who use objectives invariably use a shortened version. In this form the previous objective becomes:

Analyze the effect of maximum cycle temperature and pressure on the performance of a Rankine cycle.

This form is easier to write, focuses on content, and is more likely to be read by students. Hanna and Cashin (1987) discuss the different types of educational objectives, noting that behavioral objectives are usually written in the form of the minimal essential objective, and that these objectives focus on relatively low-level skills since such skills are easiest to measure. For the higher-level skills which practicing engineers need, behavioral indicators of achieve-

TABLE 4-1   EXAMPLES OF THERMODYNAMICS OBJECTIVES

1. The student can write the first and second laws.
2. The student can describe the first and second laws in his or her own language. (That is, describe these laws to the student's grandmother.)
3. The student can solve simple single-answer problems using the first law.
4. The student can solve problems requiring both the first and second laws either sequentially or simultaneously.
5. Given the characteristics of a standard compressor, the student can develop schemes to compress a large amount of gas to a high pressure where both the amount of gas and the required pressure increase are larger than a single compressor can handle.
6. The student can use the second law to determine fallacies in power cycles. The student can describe the fallacies clearly and logically both in memos and in oral debate.
7. The student understands the limits of his or her knowledge and knows when classical thermodynamics is not the appropriate analysis tool.
8. The student can evaluate his or her own solutions and those of others to find and correct errors.
9. The student can search appropriate data bases and the literature to find required thermodynamic data, and if the data are not available the student can select appropriate procedures and predict the values of the data.
10. Since one of the goals of this course is to help students become broadly educated, the student can appreciate the beauty of classical thermodynamics and can briefly outline the history of the field.

ment without minimum standards are more appropriate. For these objectives, student behaviors are illustrations only. Minimum standards are not given since students are encouraged to do the best they can. Conditions for performance are explicitly stated, but this may be done for an entire set of objectives and may be considered to be understood. A set of content-oriented related examples for a thermodynamics course is given in Table 4-1.

Objectives aid the instructor since they clarify the important content to be covered in readings, lectures, homework, and tests. If material is not important enough to have an objective, then it should be omitted. When developing tests, the professor can look at the list of objectives and check that the most important are included in the test questions. This procedure is discussed further in Chapter 11.

Objectives should be shared with students so that they know what material to study and what material they will be tested on (DeBrunner, 1991). Students should also be explicitly told if other skills, such as those involving a computer or communication, will be required in the course. And they should know if they are expected to become broadly educated in the field and be able to do more than just solve problems. Examples of both these areas are included in the set of thermodynamics objectives. These objectives are written at several different levels. It is important to ensure that the course objectives and hence readings, lectures, homework, and tests cover the range of levels desired. The appropriate levels and types of objectives are included in taxonomies.

4.2. TAXONOMIES OR DOMAINS OF KNOWLEDGE

Taxonomies of educational objectives were created by two very significant committee efforts in the 1950s and early 1960s. The taxonomy in the cognitive domain (Bloom et al., 1956), which includes knowledge, intellectual abilities and intellectual skills, has been widely adopted, whereas the taxonomy in the affective domain (Krathwohl et al., 1964), which includes interest, attitudes, and values, has had less influence. A third domain is the psychomotor, manipulative, or motor skills area. This area appears to be continuously decreasing in importance in engineering education, although the success of graduate students doing experimental research often depends on this domain. A problem-solving taxonomy has also been developed by engineering educators. These taxonomies are discussed in the following four sections.

4.2.1. Cognitive Domain

Since the cognitive domain is involved with thinking, knowledge, and the application of knowledge, it is the domain of most interest to engineering educators. Bloom et al. (1956) divided the domain into six major levels and each level into further subdivisions. The six major divisions appear to be sufficient for the purposes of engineering education.

1 Knowledge. Knowledge consists of facts, conventions, definitions, jargon, technical terms, classifications, categories, and criteria. It also consists of the ability to recall methodology and procedures, abstractions, principles, and theories in the field. Knowledge is necessary but not sufficient for solving problems. Examples of knowledge that might be required include knowing the values of e and π, knowing the sign conventions for heat and work in an energy balance, knowing the definition of irreversible work, knowing what a quark is, being able to list the six areas of the taxonomy of educational objectives, defining the scientific method, and recalling the Navier-Stokes or Maxwell equations. The potential danger in the knowledge level is that it is very easy to generate test questions, particularly multiple-choice questions, at this level. The ability to answer these questions correlates with a student's memorization skills but not with his or her problem-solving skills. In some areas of science such as biology, students are expected to memorize a large body of knowledge, but this is unusual in engineering. The first objective in Table 4-1 is an example of a knowledge objective.

2 Comprehension. Comprehension is the ability to understand or grasp the meaning of material, but not necessarily to solve problems or relate it to other material. An individual who comprehends something can paraphrase it in his or her own words. The information can be interpreted, as in the interpretation of experimental data, or trends and tendencies can be extended or extrapolated. Comprehension is a higher-order skill than knowledge, but knowledge is required for comprehension. Testing for comprehension includes essay questions, the interpretation of paragraphs or data (this can be done with multiple-choice) or oral exams. The second objective in Table 4-1 is an example of an objective at the comprehension level.

3 Application. Application is the use of abstract ideas in particular concrete situations. Many straightforward engineering homework problems with a single solution and a single part fit into this level. Application in engineering usually requires remembering and applying technical ideas, principles, and theories. Examples include determining the pressure for an ideal gas, determining the cost of a particular type of equipment, determining the flow in a simple pipe, determining the deviation of a beam to a load, and determining the voltage drop in a simple circuit. Objective 3 in Table 4-1 is an example.

4 Analysis. In engineering, analysis usually consists of breaking down a complex problem into parts. Each part can then be further broken down or be solved by application of engineering principles. The connections and interactions between the different parts can be determined. Objective 4 in Table 4-1 is an example of an analysis objective since it requires breaking a more complex problem into parts and then determining the relationship between the parts. Many engineering problems fall into the analysis level because very complicated engineering systems must be analyzed.

5 Synthesis. Synthesis involves taking many pieces and putting them together to make a new whole. A major part of engineering design involves synthesis. One problem for the professor in teaching synthesis is that there is no longer a single correct answer. Many students, particularly at the lower levels in Perry's scheme of intellectual development (see Chapter 14), find synthesis difficult because the process is open-ended and there is no single answer. Synthesis should be incorporated into every course and not be delayed until the "capstone" senior design course. Objective 5 in Table 4-1 is an example of a synthesis problem for a thermodynamics course.

6 Evaluation. Evaluation is a judgment about a solution, process, design, report, material, and so forth. The judgment can be based on internal criteria. Is the solution logically correct? Is the solution free of mathematical errors? Is the report grammatically correct and easy to understand? Is the computer program documented properly? Objectives 6 and 8 in Table 4-1 are examples of objectives at the evaluation level which use internal criteria. Objective 7 is also an evaluation example based on internal evidence but is easier to attain if external sources are also utilized. In this case the external sources would be some knowledge of statistical thermodynamics and irreversible thermodynamics. In many engineering problems the evaluation requires external criteria such as an analysis of both economics and environmental impact. Objective 9 in Table 4-1 requests evaluation using external criteria, and it also requests analysis.

Bloom's taxonomy is a hierarchy. Knowledge, comprehension, application, and analysis are all required before one can properly do synthesis. It can be argued that in engineering, synthesis is a higher-order activity than evaluation, since evaluation is needed to determine which of many answers is optimum. Without getting into this argument, note that students need practice and feedback on all levels of the taxonomy to become proficient. The major use of the taxonomy for professors is to ensure that objectives, lectures, homework, and tests include examples and problems at all levels. Stice (1976) noted that when he classified the test questions in one of his classes he was horrified to find that almost all of them were in the three lowest levels of Bloom's taxonomy. Since students tend to learn what they are tested for, most of the students were not developing higher-level cognitive skills in this class. If the teaching style, homework, and test questions are suitably adjusted, students can be taught content at all levels of the taxonomy.

4.2.2. Affective Domain

The affective domain is concerned with behaviors and objectives which are emotional and deal with feelings. It includes likes and dislikes, attitudes, value systems, and beliefs. Development of a taxonomy for the affective domain proceeded in a parallel but slower fashion than for the cognitive domain. There was overlap on the two development committees, and the logic in developing the taxonomies was similar. However, the taxonomy in the affective domain was much more difficult to develop because there is much less agreement on the hierarchical structure. Krathwohl et al. (1964) used the process of internalization to describe the hierarchical structure of learning and growth in the affective field. Internalization is similar to socialization and refers to the inner growth as an individual adopts attitudes, principles, and codes to guide his or her value judgments. The affective domain taxonomy has had considerably less influence in education than the cognitive domain taxonomy. This is particularly true in engineering education, perhaps because of the large number of thinking types in engineering (see Chapter 13). The five levels of the affective domain are briefly outlined below (Kibler et al., 1970; Krathwohl et al., 1964).

1 Receiving and attending. Is the individual aware of a particular phenomenon or stimulus? Is he or she willing to receive the information or does he or she automatically reject it? Does the individual choose to pay attention to a particular stimulus? Information above the individual's level of intellectual development may not be attended to because it cannot be understood.

2 Responding. The individual is willing to respond to the information. This occurs first as passive compliance when someone else initiates the behavior. Then the individual becomes willing to and desires to respond on his or her own initiative. Finally, the response leads to personal satisfaction which will motivate the individual to make additional responses.

3 Valuing. The individual decides that an object, phenomenon, or behavior has inherent worth. The individual first accepts the value, then prefers the value, and finally becomes committed to the value as a principle to guide behavior.

4 Organization. The individual needs to organize values into a system, determine how they interrelate, and establish a pecking order of values.

5 Characterization by a value. The individual's behavior becomes congruent with his or her value structure, and the individual acts in a way that allows others to see his or her underlying values. Many modes of common speech point to people who are characterized by their values: "She is a caring person." "He always puts students first." "He is very up-front."

The affective domain has not been heavily studied or discussed in engineering education, yet engineering professors do have value goals for their students. They want them to be honest, hard-working, ethical individuals who study engineering because of an intrinsic desire for knowledge. Perhaps there would be a little more movement toward these goals if professors explicitly stated some of their expectations and objectives in this domain. One example is the use of an honor code. A second example is objective 10 in Table 4-1.

4.2.3. Psychomotor Domain

The psychomotor domain includes motor skills, eye-hand coordination, fine and major muscle movements, speech, and so forth. The importance of this domain in engineering education has been continually decreasing as shop courses have been removed, calculators have replaced slide rules, and digital meters have replaced analog meters. Psychomotor skills are still useful in engineering education, particularly for graduate students doing experimental research. Examples include reading an oscilloscope, glassblowing, welding, turning a valve in the correct direction, soldering, titration, typing and keyboarding numbers on a calculator, gestures while speaking, and proper speech.

The taxonomy in the psychomotor domain includes (Kibler et al., 1970):

1 Gross body movements.
2 Finely coordinated body movements.
3 Nonverbal communication behaviors.
4 Speech behaviors.

Finely coordinated body movements include typing and keyboarding. Because of the importance of computers and calculators in the practice of engineering, these psychomotor skills have become much more important than in the past. We feel strongly that all engineers should learn these skills.

Nonverbal communication skills are very important to actors. Salespeople probably need to study these skills also. If an engineer is interested in technical sales, some development of these skills may be helpful. For other engineers, including engineering professors, nonverbal communication needs to be congruent with the spoken message. Individuals can be successful engineers with speech handicaps. However, the ability to talk clearly and distinctly and to project one's voice is a distinct aid to communication. In addition, communication can be enhanced by coordinating facial expressions, body movement, gestures, and verbal messages (see Chapter 14). Professors who desire to become outstanding lecturers need to develop their skills in speech behaviors (see Chapter 6).

4.2.4. Problem-Solving Taxonomy

A taxonomy for problem solving was developed by Plants et al. (1980). This taxonomy was published in the engineering education literature and has not been as widely distributed or adopted as the other taxonomies. Because of the importance of problem solving in engineering education, this taxonomy can be useful. Applications of the problem-solving taxonomy to engineering education are discussed in Chapter 5, by Plants (1986, 1989), and by Sears and Dean (1983). The five levels of the taxonomy are briefly discussed below.

1 Routines. Routines are operations or algorithms which can be done without making any decisions. Many mathematical operations such as solution of a quadratic equation, evaluation of an integral, analysis of variance, and long division are routines. In Bloom's taxonomy these would be considered application-level problems. Students consider these "plug-and-chug" problems.

2 Diagnosis. Diagnosis is selection of the correct routine or the correct way to use a routine. For example, many formulas can be used to determine the stress on a beam, and diagnosis is selection of the correct procedure. For complex integrations, integration by parts can be done in several different ways. Selecting the appropriate way to do the integration by parts involves diagnosis. This level obviously overlaps with the application and analysis levels in Bloom's taxonomy.

3 Strategy. Strategy is the choice of routines and the order in which to apply them when a variety of routines can be used correctly to solve problems. Strategy is part of the analysis and evaluation levels of Bloom's taxonomy. The strategy of problem solving and how to teach it are the major topics of Chapter 5.

4 Interpretation. Interpretation is real-world problem solving. It involves reducing a real-world problem to one which can be solved. This may involve assumptions and interpretations to obtain data in a useful form. Interpretation is also concerned with use of the problem solution in the real world.

5 Generation. Generation is the development of routines which are new to the user. This may involve merely stringing together known routines into a new pattern. It may also involve creativity in that the new routine is not obvious from the known information. Creativity is also a topic of Chapter 5.

4.3. THE INTERACTION OF TEACHING STYLES AND OBJECTIVES

Once the content and objectives have been chosen, appropriate teaching methods can be selected. To meet any of the objectives (including the affective objectives), students must have a chance to practice and receive feedback. If you want them to meet certain objectives, share these objectives with them and test for these objectives. Students will work to learn the stated objectives in the course. If the objectives are not stated and are unclear, they will work to learn what they think the objectives are. This is much more of a hit-and-miss proposition than stating clear objectives.

The importance of clear objectives is highlighted by research on teaching styles and student learning (Taveggia and Hedley, 1972). Student learning of subject matter content as measured by course content examinations is essentially the same regardless of the teaching style as long as the students are given clear, definite objectives and a list of materials for attaining the objectives. Note that this applies to the knowledge, comprehension, application, and perhaps analysis levels, but not to synthesis, evaluation or problem solving.

Most of the time in engineering classes is spent trying to teach cognitive content objectives. Knowledge-level objectives and content are the easiest to teach and test for. Knowledge-level material can be learned from well-written articles, books, and class notes. If the objectives are clear, students will memorize the material. For example, if a student reading this book is told to learn the six levels of the cognitive domain, he or she will memorize them. Lecture can also be used for transmission of knowledge-level material, but it is less effective than written material except for clarifying questions. Comprehension is a higher level than knowledge, and more student activity can be useful. Written material is useful, particularly if the student paraphrases the material or develops his or her own hierarchical structure. To be effective, lectures need to have reasonable amounts of discussion and/or questions so that students actively process the material. Discussion and groups can also be helpful for comprehension, as can homework and problem solving.

Applications in engineering usually means problem solving. It is useful to show some solutions in class, but there is the danger that the solutions shown may be too neat and sterile since the professor has removed all the false starts and mistakes (see Chapter 5). Watching someone else solve problems does not make a student a good problem solver: The student must solve problems. A good starting point is homework with prompt feedback and with the requirement that incorrect problems be reworked. Group problem solving both in and out of class is a good teaching method since the interactions help many students. Tutoring and teaching are also excellent methods for mastering application objectives since tutoring and teaching require one to structure the knowledge. Analysis objectives usually involve more complex problems and can be taught by the same methods used for application.

To learn to do synthesis one must do synthesis. This can be started in the freshman year in computer programming courses. Once students have mastered application and analysis, they can synthesize a large program by combining already created parts and new parts. Group work can again be valuable since it helps motivate students and increases retention (Hewitt, 1991). Synthesis in upper-division classes often involves developing a new design, whether it is an integrated circuit, a chemical plant, a nuclear reactor, or a bridge. Creativity can be encouraged by providing computer tools which will do the routine calculations. The PMI approach (see Section 5.6.3) which finds pluses, minuses, and interesting aspects of the proposed solution is useful in encouraging students to be creative in synthesis.

Evaluation is not something that only the professor should do. Students need to practice this skill since they will be expected to be able to evaluate as practicing engineers. The professor can demonstrate the skill in class, have the students practice evaluation, and provide feedback on their evaluations. One way to do this is to show an incorrect solution on an overhead transparency. After giving the students a few minutes to study the solution, the professor can grade the solution while it is on the overhead. The students can then be given several solutions to evaluate as homework. At least one of these solutions should be correct since part of evaluation involves recognizing correct solutions. The students' papers are then turned in and graded. A slight twist to this is to return student homework or tests with no marks and tell the student to evaluate and correct the paper before turning it in for a grade.

Engineering professors can help students master objectives in the affective domain by sharing the explicit objectives with them in a positive fashion. For example, an instructor might say, "Since you all expect to become practicing engineers, I expect you to demonstrate professional behavior and ethical standards in this class." This is preferable to saying, "If I catch any of you cheating I am going to prosecute you and force you out of engineering."

Short (and be sure they are short) "war stories" during lectures can be helpful in helping students socialize and internalize the engineering discipline (this socialization is usually a major unstated affective objective). Engineering experience through co-op, internships, and summer jobs is an excellent way to socialize engineering students if the experience is positive. Enjoyment of the class is one of our affective objectives. A professor who is pleasant, greets students by name, and is both fair and reasonable to them is likely to have students who enjoy the class.

Psychomotor objectives require practice of the skills. Most of these can be done in laboratory, but the professor needs to be aware that students may need instruction in some simple manual manipulations. Groups are effective since one member of the group often already possesses the psychomotor skills. Few engineering professors are trained to work with students who have major deficits in the psychomotor area. Since psychomotor problems, particularly in speech, can cause both students and practicing engineers major difficulties, engineering professors should know what resources are available for help.

4.4. DEVELOPING THE CONTENT OF THE COURSE

The content of each course in the curriculum is the topic of many faculty discussions. We do not intend to discuss disciplinary details since that is not the purpose of this book. Instead, we will briefly explore some pedagogical details. In required courses the content must make the course fit into the curriculum.

Although there is never complete unanimity, most engineering departments generally agree on the content a student must have before completing his or her degree. This content must appear somewhere in the curriculum. In addition, required courses often serve as prerequisites for other courses, and the appropriate prerequisite material must be covered. The only way to ensure that the expected content is covered is to communicate with other faculty. Find out what is needed for other courses. Discuss in detail what material the students have had in prerequisite courses and find out what they are capable of doing after they have passed the prerequisite courses. (Obviously, what a student can do is not necessarily the same as what the professor covered.) Discuss the outline with other faculty who have taught the course in the past or who might want to teach it in the future. Before making major course revisions or before changing the textbook be sure that critical material is not deleted. Talk to engineers in industry to determine what they use and do not use. Update the content you use so that it is in computer-accessible form. This will avoid discouraging students from using computers.

Once the major content for the course has been outlined, look at the hierarchy of objectives you wish to cover. The time required for each topic depends on the depth of your coverage and on the objectives, in addition to the beginning knowledge of the students. A well-thought-out textbook will have done this, but you may disagree with some of the author's decisions. Plan the level of presentations considering the students' maturity (see Chapter 14). Then you can plan the major objectives for each lecture.

We suggest that the bulk of the course be developed for the sensing types and serial learners in the class. Following a logical development makes it much easier for these students to learn the material, and this sequence does not hamper the intuitive types and the global learners. Sensing types will appreciate examples and concrete applications. At the beginning and/or end of each class include the global picture for intuitive types and global learners. Intersperse theory with applications to keep both the intuitive and sensing types interested. Include both visual and kinesthetic material. Conscious use of a learning cycle will increase student learning. This arrangement will ensure that every student has part of the course catered to his or her strengths, but that the student will also be encouraged to strengthen his or her weaknesses.

4.5. TEXTBOOKS

Textbooks are commonly used in undergraduate engineering education in the United States. Selection of a textbook is important since it can add or subtract from the course quality and because many engineers keep their textbooks and use them as a primary reference for many years. Useful discussions on textbook selection are included in Eble (1988), Johnson (1988), and Plants et al. (1973).

4.5.1. Should a Textbook Be Used?

The advantages of a well-written textbook are that it provides content at the appropriate level in a well-structured form with consistent nomenclature and includes appropriate learning aids such as example problems, objectives, figures, tables, and homework problems at a variety of levels of difficulty. The disadvantages are that a textbook usually provides only one viewpoint, may not include the content you want, may be out of date, and may not be the ideal format for helping students learn to learn on their own.

In beginning courses students often want and need the structure that a good textbook provides. Since basic knowledge is not changing rapidly, textbooks at this level do not become obsolete rapidly; and because of the numerous pressures to standardize lower division courses [e.g., transferring of credits, ABET requirements (see Section 4.6), "standard" textbooks, and movement of faculty between schools], textbooks which closely match the requirements of these courses are usually available. If an appropriate textbook is not available, then a publish-on-demand textbook can be considered (see Section 4.5.3). The result is that textbooks are usually used for required undergraduate courses, particularly lower-division courses.

The situation is often different for undergraduate elective courses and all courses at the graduate level. A book for a specialized course may not exist. Books for courses at the frontiers of knowledge can rapidly become obsolete. Since the market for these books is smaller than for required undergraduate courses, there will be fewer books to choose from and they will be more expensive. Seniors and graduate students need less structure and can better cope with varying author styles and different nomenclatures. The original literature is not as efficient a way to teach since it was not written for students, but it is a good vehicle to help students learn how to learn on their own. The original literature can often provide a sense of excitement which is missing from most textbooks. Thus, it may be appropriate to assign readings from the original literature. Is the cost of the textbook reasonable? Many engineering textbooks, particularly at advanced levels, are not reasonably priced, and this may be a reason to use readings from the original literature. However, copyright law is in flux and professors need to be cautious when making a number of copies of copyrighted material for a class. Permission must be obtained from the copyright owners before making copies. A good source of information on this issue is the National Association of College Stores (1991) brochure.

A good textbook can be a tremendous aid and save a great deal of time. By developing the book for a course, the author has already done much of the organization and presentation of content for you. But books do limit what you can do in a class. Students won't mind if you occasionally skip around in the textbook or require other readings. However, if this is done extensively, they will become annoyed and wonder why you have made them buy an expensive book and then never use it.

4.5.2. Textbook Selection

To some students textbooks develop an almost mystical importance. The book is treated as if it contains **The Truth**. Perhaps this is a carry over from the monastic beginnings of universities where students studied "sacred texts" (Palmer, 1983). Because of this devotion of many students, textbook selection is important. An unnecessarily difficult textbook will discourage, excessive errors can lead to a loss of faith, and an obsolete textbook serves students poorly.

Intelligent choice of a textbook requires a significant amount of effort. If nomenclature and jargon are standard in your area, you can obtain a good feel for content coverage by looking at the table of contents. The most recent copyright date can tell if recent advances might be included, but not all authors of undergraduate textbooks are up-to-date with research. You'll have to delve into a few chapters to make sure the ideas are current. A convenient way of comparing a number of books is to check a few key items that you will cover in your course.

You'll also need to read some sections to see if the author's writing style will be understandable to your students. Although you can assume that most authors of engineering textbooks understand the content, you cannot assume that they understand how students learn. An inductive approach starting with specifics and leading to generalities is much more appropriate in an introductory textbook than an deductive approach. For introductory courses, books written in a concrete instead of an abstract style are also easier for most students to understand. Sensing students particularly appreciate detailed examples. Explicitly listing objectives is also helpful to tell students what they are expected to be able to do. The writing should be at a level appropriate for the students, and new jargon should be carefully defined. Figures and tables should be clearly labeled so that nothing needs to be assumed to understand them. Relatively short sections are easier for most students since there is a sense of accomplishment when each section is completed. Intuitive students may use the section headings and subheadings to obtain an overview of the chapter contents, so it is important that these give a true picture of the organization of the content. Homework problems should be clear and unambigious. It is also helpful if the level of difficulty of the problems is indicated. If possible, examine the solutions manual since it is often a good guide to how carefully the homework problems have been crafted. The absence of a solutions manual may indicate that the author did not spend much time on the homework problems. Books using a deductive approach or written in an abstract style with few examples may be appropriate for advanced-level classes where students are seeing the material for a second time.

A careful check of content versus your preferred course outline is necessary. Does the sequence of material make sense? Skipping around in the book is often confusing to students. If the book has light coverage of some topics, you may have to supplement it with course notes and/or outside reading. If some topics are explained in insufficient detail, you may be able to compensate in lecture. And if the book has extra material that the course will not cover, you need to determine how easy it will be to skip sections. Some authors clearly state the prerequisite chapters for each chapter so that users know which sections can be skipped. Other authors provide supplemental sections of optional material. While looking at the content, check for typographical errors and fundamental mistakes. Not all books are created equal with respect to accuracy. Typographical errors, particularly in example problems, can be extremely confusing to students who have not yet learned how to evaluate the material for correctness. Such errors may also undermine the book's credibility with students.

Is there supplemental material which will help you teach the course? If you will be teaching a course that is not your major interest, a solutions manual which correctly solves problems will be helpful. If the course is in your area of primary interest, you may choose not to use a solutions manual. Computer software bundled with the adoption of a textbook can be very advantageous if the software is compatible with the school's computer system. Some engineering textbooks integrate software into the homework assignments and the teaching of the content.

Will the book be useful to the students in later courses or as a reference after they graduate? An excellent index is not necessary when a book is used as a textbook, but it is essential for reference use. Proper referencing of appropriate source materials is also important for reference use of the book. If students will keep the book for a long period, it needs to be printed on good quality paper and be durably bound. A laboratory workbook which will probably be discarded does not need this kind of quality.

Textbook adoptions should be considered to be tentative. After a semester's use, the book can be reevaluated. Ask the students for feedback on the book. Consider how well it worked on a line-by-line and day-by-day basis. If the book does not work out or a better book becomes available, you can switch.

4.5.3. Publish-on-Demand Textbooks

The future of textbook publishing may well be publish-on-demand textbooks. McGraw-Hill, Inc., has been the innovator in this area. A fairly large book is stored on a computer. The user selects the chapters that he or she wants to use and the order in which they should appear. The computer software automatically renumbers all chapters, figure and table numbers, equation numbers, and so forth. The new book is printed out in the desired order, and the books are bound and shipped to the school. The cost is proportional to the size of the book created.

With this technology, chapters from different books and even chapters written by the professor can be included in the made-to-order book. The publisher takes care of obtaining permissions and paying appropriate royalties to the authors. Since professors customize the books, the actual number of pages each student purchases will be less and the cost will probably be less. However, there is likely to be a smaller market in used books since customized books change more often and are much less transferable from school to school. Thus, the publisher would probably sell more new copies.

However, this is a new technology and not all the problems have been resolved. The technology for ensuring that the nomenclatures of different chapters are compatible if the chapters are from different sources has not been developed. Of course, there's no guarantee that a single author will be consistent in the use of nomenclature either. Methods of combining chapters from books published by different companies have not been completely developed. Acceptance by the professoriate is also not assured.

4.5.4. Writing Textbooks

"There are bad texts—which someone else writes—good texts—which we write—and perfect texts—which we plan to write some day" (Eble, 1988). Many young professors have the goal of someday writing the classical textbook in their area of expertise. The motivation to write a textbook in engineering often arises from dissatisfaction with the textbooks which are available or the total unavailability of any textbook in a new field. Writing a textbook is difficult but rewarding work. There is personal satisfaction from having done a difficult task well, a good textbook can help an engineering professor become well known, and a successful textbook can be financially rewarding. In addition, while writing the textbook, the professor is likely to be vitally interested in the class and will probably do a good job of teaching the course.

The common wisdom is that engineering professors should wait to write a textbook until they have tenure. Sykes (1991) quotes an unidentified professor: "Nobody seeking tenure can possibly have the time to write a textbook!" The professor should have several years of teaching experience, which will be helpful in writing the textbook, and should probably be an expert (see Section 9.2). And, most importantly, because of the period of time required to complete and publish a textbook, writing one is risky for an assistant professor.

Engineering professors are not trained in all the various aspects of writing textbooks, and a certain amount of on-the-job training takes place. Fortunately, successful authors enjoy writing about writing, and there are a variety of sources of advice both for writing on engineering in particular (Beakley, 1988; Bird, 1983; Roden, 1987) and for writing books in general (Levine, 1988; Krull, 1989; Mueller, 1978). The new author can also join the Textbook Author's Association (TAA, P.O. Box 535, Orange Springs, FL 32182-0535) and benefit from the *TAA Report.* Joining TAA is particularly helpful for learning about contracts and what publishers do but not for deciding upon appropriate content. A little knowledge (such as that a 15 percent royalty on a publisher's net receipts is common for college textbooks) is very helpful when a contract is negotiated. Our advice to potential authors is simple. If writing a book is the right thing to do, do it. Signs that it is the right thing to do include:

• You've taught the course for several years, and none of the available books is really satisfactory.
• You know you can write a better book.
• You feel compelled to write a book.
• You have already written extensive handouts for the class to supplement the book you are using.
• Students ask why you haven't written a book since they are sure you can do a better job.
• You have sufficient energy and time so that the thought of another project does not make you cringe.

4.6.   ACCREDITATION CONSTRAINTS ON UNDERGRADUATE PROGRAMS

Most engineering programs in the United States are accredited by the Accreditation Board for Engineering and Technology (ABET). Accreditation is considered desirable since it allows graduates to take the appropriate examinations to become a professional engineer, makes the transfer of credits to other universities easier, makes it easier for graduates to get admitted into graduate school, and serves as a stamp of approval on the quality of the program. However, accreditation does put some constraints on undergraduate engineering programs. These constraints have been the focus of considerable debate since many engineering educators believe they stifle educational innovation.

ABET's policy is to accredit individual engineering or technology programs, not an entire school. It is not unusual to have both accredited and unaccredited programs at the same university. The unaccredited programs are not necessarily poorer; instead, they may represent innovative programs that do not fit within ABET's constraints.

The ABET criteria are delineated in an ABET publication (ABET, 1989) and summarized in Table 4-2. These are minimum requirements, and individual engineering disciplines may impose additional requirements. The mathematical studies must include differential and integral calculus and differential equations. The basic sciences must include both general chemistry and general physics and may include other sciences. The engineering sciences include mechanics, thermodynamics, electrical circuits, materials science, transport phenomena, and computer science (but not programming courses). Engineering design has proven to be a controversial area and is discussed separately below. The humanities and social sciences include anthropology, economics (but not engineering economics), fine arts (but not practice-oriented courses), history, literature, political science, psychology, sociology, and foreign languages (but not speaking courses in the student's native language). The laboratory experience should be appropriate to combine elements of theory and practice. Kersten (1989) discusses the laboratory requirement in more detail. The computer-based experience should be sufficient enough so that the student can demonstrate efficiency in application and use of digital computers. Competency in written and oral communication is expected. The semester credits listed in Table 4-2 are based on a total of 128 for graduation. The requirements are adjusted if more or fewer hours are required for graduation, and the numbers are adjusted for schools on a quarter system.

Engineering design has been the most controversial area of the ABET criteria. There is no consensus on exactly what is and what is not design. Schools that see their mission as producing candidates for graduate schools or broadly educated individuals tend to want to decrease the design requirement, whereas schools producing graduates for industry want to increase the design requirement. An additional problem is that many faculty do not have industrial design experience and have difficulty teaching design.

The ABET (1989) document states that design produces a system, component, or process for specific needs. The design process is often iterative and includes decision making normally with economic and other constraints. Appropriate mathematics, science, and engineering

TABLE 4-2   SUMMARY OF ABET CRITERIA FOR ACCREDITATION OF ENGINEERING PROGRAMS

Mathematics past Trigonometry	16	
Basic science	16	
Engineering science	32	
Engineering design	16	
Humanities and social science	13	
Laboratory experience	*	
Computer-based experience	*	
Written and oral communication	*	
Other	*	
	Base	128

* Credit hours not specified.

principles should be employed in the design process. The fundamental elements often include setting objectives and criteria, synthesis, analysis, construction, testing, evaluation, and communication of results. Student problems should include some of the following features: creativity, open-ended problems, design methodology, formulation of problem statements, alternate solutions, feasibility, and design of system details in addition to economics. Drafting skill courses cannot be used to satisfy the design requirements. ABET states that normally at least one course must be primarily design at the senior level and draw upon material from other courses. This is often interpreted as the need for a "capstone" course, although ABET (1989) does not use this wording. Proposed changes in the accreditation criteria for design are discussed by Christian (1991). If approved, these changes will strengthen the requirement for a "meaningful, major engineering design experience," but the engineering science and design categories will be combined. This latter provision might reduce the amount of design at some schools, and the proposal is controversial.

Engineering courses do not need to be listed as entirely engineering science or engineering design but can be split between the two. When a program is accredited, the choice of split may have to be justified. Thus, a professor teaching an undergraduate course does not have complete freedom of content, but must take care to follow the split between engineering science and design that the department has designated.

The ABET accreditation procedure starts with a letter to the dean who responds that reaccreditation is desired. The school then fills out very detailed questionnaires for each program to be accredited. One volume of general information and a second volume with detailed information on each accredited engineering program are prepared. Resumes for all faculty members in the programs are included. An ABET team, which consists of the team captain and one member for each program to be accredited, visits the school for three days. The team members speak with faculty and students, study course notebooks prepared by the faculty, investigate student transcripts, tour the facilities, and ask for any information they consider to be pertinent. ABET examiners typically ask about class size, teaching loads, space availability, course work, and the quality and morale of faculty and students. A weakest-link theory is used to determine whether students have met the minimum ABET requirements. That is, it must be impossible for a student to graduate without satisfying the ABET requirements. Accreditation visits are considered extremely important, and considerable time is spent preparing for them.

The accrediting team has several choices of outcome in their report. They can accredit the program for a full six-year term or for an interim three-year period with a report to justify the additional three years. Or, the accreditation can be for three years with both a report and an additional visit required before the next three years will be accredited. For unsatisfactory programs a show cause might be given. A *show cause* means that the school must show why ABET should not remove accreditation. Finally, the visiting team may decide not to accredit the program. Note that an accreditation report that gives less than complete accreditation is often used to obtain needed additional resources from the university.

4.7. CHAPTER COMMENTS

Writing objectives may be like many other things that are good for you but are not particularly pleasant. Prepare them once for one course. The experience will sharpen your teaching both in that course and in other courses, even if you do not formally write objectives for other courses. Bloom's taxonomy is extremely helpful in ensuring the proper distribution of class time, student effort, and quiz questions. Carefully classifying objectives and test questions as to the level on the taxonomy is also a very useful exercise to do for at least one class. Then in later classes the level will usually be obvious.

The ABET requirements may not be high on your list of interesting reading. However, if new faculty are unaware of the ABET requirements, it is unlikely that their courses will meet the spirit of these criteria. This is particularly true of including design as some fraction of a course. In addition, to be informed participants in the current debate on accreditation requirements, faculty must understand the current requirements.

4.8. SUMMARY AND OBJECTIVES

After reading this chapter, you should be able to:

• Write objectives at specified levels of both the cognitive and the affective taxonomies.
• Develop a teaching approach to satisfy a particular objective.
• Decide whether to use a textbook in a course and select an appropriate textbook.
• List and discuss the ABET requirements for accreditation of an undergraduate engineering program.

HOMEWORK

1 Pick a required undergraduate engineering course. Write six cognitive objectives for this course with one at each level of Bloom's taxonomy.
2 Write two objectives in the affective domain for the course selected in problem 1.
3 Pick an undergraduate laboratory course. Write two objectives in the psychomotor domain.
4 Objective 10 in Table 4-1 includes a cognitive and an affective domain objective. Classify each of these.
5 For the course selected in problem 1 decide whether a textbook should be used. Explain your answer.
6 The following statement can be debated. "ABET accreditation has strengthened engineering education in the United States."
   a Take the affirmative side and discuss this statement.
   b Take the negative side and discuss this statement.

REFERENCES

ABET, *Criteria for Accrediting Programs in Engineering in the United States,* Accreditation Board for Engineering and Technology, New York, 1989.

Beakley, G. C., "Publishing a textbook? A how-to-do-it kit of ideas," *Eng. Educ.,* 299 (Feb. 1988).

Bird, R. B., "Book writing and chemical engineering education: Rites, rewards and responsibilities," *Chem. Eng. Educ.,* 17, 184 (Fall 1983).

Bloom, B. S., Engelhart, M. D., Furst, E. J., Hill, W. H., and Krathwohl, D. R., *Taxonomy of Educational Objectives: The Classification of Educational Objectives. Handbook I: Cognitive Domain,* David McKay, New York, 1956. (This book has many examples from a variety of areas.)

Christian, J. T., "Current ABET accreditation issues involving design," *Proceedings ASEE Annual Conference*, ASEE, Washington, DC,1519 (1991).

DeBrunner, V., "Performance-based instruction in electrical engineering," *Proceedings ASEE Annual Conference,* ASEE, Washington, DC, 1589, 1991.

Eble, K. E., *The Craft of Teaching*, 2nd ed., Jossey-Bass, San Francisco, 1988.

Hanna, G. S. and Cashin, W. E., "Matching instructional objectives, subject matter, tests, and score interpretations," Idea Paper No. 18, Center for Faculty Education and Development, Kansas State University, Manhattan, KS, 1987.

Hewitt, G. F., "Chemical engineering in the British Isles: The academic sector," *Chem. Engr. Rsch. Des.,* 69 (A1), 79 (Jan. 1991).

Johnson, G. R., *Taking Teaching Seriously: A Faculty Handbook,* Texas A&M University Center for Teaching Excellence, College Station, TX , 1988.

Kersten, R. D., "ABET criteria for engineering laboratories," *Proceedings ASEE Annual Conference*, ASEE, Washington, DC, 1043, 1989.

Kibler, R. J., Barker, L. L., and Miles, D. T., *Behavioral Objectives and Instruction*, Allyn and Bacon, Boston, 1970.

Krathwohl, D. R., Bloom, B. S., and Masia, B., *Taxonomy of Educational Objectives: The Classification of Educational Goals. Handbook II: The Affective Domain*, David McKay, New York, 1964.

Krull, K., *Twelve Keys to Writing Books That Sell,* Writer's Digest, Cincinnati, 1989.

Levine, M. L., *Negotiating a Book Contract: A Guide for Authors, Agents and Lawyers*, Moyer Bell, New York, 1988.

Mager, R. F., *Preparing Instructional Objectives*, Fearon Publishers, Palo Alto, CA, 1962.

Mueller, L. W., *How to Publish Your Own Book*, Harlo Press, Detroit, 1978.

National Association of College Stores, *Questions and Answers on Copyright for the Campus Community*, NACS, 199. (For copies write to NACS, 500 East Lorain St., Oberlin, OH, 44074-1294.)

Palmer, P. J., *To Know As We are Known: A Spirituality of Education*, Harper, San Francisco, 1983.

Plants, H. L., "Content comes first," *Eng. Educ.,* 533 (March 1972).

Plants, H. L., "Basic problem-solving skills," *Proceedings ASEE Annual Conference,* ASEE, Washington, DC, 210, 1986.

Plants, H. L., "Teaching models for teaching problem solving," *Proceedings ASEE Annual Conference*, ASEE, Washington, DC, 983, 1989.

Plants, H. L., Dean, R. K., Sears, J. T., and Venable, W. S., "A taxonomy of problem-solving activities and its implications for teaching," In Lubkin, J. L. (Ed.), *The Teaching of Elementary Problem Solving in Engineering and Related Fields*, ASEE, Washington, DC, 21-34, 1980.

Plants, H. L., Sears, J. T., and Venable, W. S., "Making tools work," *Eng. Educ.,* 410 (March 1973).

Roden, M. S., "How to make more than $.25 per hour as a textbook author," *Proceedings ASEE Annual Conference,* ASEE, Washington, DC, 52 1987.

Sears, J. T. and Dean, R. K., "Chemical engineering applications of a problem-solving taxonomy," *AIChE Symp. Ser.,*79(228), 1 (1983).

Stice, J. E., "A first step toward improved teaching," *Eng. Educ.,* 394 (Feb. 1976).

Sykes, T., "Textbooks as scholarships?" *TAA Report* 5(4), 5 (Oct. 1991).

Taveggia, T. C., and Hedley, R. A., "Teaching really matters, or does it?" *Eng. Educ.,* 546 (March 1972).

CHAPTER 5

PROBLEM SOLVING AND CREATIVITY

Engineering education focuses heavily on problem solving, but many professors teach content and then expect students to solve problems automatically without being shown the process involved. Our position is that an explicit discussion of problem-solving methods and problem-solving hints should be included in every engineering class. A problem-solving taxonomy was briefly discussed in Section 4.2.4.

Most engineering schools are very good at teaching routines, and most engineering students become very proficient at them. And since diagnosis is required for many problems, particularly in upper-division courses, most students become reasonably proficient at it also. Students in general are not proficient at strategy, interpretation, and generation. These three areas of the problem-solving taxonomy will be discussed throughout this chapter.

In this chapter we will first briefly discuss some of the basic ideas about problem solving and compare the differences between novices and experts. Then a strategy for problem solving which works well for well-understood problems will be presented, and methods (heuristics) for getting unstuck will be discussed. The teaching of problem solving will be covered with a number of hints that can be used in class. Finally, creativity will be discussed.

5.1. PROBLEM SOLVING—AN OVERVIEW

Extensive studies have shown that problem solving is a complicated process. The concept map shown in Figure 5-1 gives some idea of the interactions and complexities involved (this figure is modified from the one in Chorneyko et al., 1979). An entire book would be required to explain the information on this map fully. Readers who feel a need to understand parts of this map which are not explained in this chapter are referred to the extensive list of references at the end of the chapter.

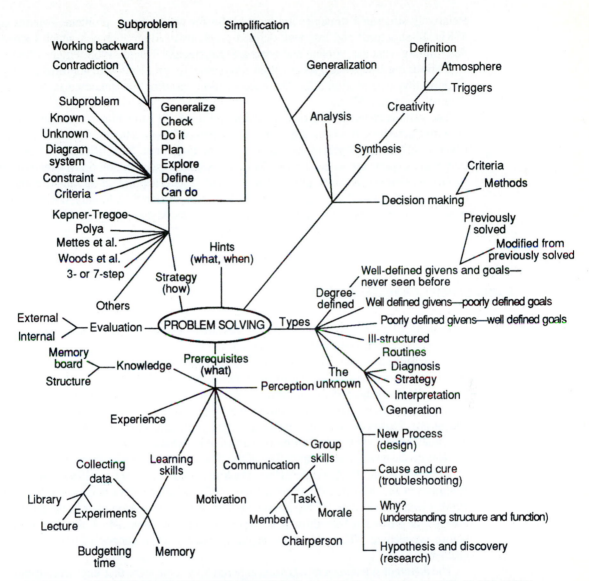

FIGURE 5-1 CONCEPT MAP OF PROBLEM SOLVING. Reprinted with permission of *CEE, 13*, 132, (1979).

Cognitive psychologists are in general agreement that there are generalizable problem-solving skills, but that problem solving is also very dependent upon the knowledge required to solve the problem [see Chapter 14 and Kurfiss (1988) for a review]. Of the prerequisites shown in Figure 5-1, knowledge and motivation are the most important.

Problem solving can be classified by the type of problem which must be solved. Three different classification schemes are shown in Figure 5-1. A scheme based on the degree of definition of the problem (Cox, 1987) is useful since it ties in closely with the strategy required.

Relatively structured strategies are most useful for well-defined problems (Mettes et al., 1981). Ill-structured and less well-defined problems need an approach which focuses on determining what the problem and goals are (Kepner and Tregoe, 1965; Fogler, 1983).

Various multistep strategies are often appropriate for problems with intermediate degrees of definition (see Section 5.3). The classification based on the unknown is discussed by Chorneyko et al. (1979).

The various elements of problem solving in Figure 5-1 show how it interacts with other cognitive activities. Analysis and synthesis are part of Bloom's taxonomy while generalization is a seldom taught part of the problem-solving taxonomy. Simplification is a procedure that many experts use to get a rapid fix on the solution (see Section 5.2.). Creativity is an extensively studied, but not really well understood, adjunct to problem solving. Creativity can be enhanced in individuals with proper coaching (see section 5.6). Finally, decision making is often a part of problem solving which connects it to the Myers-Briggs analysis (see Chapter 13) and is a major part of the Kepner and Tregoe (1965) approach.

5.2. NOVICE AND EXPERT PROBLEM SOLVERS

Experts have about 50,000 "chunks" of specialized knowledge and patterns stored in their brains in a readily accessible fashion (Simon, 1979). The expert has the knowledge linked in some form and does not store disconnected facts. Exercises which require students to develop trees or networks can help them form appropriate linkages (Staiger, 1984). Accumulation of this linked knowledge requires about ten years. Since it is not feasible to accumulate this much information in four or five years, producing experts is not a realistic goal for engineering education. However, it is reasonable to mold proficient problem solvers who have the potential to become experts after more seasoning in industry.

How do the novices who start college differ from experts? This has been the topic of many studies (Dansereau, 1986; Fogler, 1983; Hankins, 1986; Larkin et al., 1980; Lochhead and Whimbey, 1987; Mayer, 1992; Smith, 1986, 1987; Whimbey and Lochhead, 1982; Woods, 1980, 1983; Woods et al., 1979; Yokomoto and Ware, 1990). A number of observations on how novice problem solvers differ from experts are listed in Table 5-1. Read through it briefly before proceeding. The table is arranged in roughly the sequence in which one solves problems.

The differences between novices and experts show some areas that engineering educators can work on to improve the problem-solving ability of students. In the category of prerequisites, students should be encouraged to learn the fundamentals and do deep processing. Knowledge should be structured so that patterns, instead of single facts, can be recalled. Motivation and confidence are important, so professors should encourage students and serve as models of persistence in solving problems.

In working problems, students need to practice defining problems and drawing sketches. The differences between a student's sketch and that of an expert should be delineated, and the student should be required to redraw the sketch. Students also need to practice paraphrasing the problem statement and looking at different ways to interpret the problem. A distinct

TABLE 5-1   COMPARISON OF NOVICE AND EXPERT PROBLEM SOLVERS

Characteristic	Novices	Experts
Memory	Small pieces Few items	"Chunks" or pattern ~ 50,000 items
Attitude	Try once and then give up Anxious	Can-do if persist Confident
Categorize	Superficial details	Fundamentals
Problem statement	Difficulty redescribing Slow and inaccurate Jump to conclusion	Many techniques to redescribe Fast and accurate Take time defining tentative problem May redefine several times
Simple well-defined problems	Slow Work backward	~ 4 times faster Work forwards with known procedures
Strategy	Trial and error	Use a strategy
Information	Don't know what is relevant Cannot draw inferences from incomplete data	Recognize relevant information Can draw inferences
Parts (harder problems)	Do NOT analyze into parts	Analyze parts Proceed in steps Look for patterns
First step done (harder problems)	Try to calculate (Do It step)	Define and draw Sketch Explore
Sketching	Often not done	Considerable time Abstract principles Show motion

strategy should be used (see the next section). Students should also practice analyzing problems to break the problem into parts, and they need to be encouraged to perform the explore step. A chug-and-plug mentality should be discouraged, and students should be encouraged to return to the fundamentals.

Once students know a strategy, they should be encouraged to monitor their progress. Methods for getting unstuck should be taught (see Section 5.4). Then once the problem has been completed, students should be required to check their results and evaluate them versus internal and external criteria. After the problems have been graded, some mechanism for ensuring that students learn from their mistakes is required. Throughout the process students should be encouraged to be accurate and active. Specifics of methods for teaching problem solving are discussed in more detail in Section 5.5.

TABLE 5-1(CONT)   COMPARISON OF NOVICE AND EXPERT PROBLEM SOLVERS

Characteristic	Novices	Experts
Limits	Do not calculate	May calculate to get quick fix on solution
Equations	Memorize or look up detailed equations for each circumstance	Use fundamental relations to derive needed result
Solution procedures	"Uncompiled" Decide how to solve after writing equation	"Compiled" procedures Equation and solution method are single procedure
Monitoring solution progress	Do not do	Keep track Check off versus strategy
If stuck	Guess Quit	Use Heuristics Persevere Brainstorm
Accuracy	Not concerned DO NOT Check	Very accurate Check and recheck
Evaluation of result	Do not do	Do from broad experience
Mistakes or failure to solve problems	Ignore it	Learn what should have done Develop new problem solving methods
Actions	Sit and think Inactive Quiet	Use paper and pencil Very active Sketch, write questions, flow paths.Subvocalize (talk to selves)
Decisions	Do NOT understand process No clear criterion	Understand decision process Clear criterion

5.3. PROBLEM-SOLVING STRATEGIES

Many experts have difficulty verbalizing the problem-solving strategy they are using since the strategy has become automatic. When an expert does verbalize how he or she solves a problem, it is clear that a distinct strategy has been used. Novices have a strategy also—it is a trial-and-error strategy. It is not very effective and does not help the novice become a better problem solver.

A distinct problem-solving strategy should be demonstrated and then required from students. The exact strategy used is not important; what is important is that the strategy be used consistently and that students be required to use it. Woods et al. (1979) suggest that the strategy have between four and fifteen steps. If shorter than four steps the strategy is probably too short and not detailed enough to be useful; if longer than fifteen steps it is too long to remember and use.

The strategy that we have used is based on the work of Don Woods and his coworkers at McMaster University (Woods et al., 1975, 1977, 1979, 1984; Woods, 1977, 1983, 1987; Leibold et al., 1976). Through the years their strategy has changed slightly. We have settled on a strategy with six operational steps and a prestep which focuses on motivation:

0 I can.
1 Define.
2 Explore.
3 Plan.
4 Do it.
5 Check.
6 Generalize.

Step 0 is a motivation step. Since anxiety can be a major detriment to problem solving, it is useful to work on the student's self-confidence (Scarl, 1990; Richardson and Noble, 1983). The professor may want to avoid being subtle when first working on this step. It is also useful to teach students a few simple relaxation exercises (Richardson and Noble, 1983; also, see Section 2.7 on handling stress).

Step 1, the define step, is often given very little attention by novices. They need to list the knowns and the unknowns, draw a figure, and perhaps draw an abstract figure which shows the fundamental relationships (remember that most people prefer visual learning). The figures are critical since an incorrect figure almost guarantees an incorrect solution. The constraints and the criteria for a solution should be clearly identified.

Step 2 is the explore step. This step was originally missing from the strategy but was added when its importance to expert problem solvers became clear (Woods et al., 1979). This step can also be called "Think about it," or "Ponder." During this step the expert asks questions and explores all dimensions of the problem. Is it a routine problem? If so, the expert will solve the problem quickly in a forward direction. If it is not routine, what parts are present? Which of these parts are routine? What unavailable data are likely to be required? What basis is most likely to be convenient? What are the alternative solution methods and which is likely to be most convenient and accurate? What control envelope should be used? Does this problem really need to be solved, or is it a smoke screen for a more important problem? Many experts determine limiting solutions to see if a more detailed solution is really needed. Since novices are often unaware of this step, they need encouragement to add it to their repertoire.

In the plan step, formal logic is used to set up the steps of the problem. For long problems a flowchart of the steps may be useful. The appropriate equations can be written and solved without numbers. This is extremely difficult for students in Piaget's concrete operational stage. This step is easier for global thinkers and intuitives, which means that serial thinkers and sensing individuals need more practice.

Do it, step 4, involves actually putting in values and calculating an answer. This is the step which novices want to do first. Even fairly skilled problem solvers often want to combine steps 3 and 4 and not develop a solution in symbolic form. The separation of the plan and do it stages makes for better problem solvers in the long run. Separating these stages makes it easier to check the results and to generalize them since putting in new values is easier. Sensing students tend to be better at doing the actual calculations.

Checking the results should be an automatic part of the problem-solving strategy. Checking requires internal checks for errors in both mathematical manipulations and number crunching, and it involves evaluation with external criteria. A very useful ploy of expert problem solvers is to compare the answer to the limits determined in the explore step. The answer should also be compared to "common sense." This step requires evaluation and many students will not be adept at it.

The last step, generalize, is almost never done by novices unless they are explicitly told to do it. What has been learned about the content? How could the problem be solved much more efficiently in the future? For example, was one term very small so that in the future it can be safely ignored? Were trends linear so that in the future very few points need to be calculated? If the problem was not solved correctly, what should have been done? Students need to be strongly encouraged to study feedback and then resolve incorrect problems.

Note that problem solvers who use this strategy consistently will use all levels of both the Bloom and the problem-solving taxonomies. However, students will rebel against using this or any other structured approach to solving problems. The method is unfamiliar and will feel awkward at first. Since many aspects of problem solving are automatic, making them conscious is uncomfortable at first and may inhibit the student for a period. An analogy is the self-taught golfer or tennis player who starts taking lessons. Thinking about the swing so that it can be improved makes it difficult to swing effortlessly. However, in the long run the person with training will become a better golfer or problem solver. (Note that an expert golfer will also be an expert problem solver in this narrow domain.)

Many other problem-solving strategies can be used. Polya (1971) originated a four step approach which is a predecessor of the approach shown here. Since Woods (1977) has published an extensive review of problem-solving strategies, these older papers will not be reviewed here. Scarl (1990) also describes a procedure very similar to that presented here, and in addition he is very directive of what students should do when. Mettes et al. (1981) describe a systematic flow sheet approach that is quite different from the method illustrated here. Their structured approach was developed specifically for solving thermodynamics problems and may not be generalizable. Smith (1986, 1987) discusses expert system models for problem solving. Kepner and Tregoe (1965) developed procedures that are most applicable to determining what the problem is (troubleshooting) and for decision making. Guided design is a method for guiding groups of students through a structured problem-solving procedure (Wales and Stager, 1977; Wales et al., 1986). This method will be discussed further in Chapter 9. In a book used to teach problem solving, Rubenstein (1975) discusses five models of problem solving.

5.4. GETTING STARTED OR GETTING UNSTUCK

A problem-solving strategy is not much help if you just cannot get started on a problem or are completely stuck. What do you do then? Novice problem solvers tend to give up or make wild guesses, whereas experts persist, recycle back through the define step, and use heuristics.

The first step for a professor is to encourage students. Remember those high school football slogans, "When the going gets tough, the tough get going," and "Winners never quit, and quitters never win," and so forth? A short pep talk is not out of order, particularly for students who have the prerequisites to be successful. Nothing makes a student more confident in her or his ability to solve problems than successfully solving difficult problems.

The second step is to encourage the student to recycle in whatever problem-solving strategy the class is using. Ask, "Have you reread the problem statement to be sure you are solving the right problem?" "Have you rechecked your figures for accuracy?" "Have you thought about whether your plan of attack still seems reasonable? Novices want to apply a strategy once through, while experts apply a strategy in a series of loops. One advantage of having an explicit strategy is that you can easily refer the student to a particular stage of the process, and both of you will have a common language.

If recycling through the strategy does not work, suggest that the student identify his or her difficulty with the problem (Woods, 1983). Where is the student stuck? What is the obstacle? Where does the student want to be? Are there alternatives that can be used? Sometimes this process will lead the student to a productive path.

If still stuck, it is now time for the problem solver to use heuristics. Heuristics are methods which might, but are not guaranteed to, work. A large number of heuristic methods have been suggested to aid a problem solver who is stuck (Adams, 1978; Cox, 1987; Koen, 1984, 1985; Polya, 1971; Rubenstein and Firstenberg, 1987; Scarl, 1990; Smith, 1986, 1987; Starfield et al., 1990; Wankat, 1982; Woods et al., 1979). A very large number of heuristics can be listed; however, it probably does not matter which ones students are taught as long as they use them. For any given obstacle many different heuristics will work, since what the heuristic does is to get the problem solver thinking productively on a new path. (Students need to realize this also—and it can be called another heuristic.) Readers interested in the use of heuristics for problem-solving should consult the short book by Starfield et al. (1990).

The second and third suggestions in this section (recycle and find the obstacle) can be considered either heuristics or parts of the problem-solving strategy. We will list a variety of other heuristics. Select from these the ones that you will teach to the students, remembering that they will need to practice using the heuristics and will need feedback. Particularly with novices, it is preferable to keep the list short so that they can remember and use the heuristics.

1 Simplify the problem and solve limiting cases.

This is a procedure often used by experts. Another closely related heuristic involves solving special cases.

2 Check to see that the problem is not under- or overspecified.

Problems that are under- or overspecified need interpretation before they can be solved.

3 Relate the problem to a similar problem which you know how to solve.

Solutions to similar problems can give a useful outline of how to solve the current problem. A closely related technique uses analogies to give hints about the problem solution.

4 Generalize the problem.

Sometimes the problem is easier to understand and solve in a very general form.

5 Try substituting in numbers.

Sometimes the problem will be clearer with numbers inserted because it will appear more concrete.

6 Try solving for ratios.

Often a problem can be solved for ratios, but not for individual numbers.

7 Get the facts and be sure there actually is a problem.

Another way to say this is, "If it's not broke, don't fix it." This heuristic can be taught and reinforced in the laboratory.

8 Change the representation of the problem.

If the first representation of the problem is too difficult, change it. This is often called education or advertising.

9 Ask questions about the problem.

Specifications are often set arbitrarily but may make the problem extremely difficult to solve. Question them. Does the purity really have to be so high? Do the tolerances really have to be so tight?

10 Concentrate on the parts of the problem that can be solved.

Very often parts that seem unsolvable become solvable when other parts of the problem have been solved. This is partly a confidence factor.

11 In groups, be a good listener and maintain group harmony.

Groups can be synergistic in solving problems, but only if people listen and there is some group harmony.

12 Use a plus-minus-interesting (PMI) approach when presented with possible solutions (deBono, 1985; Gleeson, 1980).

The plus helps the morale of the person suggesting the solution. Minuses are why the solution is not yet complete. Interesting are the ideas that can be adapted.

13 Use a mixed scanning strategy (Rubenstein and Firstenberg, 1987).

A mixed scanning strategy alternates a broad look at the entire problem with in-depth looks at small parts of the problem.

14 Alternate working forward and backward.

Although experts work forward on simple problems, they alternate working forward and backward on difficult problems.

15 Take a break.

This is not quitting but is a break allowing you to do something else before returning to the problem with a fresh view.

16 Ask what the hidden assumptions are or what you have forgotten to use.

Novice problem solvers often limit their solutions by assuming constraints which are not part of the problem.

17 Apply a control strategy.

Experts keep track of where they are in solving a problem with a metacognitive control strategy.

Schoenfield (1985) suggests that you ask yourself three questions: What are you doing? (Be exact in the description.) Why are you doing it? How will it help you solve the problem?

18 Refocus on the fundamentals.

Sometimes asking what is fundamental will break the log jam.

19 Guess the solution and then check the answer.

Yes, guessing is a novice approach. However, sometimes when we are stuck, we have strong hunches. If we guess the answer, it may be easy to prove whether it is correct or incorrect. The differences between novice and expert behavior here are that the expert makes her or his guess after working on the problem for a period and always checks the guess.

20 Ask for a little help.

Even experts ask for help. The key is to get only a little help and not to let the helper solve the problem for you.

To close this section it may be useful to consider the six categories of blocks which Adams (1978) has identified. *Perceptual blocks* are difficulties in seeing various aspects or ramifications of the problem. *Cultural blocks* lead to inadvertent assumptions about the solution method or the solution path. In particular, in engineering there is a cultural bias toward convergent (logical) thinking and away from divergent (lateral or creative) thinking. *Environmental blocks* are due to the problem solver's surroundings, including people. For students this means the professor and other students. A lack of acceptance of novel ideas can be a major environmental block. *Emotional blocks* such as anxiety or fear of failure can make problem solvers much less effective. *Intellectual blocks* can include a lack of knowledge or trying to use inappropriate knowledge. The use of unannounced review questions on homework can help overcome this block. *Expressive blocks* involve the use of inappropriate problem-solving languages or inappropriate paths. For example, trying to solve a problem without an appropriately drawn figure can be an expressive block. An additional heuristic is: Determine the blocks which are preventing you from solving the problem.

5.5. TEACHING PROBLEM SOLVING

Many excellent papers and books have been written on how to improve the problem-solving abilities of students. In this section we have distilled many of these ideas. Readers interested in more ideas and applications are referred to the literature (Goodson, 1981; Greenfield, 1987; Kurfiss, 1988; Lochhead and Whimbey, 1987; Plants, 1986; Scarl, 1990; Starfield et al., 1990; Stice, 1987; Wales and Stager, 1977; Wales et al., 1986; Whimbey and Lochhead, 1982; Woods, 1983, 1987; Woods et al., 1975; Yokomoto, 1988). With a little creativity you can adapt the ideas in this chapter and invent new methods to improve your teaching of problem solving.

Lumsdaine and Lumsdaine (1991) and Rubenstein (1975) recommend a separate course in problem solving. However, specific knowledge in the problem domain is essential for solving problems. Thus, we suggest embedding problem solving into existing engineering courses.

Then, the problem solving and specific knowledge can reinforce each other. It is helpful if the knowledge is organized by students in a hierarchical structure since this is what most expert problem solvers do. Some information at the knowledge level of Bloom's taxonomy is essential, and professors should not hesitate to require memorization of certain crucial numbers. Most problems in lower-level engineering classes require facility with algebraic manipulations. Thus it is essential that students master algebra. Obviously, other mathematical skills are important, but algebra appears to be the lowest common denominator.

Problem solving should be taught throughout the student's college career. This can often be done in the form of little hints or suggestions of a heuristic to try while students are struggling with problems. Ideally, the same strategy would be used in all science and engineering classes and in textbooks. However, since most strategies are similar, students will not be hopelessly confused if the strategy changes. Illustrate the strategy when solving problems in class and in handouts. This includes solutions to homework and test problems. Many students will learn to use a strategy on their own, but students most in need of help in problem solving will not learn without help. Encourage and perhaps even force students to use the strategy on homework and test problems.

Although no student can become an accomplished problem solver merely by watching a professor solve examples, example problems are an important learning device, particularly for sensing students. Unfortunately, most professors inadvertently foster the idea that problem solving is a neat process and thereby do damage to the student's confidence. Using a routine to determine an answer is a neat process once the problem has been interpreted, a strategy chosen, the problem diagnosed, and the routine selected. These other steps are messy but represent the real heart of problem solving. Suppose that solving a problem took fifteen minutes and resulted in two dead ends and a page of scrap paper. Your typical approach may be to clean this up and show it to the students in five minutes with no mistakes and no dead ends. What the students see is a process that they cannot duplicate. Then when they are unable to solve problems in this way, they begin to doubt their abilities. Occasionally show a messy solution. Solve a problem in front of the class that you have not seen before and verbalize as you solve it. This can be done by having students select a problem from the textbook for you to solve. Yes, this is scary since you may fail. However, it does demonstrate the process that one goes through when solving novel problems, including step 0, the motivation or confidence-bolstering step.

Students need to solve problems to learn how to solve problems. Unfortunately, rote learning and drill will at most teach how to do routines which are necessary but not sufficient to becoming a good problem solver. Students need to solve more challenging problems requiring all levels of the problem-solving taxonomy. The good news is that all the professor has to do with the better students is to challenge them with good problems and provide feedback. The bad news is that this classical procedure does not work with the poorer students. Yet, these poorer students have the potential to become excellent engineers. How can one teach problem solving to make it accessible to them?

Particularly for beginning students, requiring a neat regular structure is useful. Tell students to lay out the problem solution in the same format for all homework problems. Require separate labeling of steps in the problem-solving strategy. Make students work down one side of the paper in regular columns. Let, and in fact encourage, students to doodle, try out ideas, and play with the problem on a separate piece of scrap paper. It is important to encourage them to write things down since this external memory is often more effective than trying to store ideas internally. (Paper is much cheaper than time.) Require a sketch even for students who can solve the problem without a sketch. Students should define all symbols even if they are the same ones as in the book. Before plugging in numbers, they should obtain an algebraic solution in symbolic form. Until an individual student has proven that he or she can skip algebraic steps, all algebraic steps should be shown. A separate equation line with all numbers substituted into the equation should be shown before the student calculates the answer. Obviously, students will resist this degree of regimentation. They will truthfully say that they are now slower and poorer problem solvers. In the long run such a procedure will produce better, neater, faster, more accurate problem solvers, and in the short run troubleshooting their solutions will be much easier. Since there is no reason why creative solutions cannot be neat and understandable, this procedure will not deaden creativity as long as the professor grades solutions with an open mind when the solutions are different.

Give a combination of application, analysis, synthesis, and evaluation problems. Be sure that the homework problems bracket the test problems in terms of difficulty, or students will think you are unfair. Be sure that some problems require the simultaneous solution of equations, or students will believe that all problems can be solved sequentially. Some problems should be open-ended, and synthesis should be required. Often students who excel in these problems are not the same students who excel in doing routines. Require students to evaluate solutions.

Separately cover all steps of the problem-solving strategy. For example, for one problem the students might do the define, explore, and plan steps only. Give multipart problems where students first have to define and draw a sketch; then after the entire class has received feedback and has that step correct, they would do the next step, and so on. Require students to completely check their solutions by solving the problem with a completely different method. Then note that the answer is still wrong if incorrect values for physical parameters are used. For another problem a complete solution with all numbers checked and no errors should be required. If accuracy is important for practicing engineers, then students must practice this level of accuracy. For these few problems where accuracy is being stressed, return the problem to the student for a corrected solution if there are any errors.

Try to cover all aspects of the problem-solving taxonomy. Give a few problems which are carefully worded to be ambiguous so that students can practice interpretation. Require students to find or estimate some of the physical constants they need (and be prepared for a variety of solutions). Give them the assignment of making up a problem so that they have practice in defining problems. Give them real cases where a clearly defined problem is not laid out in front of them. These can include troubleshooting or debottlenecking problems.

Students can be made more aware of their problem-solving procedures by verbalizing what they are doing while solving problems. This can be done conveniently in class with the Whimbey-Lochhead pair method (Whimbey and Lochhead, 1982; Lochhead and Whimbey, 1987). The class is divided into pairs, and one member of each pair is designated the problem solver. His or her job is to solve the problem and to say out loud everything he or she is thinking while solving the problem. The other person in the pair is the recorder-encourager whose job is to take notes on what the person is doing and to encourage the problem solver to keep verbalizing. As the the encourager he or she can say things such as, "What are you thinking now," or "Tell me what you're thinking." As the the recorder he or she needs to try to understand every step, diversion, and error made by the problem solver. When the reasons for a step are unclear, the recorder asks what the problem solver is doing and why. The recorder can point out algebraic or numerical errors, but he or she should not be specific as to where the error is. The two cardinal rules for the recorder are to avoid solving the problem and to not lead the solver toward a solution.

After explaining the roles to students, give the problem solver a short written problem statement. Then, as students start to read this to themselves, remind them they have to read out loud. Encourage them to verbalize their anxiety as they read a new problem and encourage them to verbalize self-encouragement. Encourage the problem solver to use a pencil and paper while solving the problem. During the remainder of the solution of the problem, visit various pairs and reinforce the role of each student.

Once the problem has been completed, either correctly or incorrectly, the recorder and the problem solver should discuss what the problem solver did while solving the problem. Remind students that learning how one solves problems is the purpose of the exercise, not correctly solving the problem. Students can then switch roles and solve a new problem.

To be effective, this procedure needs to be used several times during the semester. Note that it can be used in quite large classes. It keeps students active and simultaneously teaches both content (the problems chosen) and problem solving. This type of activity is a nice break from excessive lecturing. Professors can also learn from the Whimbey-Lochhead pair method and verbalize while they solve example problems.

Problem solving can also be taught with discovery methods of instruction (Canelos, 1988). These approaches include simulation, case study, guided design, and discussion. In all these methods students should work on real, or at least realistic, engineering problems. They should help define the problem and then work at developing a solution. Then the professor should push the students to evaluate their solution and look for a better one. When the process is completed, the professor helps the students describe the problem-solving process so that they discover the method. These methods are suitable for either individual or group work. Further details of these methods are given in Chapters 7 through 9.

Student work in groups is particularly conducive to learning problem solving. Being in a group of one's peers can help reduce a student's anxiety if it is clear that no one has all the solutions. Extroverts and field-sensitive individuals will benefit from the group support. The verbalization that occurs in a group is a good way to obtain internal feedback. Groups are excellent for clarifying difficult-to-interpret problems since each group member will look at the problem in a slightly different way. Brainstorming during the explore step is easily done in groups. From the professor's viewpoint it is more efficient to work with groups of three to five students rather than individual students since the number of questions is reduced by a factor of from three to five. Finally, when they graduate, the new engineers will be expected to work in teams in industry. Providing practice in teamwork while they are students will help their transition to industry.

Your goal while working with problem-solving groups or individuals is not to give students what they want. Students want the solution. As a professor you want the students to find a solution essentially on their own and to improve their problem-solving skills in the process.

Encourage them to verbalize and refuse to let them quit prematurely. You can check to see if the students' knowledge base is correct and can help them see the hierarchical structure of the knowledge. You can also focus their activities on problem-solving methods. For example, if they are stuck, you can ask, "What heuristics have you tried?" and "What other heuristics can you try?" If students are stuck on a clearly incorrect approach, show them why they are incorrect but without showing them a correct approach. Guided design suggests that the entire problem and responses be developed in advance so that students are automatically guided through the steps of problem solving (Wales and Stager, 1977; Wales et al., 1986). (See Section 9.1.) Although many professors do not have time for this much detail, it is helpful to have a brief outline or script of how you want to proceed. This will help you to remember to cover all important points.

5.6. CREATIVITY

Creativity is a novel and unexpected way of defining or solving a problem which leads the observer to ask, "How did you think of that?" Creativity can be a part of problem solving, but many successful solutions do not illustrate creativity. Creativity requires divergent thinking and usually appears at the define or explore step in problem solving if it is present. Note that creativity is only part of the entire problem-solving step. The creative idea must be proven to be a valid solution by a logical analysis during the plan, do it, and check steps. The generalize step can be used to further develop the creative idea and to look for other applications. The importance of creativity in engineering is summarized by Florman (1987, p. 75): "Engineering is an art as well as a science, and good engineering depends upon leaps of imagination as well as painstaking care."

Everyone is born with creative abilities. According to Hueter (1990) these abilities increase in elementary school up to an age of about eight and then steadily decrease with further schooling. At about eight years old children become very aware of the opinions of other people. It becomes important for them to fit in and to use objects for "what they are supposed to be used for." The result is a decline of creativity that continues through college. If Hueter is correct, then engineers are in a paradoxical situation. The very education which makes an engineer more capable of solving difficult problems decreases the likelihood that he or she will invent a creative solution. However, creativity can be enhanced in class with a positive attitude and suitable exercises (Christenson, 1988).

The professor's job is to nuture the creative abilities which everyone possesses and to stem any decline in creativity . What can professors do to encourage the latent creativity of every student? We will discuss three things that professors can do in engineering classes.

1 Tell students to be creative.
2 Teach students some creativity methods.
3 Accept the results of creative exercises.

5.6.1. Tell Students to be Creative

People are more creative when they are told to be creative. More creative solutions are generated when people are told to generate many possible solutions. There appears to be a bias, particularly among college students, toward producing a single solution unless they are explicitly told to produce multiple solutions. Thus, the first step professors can take is surprisingly simple. Ask for many solutions. Examples of ways to do this are:

"Develop some creative solutions for this problem."
"Give some different ways to interpret this problem statement."
"List twenty (or fifty) possible solutions to this problem."

Once a large number of possibilities have been generated, you can ask students to further develop two or three of these ideas. For example, in a design class the assignment could be to develop a folding cane. Students are asked to generate twenty different possibilities and then to do detailed designs for two of these ideas. Note that when doing this you need to accept ideas positively even if they probably would not work. The second part of this assignment asks students to do the necessary work and logical analysis to make the creative idea work. A second example which is applicable to any class is to require students to write homework or test questions with answers (Felder, 1987). This is a useful problem-solving exercise, and it becomes a useful creativity exercise also if students are told that grading will depend upon the novelty of their questions. However, remember that this will be quite time-consuming for students. A third exercise is to ask students to identify as many uses for a common object (e.g., a brick or a pencil) as possible (Christenson, 1988).

5.6.2. Creativity Techniques

Knowledge is necessary to be creative. An interesting question is, What knowledge? Prausnitz (1985) strongly makes the point that cross-fertilization of knowledge is required. Students should take a variety of courses in many areas and not overspecialize. This should include advanced-level courses in different disciplines. People who hunt or fish know that the edges are by the far the most productive areas. The edges between disciplines are often the most productive areas for creative ideas.

A variety of creativity techniques have been developed and can easily be adopted for use in engineering classes. Brainstorming was invented by Osborn (1948, 1963) and has since become so common that the term is now part of common usage. The technique is easy to use in class. The professor tells the class to think up any possible solution to a specific problem and say it out loud. One or two students who are selected as recorders write the ideas down. The one cardinal rule is there can be no judgment of any of the ideas during the brainstorming session. The professor acts as a facilitator to encourage students to generate more ideas and ensures that there is no criticism of ideas during this stage. Students are encouraged to build on the ideas of others. After the session, the ideas are collected and a list of possible solutions is generated. Then, in a separate session the ideas are criticized and possible workable solutions based on one or a combination of several of the ideas are chosen. In a design class different design teams can then be assigned to further develop these ideas. The principles of brainstorming are easily distilled.

1 Develop a lot of ideas.
2 Build on the ideas of others.
3 Make no criticism during the development phase.
4 Evaluate the ideas afterward.
5 Further develop promising ideas.

These principles can be applied to other creative exercises. For example, individuals can brainstorm by themselves. Groups can brainstorm in a conference telephone call or by electronic mail. In all cases the idea generation and evaluation stages must be separated; otherwise, the evaluation will inhibit idea generation.

Lateral thinking is an approach suggested by de Bono (1971, 1973, 1985). It involves restructuring patterns, changing viewpoints, jumping around, deliberately trying to change things, changing the problem statement, and avoiding logical (vertical thinking) analysis. Lateral thinking, unlike logical analysis, does not have to be sequential, does not have to be correct at each stage, does not have to use relevant information, and is not restricted to the problem as posed. Since lateral thinking is used only in the define and explore stages, it is completely checked by logical analysis in the later stages. Essentially, lateral thinking is more an attitude than a method.

A few examples will help illustrate. The same amount of money can be collected in tolls at less cost and with less disruption of traffic by closing half the toll booths. Stop for a minute and think about how this could be done. If all the toll booths going onto an island are closed, the toll can be doubled for cars leaving the island. This solves the problem.

A process called reversal can be illustrated with the following problem. The occupants of a new office building complained that the elevators were too slow and that the wait for elevators was too long. The straightforward logical solution was to speed up the elevators. What was the reversed problem? Reversal suggested slowing down the people. Mirrors were installed next to the elevators so that people could watch themselves (and others) while waiting for the elevators. Complaints plummeted afterward.

Dieting is a problem for many people. The straightforward solution is to tell people to eat less. A great deal of money has been spent on variations of this straightforward solution. What is the reversal solution? Go ahead and think up some ideas—none of them are wrong. One possible reversal solution is to tell people to eat as much of anything they want, whenever they want but with one simple rule. When they eat, that is all they can do. No television, no conversation, no thinking about problems, no radio, no music, no reading, and so forth. They eat, and while they eat they think about what they are eating (Smith, 1975). One of the authors (PCW) can attest that this wonder diet does work. It apparently works because the body gives a signal that it is full. When people do nothing but eat, they are much less likely to ignore this signal, and in addition, when on this diet there is little worry about going hungry later. However, the diet is not necessarily simple since it requires changing habits, but it is a different solution. Many of the heuristics, challenges to students and exercises discussed in the remainder of this section can be considered part of lateral thinking.

Writing can be a very useful method to get students to think about thinking and to think creatively in engineering (Hoerger and Bean, 1988; Elbow, 1986). Writing in a journal can be a useful method for getting students to think creatively, whether as freewriting or as fast exploratory writing where the student just writes without worrying about grammar or spelling. For example, he or she might write a page about uses for a screwdriver. One can also have students develop an idea map which can be considered a less sophisticated version of a concept map (see Figure 5-1). A journal is also useful for improving the writing skills of students.

Challenging students with creative games, questions, and exercises is a good way to increase their creativity (Felder, 1988). These do not have to be tied to engineering content. For example, have them brainstorm 100 possible uses for a brick. Ask them the meaning of word games such as what is "12safety34," or what is "milonelion"? There are also many mathematical exercises which require creativity to solve rapidly. For example, if there will be a single elimination tennis tournament with 360 players, how many matches need to be held to determine the winner. This can be done laboriously, but a rapid creative solution can be obtained. Since every player except one must lose a match, there must be 359 matches (Gardner, 1978). There are several books which specialize in this type of question and Gardner (1978) is a good source of both problems and references for additional problems. Open-ended creative questions can be invented by the instructor, and it is not necessary to have an answer. For example, Why do bridges freeze before the road surface? How could this be prevented economically? or, What is a good economical use for snow?

Heuristics were discussed extensively in Section 5.4, and many of them are useful for the generation of creative ideas. A few of many possible new heuristics developed specifically for creativity are listed below.

1 Have many ideas. The more ideas, the more likely that one will be good (Christenson, 1988).

2 Reverse the problem.

3 Build on a random stimulus (deBono, 1971). For example, pick a word at random from the dictionary and see if it leads to any possible solutions.

4 Think of something funny about the problem. Make a pun out of the problem. Humor and wit can often lead to original solutions.

5 Think of analogous solutions in nature to similar problems. This is a key part of the synectics approach to creativity (Gordon, 1961).

6 Develop word lists of stimulus words, properties, or key concepts (Staiger, 1984).

7 Use checklists or keywords to trigger different ways of looking at a problem. For example, the word creativity can be used (Sadowski, 1987):

C - combine
R - reverse
E - expand
A - alter
T - tinier
I - instead of
V - viewpoint change
I - in another sequence
T - to other uses
Y - yes! yes! (affirm new ideas)

Most engineers tend to be heavily left-brain-oriented. Their creativity can be enhanced by having them learn how to shut off the left brain and use the right brain. Following the pioneering work of Roger W. Sperry, it is now clear that the left hemisphere of the brain is mainly involved in verbal analytical thinking. The right hemisphere mainly processes visual and perceptual thinking, and its mode of processing involves intuition and leaps of insight. Clearly, engineering education is heavily left-hemisphere oriented.

People can learn how to consciously shut off the left brain and use the right brain. The subdominant right brain can be engaged by giving the whole brain a job which the left brain will refuse to do (Edwards, 1989). One way to practice the shift from left to right brain is to look at perceptual illusion drawings and consciously force yourself to see one part of the illusion and then another. Examples of this type of drawing are the vase that becomes two faces and the many drawings by M. C. Escher. A second exercise is to look at photographs of familiar faces, but with the photographs upside down. This exercise requires a shift in pattern recognition. A third exercise is to draw by using the right side of the brain without using words to name parts. Edwards' (1989) book has detailed exercises for learning how to do this. While doing these exercises you may want to quietly reassure the left brain that you will return to it shortly. In order to be able to shift at will to right-brain thinking, you must monitor the activity so that you know when the shift has occurred. Remember that the purpose of teaching engineering students how to shift to the right brain is to provide them with an alternative way of looking at things since this may produce creative ideas for solutions. Once the ideas are generated, the left brain takes over to prove or disprove the ideas. (A personal note: We find that our most creative ideas often come when we are tired. Apparently being tired relaxes the control of the left brain and the right brain has the chance to generate ideas. This can happen only if the problem has been thoroughly thought about previously.)

How do you incorporate creativity successfully into a class? Flowers (1987) has some useful suggestions based on his experience at MIT. One needs willing students, an enthusiastic instructor, "good" problems, and appropriate feedback. Most students are willing to try something new, and creativity is usually new. Instructors who voluntarily add creativity exercises to their courses will usually be enthusiastic. But picking good problems can be difficult: The instructor needs to know enough about the problem to know that it cries out for a creative solution, but he or she does not want to know the solution. (Instructors who know the solution have a very difficult time not teaching toward that solution.) Since pressure is real in the engineering profession, a project needs deadlines. For motivational purposes it is important to have successes. In the feedback, celebrate the successes which occur in the details. Such things as a clever mechanism, a trick circuit, and a clever coupling of processes need to be celebrated as creative accomplishments. It is these detailed ideas that most often delineate commercial successes since the development of a Xerox machine or the first introduction of a hand calculator occurs rarely. Flowers (1987) suggests individual exercises before group exercises since group exercises introduce a whole new area of group dynamics.

5.6.3. Acceptance of Ideas

A very important part of fostering creativity in students is to accept ideas and help students build on ideas. This acceptance is an inherent part of brainstorming. In working with students both on class projects and as a research advisor, a professor who accepts ideas will foster creativity. But acceptance does not mean stopping the search for more ideas; instead, it means ideas are not turned down.

There are many ways to accept ideas. One way is never to criticize an idea (Hueter, 1990). Instead, suggest to the student that he or she work on it and report back to you on the result. If the idea works, then all is fine and good. If the idea does not work, the student will learn from the evaluation process. Your hope is that in the future he or she will test ideas a little more carefully before talking to you. In either case the student will not be inhibited from generating new ideas.

A second method is to consciously use the PMI approach (deBono, 1985; Gleeson, 1980). First, note the *plus* (P) aspects of the idea. Then note the *minuses* (M) in the idea. Finally, note the *interesting* (I) aspects that can be built on. Encourage the student to build on the idea to retain the pluses while eliminating the minuses.

Building on ideas can be practiced in class. You can outline an interesting, creative idea for the class. Then assign students homework building on this idea. A second approach is to have small groups work on an idea and have each student in turn add to the idea. When this is done, the rules of brainstorming (no criticism) apply.

A third approach is to watch for creative solutions in homework assignments and tests (Felder, 1988). When one occurs, praise the student even if the final result is incorrect. Calling the student into your office and discussing the solution is one way to praise the student and to start building a relationship.

5.7. CHAPTER COMMENTS

This chapter could easily be turned into a book or two. We have tried to keep the information within the bounds of a chapter and at the same time to provide some concrete examples of what a professor can do to foster the creativity of students as well as to help improve their problem-solving skills. A large number of references are included for readers who want more information. If each professor spent five to ten minutes in class about once a week, we believe that students would become both better problem solvers and more creative engineers—certainly two goals worth striving for.

5.8. SUMMARY AND OBJECTIVES

After reading this chapter, you should be able to:

- Discuss and modify Figure 5-1 to fit your understanding of problem solving.
- Delineate the differences between novices and experts. Use these differences to outline how to teach novices to be better problem solvers.
- Discuss the steps in a problem-solving strategy (one different from the one discussed here can be used as a substitute) and use this strategy to help students solve problems.
- List and help students use some of the methods for getting unstuck.
- Develop a plan to incorporate both problem-solving and creativity exercises in an engineering course.
- Explain the three steps which can foster creativity and use some of the techniques.

HOMEWORK

1 Develop several five- to ten-minute problem-solving exercises for an undergraduate engineering course.

2 Develop several five- to ten-minute creativity exercises for an undergraduate engineering course.

3 List thirty open-ended questions which are appropriate for a specific engineering course.

4 For a specific engineering class set up some example problems in the format of the strategy you are using.

5 Write a script for a brainstorming session in an engineering class.

REFERENCES

Adams, J. L., *Conceptual Blockbusting*, Norton, New York, 1978.
Canelos, J., "The psychology of problem solving. What the research tells us," *Proceedings ASEE Annual Conference,* ASEE, Washington, DC, 2091, 1988.
Chorneyko, D. M., Christmas, R. J., Cosk, S., Dibbs, S. E., Hamielec, C. M., MacLeod, L. K., Moore, R. F., Norman, S. L., Stoankovich, R. J., Tyne, S. C., Wong, L. K., and Woods, D. R., "What is

problem solving?" *Chem. Eng. Educ.*, 132 (Summer 1979).

Christensen, J. J., "Reflections on teaching creativity," *Chem. Eng. Educ.*, 22, 170 (Fall 1988).

Cox, V. G., "An application of cognitive science to understanding problem solving activity for well structured problems: Cognition, algorithms, metacognition, and heuristics," *Proceedings ASEE/IEEE Frontiers in Education Conference*, IEEE, New York, 664, 1987.

Dansereau, D. F., "Technical learning strategies," *Proceedings ASEE/IEEE Frontiers in Education Conference*, IEEE, New York, 165, 1986.

de Bono, E., *Lateral Thinking: Creativity Step by Step*, Harper and Row, New York, 1973.

de Bono, E., "The virtues of zigzag thinking," *Chem. Technol.*, 10 (Jan. 1971).

de Bono, E., *DeBono's Thinking Course*, Facts on File, New York, 1985.

Edwards, B., *Drawing on the Right Side of the Brain*, rev. ed., Jeremy P. Tarcher, Los Angeles, 1989.

Elbow, P., *Embracing Contraries: Explorations in Learning and Teaching*, Oxford University Press, New York, chap. 3, 1986.

Felder, R. M., "Creativity in engineering education," *Chem. Eng. Educ.*, 22, 120 (Summer 1988).

Felder, R. M., "On creating creative engineers," *Eng. Educ.*, 222 (Jan. 1987).

Florman, S. C., *The Civilized Engineer*, St. Martin's Press, New York, 1987.

Flowers, W.C., "On engineering students' creativity and academics," *Proceedings ASEE Annual Conference*, ASEE, Washington, DC, 227, 1987.

Fogler, H. S., "The design of a course in problem solving," *AIChE Symp. Ser.*, 79 (228), 40 (1983).

Gardner, M., *aha! Insight*, Scientific American, W. H. Freeman, New York, 1978.

Gleeson, A. M., "Think PMI," *Chem. Eng.*, 131 (Sept. 8, 1980).

Goodson, C. E., "An approach to the development of abstract thinking," *Proceedings ASEE Annual Conference*, ASEE, Washington, DC, 187, 1981.

Gordon, W. J. J., *Synectics*, Harper and Row, New York, 1961.

Greenfield, L. B., "Teaching thinking through problem solving," in Stice, J. E. (ed.), *Developing Critical Thinking and Problem-Solving Abilities: New Directions for Teaching and Learning, No. 30*, Jossey-Bass, San Francisco, 5, 1987.

Hankins, G., "Expert systems and the learning process," *Proceedings ASEE/IEEE Frontiers in Education Conference*, IEEE, New York, 1972, 1986.

Hoerger, C. R. B. and Bean, J. C., "Design journals to improve thinking and writing skills in engineering," *Proceedings ASEE Annual Conference*, ASEE, Washington, DC, 1242, 1988.

Hueter, J. M., "Innovation and creativity: A critical linkage," *Proceedings ASEE Annual Conference*, ASEE, Washington, DC, 1634, 1990.

Kepner, C. H. and Tregoe, B. B., *The Rational Manager*, McGraw-Hill, New York, 1965.

Koen, B. V., *Definition of the Engineering Method*, ASEE, Washington, DC, 1985.

Koen, B. V., "Towards a definition of the engineering method," *Eng. Educ.*, 150 (Dec. 1984).

Kurfiss, J. G., *Critical Thinking: Theory, Research, Practice, and Possibilities*, ASHE-ERIC Higher Education Report No. 2, Washington, DC, Association for the Study of Higher Education, 1988.

Larkin, J., McDermott, J., Simon, D. P. and Simon, H. A., "Expert and Novice Performance in Solving Physics Problems," *Science*, 208, 1335 (June 20, 1980).

Leibold, B. G., Moreland, J. L. C., Ross, D. C., Butko, J. A., et al., "Problem-solving: A freshman experience," *Eng. Educ.*, 172 (Nov. 1976).

Lochhead, J. and Whimbey, A., "Teaching analytical reasoning through thinking aloud pair problem solving," in Stice, J. E. (ed.), *Developing Critical Thinking and Problem-Solving Abilities, New Directions for Teaching and Learning, No. 30*, Jossey-Bass, San Francisco, 73, 1987.

Lumsdaine, E. and Lumsdaine, M., "Full implementation of a first-year course in creativity and problem solving," *Proceedings ASEE Annual Conference*, ASEE, Washington, DC, 1572, 1991.

Mayer, R. E., Thinking, Problem Solving, Cognition, W. H. Freeman, New York, 1992.

Mettes, C. T. C. W., Pilot, A., Roossink, H. J., and Kramers-Pals, H., "Teaching and learning problem solving in science: Part II: Learning problem solving in a thermodynamics course," *J. Chem. Educ.,* 58 (1), 51 (Jan. 1981).

Osborn, A. F., *Applied Imagination,* 3rd ed., Scribner, New York, 1963.

Osborn, A. F., Your Creative *Power,* Scribner, New York, 1948.

Plants, H. L., "Basic problem-solving skills," *Proceedings ASEE Annual Conference*, ASEE, Washington, DC, 210, 1986.

Polya, G., *How to Solve It,* Princeton University Press, Princeton, NJ, 1971.

Prausnitz, J. M., "Towards encouraging creativity in students," *Chem. Eng. Educ.,* 22 (Winter 1985).

Richardson, S. A. and Noble, R. D., "Anxiety: Another aspect of problem solving," *AIChE Symp. Ser.,* 79(228), 28 (1983).

Rubenstein, M., *Patterns of Problem Solving*, Prentice-Hall, Englewood Cliffs, NJ, 1975.

Rubenstein, M. F. and Firstenberg, I. R., "Tools for thinking," in Stice, J. E., (ed.), *Developing Critical Thinking and Problem-Solving Abilities, New Directions for Teaching and Learning, No. 30*, Jossey-Bass, San Francisco, 23, 1987.

Sadowski, M.A., "Thinking creatively," *Proceedings ASEE/IEEE Frontiers in Education Conference*, IEEE, New York, 332, 1987.

Scarl, D., *How to Solve Problems. For Success in Freshman Physics, Engineering, and Beyond*, 2nd ed., Dosoris Press, Glen Cove, NY, 1990.

Schoenfield, A., M*athematical Problem Solving*, Academic Press, New York, 1985.

Simon, H. A., "Information processing models of cognition," *Ann. Rev. Psychol.*, 30, 363 (1979).

Smith, A., *Powers of Mind*, Random House, New York, 1975.

Smith, K. A., "Artificial intelligence and learning how to learn," *Proceedings ASEE/IEEE Frontiers in Education Conference*, IEEE, New York, 177, 1986.

Smith, K.A., "Educational engineering: Heuristics for improving learning effectiveness and efficiency," *Eng. Educ.* , 274 (Feb. 1987).

Staiger, E. H., "Probing more of the mind," *Eng. Educ.*, 65 (April 1984).

Starfield, A. M., Smith, K. A., and Bleloch, A. L., *How to Model It: Problem Solving for the Computer Age*, McGraw-Hill, New York, 1990.

Stice, J. E., "Learning how to think: Being earnest is important, but it's not enough," p. 93, in Stice, J. E., (ed.), *Developing Critical Thinking and Problem-Solving Abilities, New Directions for Teaching and Learning, No. 30*, Jossey-Bass, San Francisco, (1987).

Stice, J. E., "Further reflections: Useful Resources," p. 101, in Stice, J. E., (ed.), *Developing Critical Thinking and Problem-Solving Abilities, New Directions for Teaching and Learning, No. 30*, Jossey-Bass, San Francisco, (1987).

Wales, C. E., Nardi, A. H., and Stager, R. A., *Professional Decision-Making*, Center for Guided Design, West Virginia University, Morgantown, WV, 1986.

Wales, C. F. and Stager, R. A., *Guided Design*, Center for Guided Design, University of West Virginia, Morgantown, WV, 1977.

Wankat, P. C., "How to overcome energy barriers in problem solving," Chem. Eng. Ed*uc..,* 121 *(Oct. 18, 1982).*

Wankat, P. C., "Analysis of student mistakes, and improvement of problem solving on McCabe-Thiele binary distillation tests," *AIChE Symp. Ser.,* 79, (228), 33 (1983).

Whimbey, A. and Lochhead, J., Problem Solving and Comprehension: A *Short Course in Analytical Reasoning,* 3rd ed., Franklin Institute Press, Philadelphia, 1982.

Woods, D. R., "On teaching problem solving. Part II: The challenges," *Chem. Eng. Educ.*, 140 (Summer

1977).

Woods, D. R., Crowe, C. M., Hoffman, T. W., and Wright, J. D., "Major challenges to teaching problem-solving skills," *Eng. Educ.*, 277 (Dec. 1979).

Woods, D. R., "Problem solving and chemical engineering," AIChE Symp. Ser. *79* (228) 22 (1983).

Woods, D. R., "How might I teach problem solving?," in Stice, J. E. (ed.), *Developing Critical Thinking and Problem-Solving Abilities, New Directions for Teaching and Learning, No. 30,* Jossey-Bass, San Francisco, 55, 1987.

Woods, D. R., Crowe, C. M. Taylor, P. T., and Wood, P. E., "The MPS Program," *Proceedings ASEE Annual Conference*, ASEE, Washington, DC, 1984.

Woods, D. R., Wright, J. D., Hoffman, T. W., Swartman, R. K., and Doig, I. D., "Teaching problem solving skills," *Eng. Educ.*, 238 (Dec. 1975).

Yokomoto, C. F., "Writing homework assignments to evoke intellectual processes," *Proceedings ASEE Annual Conference*, ASEE, Washington, DC, 579, 1988.

Yokomoto, C. F. and Ware, R., "Facilitating student awareness of the complexity of expertise," *Proceedings ASEE Annual Conference*, ASEE, Washington, DC, 1962, 1990.

CHAPTER 6

LECTURES

The most common form of teaching in engineering classes in the United States is undoubtedly lecturing, and for many professors lecturing is synonymous with teaching. Lecturing can be an effective, efficient, and satisfying method for both professors and students. Yet many lectures do not satisfy learning principles and are not conducive to student learning. Despite common misuse of the lecture method, a perusal of *Engineering Education*, of *ASEE Prism*, or of the ASEE annual conference program shows few articles or presentations on lecturing.

One of the fundamental principles of engineering is to attack the critical problem which can make the most difference. In engineering education, improving lecturing is arguably the critical problem; and the first focus on improving engineering education at all schools should be on improving lecturing.

We will first consider the advantages and disadvantages of lectures, and then methods for improving content, organization, performance aspects, and interpersonal rapport in lectures. Next, special lecture techniques and special problems for large classes will be explored. Finally, the lecture will be looked at as one component of an entire course.

6.1. ADVANTAGES AND DISADVANTAGES OF LECTURES

Lecturing is a two-sided coin. An aspect of lecturing which is advantageous for an excellent lecturer can be a disadvantage for a poor lecturer. However, practically every disadvantage can be overcome if the professor makes an effort to overcome the problems. The following advantages and disadvantages are gleaned from our experience and from Alexander and Davis (1977), Cashin (1985), Eble (1988), and Johnson (1988). Some of the advantages of lectures include the following:

Audience focus. The lecturer can be aware and responsive to a specific audience so that each student feels that he or she is being talked to as an individual.

Versatile and flexible. There are many variants of lectures, and other teaching methods can be included within the lecture format.

Easily updated. Unlike some other teaching methods, changing lectures is easy and inexpensive. Material which is not otherwise available can easily be included in a lecture.

Low technology. Little can go wrong other than the lecturer becoming ill.

Acceptable and familiar. Some students like lecture because it is usually nonthreatening and they can hide in the multitudes.

Can incorporate learning principles. Those learning principles which are not incorporated into the lecture itself can easily be included in the entire course package.

Live contact. Rapport and immediate feedback to the student are possible.

Professor-efficient. Preparation time can be kept within reasonable limits.

Time-efficient. Can be presented to a large number of students which is an efficient use of the professor's time.

Instructor control. Many professors prefer a teaching style which allows them to have direct control rather than the semichaos which can occur in discussion or self-paced courses.

Anyone can lecture. All new professors have taken lecture classes, and they can copy the procedures. The special knowledge required to lecture is low.

Potentially outstanding for motivation and for conveying information. A professor can convey the interest and enthusiasm that he or she has in the topic. Information can be presented and rearranged in a variety of ways to help students learn.

Lecturing can be **exhilarating** for the professor.

Student learning can be high. If clear objectives are given to students and good support materials are available, research shows that student learning in a lecture course as measured by a content examination is equal to that of other teaching methods (Taveggia and Hedley, 1972). This result refers to the knowledge, comprehension, and application levels of Bloom's taxonomy.

Although anyone can lecture after a fashion, becoming an outstanding lecturer is difficult. If a professor does not know how to appropriately adjust lectures, then each of the advantages listed previously can become disadvantages:

Audience ignored. Poor lecturers push on despite the pain and suffering which is obvious to all but the lecturer.

Inappropriate lecture form may be used. Many professors are unfamiliar with the many variants of lectures and try to force-fit one form onto all circumstances.

Stagnation. Although lectures are easy to change and up-date every semester, many professors don't bother. This is obviously a teacher problem and not the fault of the technique.

Murphy's law. Although less technologically dependent than some other techniques, things can go wrong during a lecture. For instance, the bulb in the overhead projector can burn out or the microphone can malfunction—almost always at the most inopportune moment.

Passivity. Like stagnation of material, acceptability of the method may lead the professor to ignore looking for ways to improve.

Few learning principles may be satisfied. This is often the case in lectures with lots of content and little professor-student interaction. The worst problem is usually the passivity of students in lectures unless special efforts are made to keep them active.

Boredom. A "live" presentation where the professor is boring, speaks in a monotone, makes no eye contact, pays no attention to the students, receives no student feedback, gives no feedback to the students, and is impersonal is "dead."

Inadequate preparation or overpreparation. Inexperienced professors often spend too much time preparing for lectures, and experienced professors who no longer care may not prepare. One of the problems of lecturing is that there is no mechanism which forces adequate preparation.

False economy. The economic efficiency of large lectures is abused by many universities. Student learning of higher-level cognitive functions would be significantly enhanced in smaller classes with more interactions.

Lack of individualization. Since the instructor controls the pace, it will necessarily be too fast for some students and too slow for others.

Anyone can lecture? Unfortunately, the apparent ease of lecturing hides the fact that lecturing is one of the hardest teaching methods to truly master. In addition, what many professors have seen and are cloning are inferior lecture classes.

When it's bad, it's really.... Although they can be outstanding, lectures can also be absymally bad. In addition, although lecturing is a good teaching method for conveying information, it is not as well suited for some higher-level cognitive tasks such as analysis, synthesis, evaluation, and problem solving.

Extremely stressful. Lecturing can be an emotional trial for some professors. In extreme cases these professors need to find alternate teaching methods which are less stressful for them.

Lack of supporting material. If clear objectives are not given to students and good supporting material is not available, then student learning will be less than with an alternate teaching method which provides these.

Probably more than any other teaching technique, lecturing is teacher-dependent. In short, lectures represent the best and the worst of teaching.

6.2. CONTENT SELECTION AND ORGANIZATION

What content should be included in the lecture and how should it be organized? The experts (such as Davis and Alexander, 1977; Eble, 1988; Lowman, 1985; and McKeachie, 1986) are in surprising agreement about both content and organization. The lecturer should never try to cover everything in the lectures—a major mistake made by inexperienced professors. Remember that students are supposed to spend two or three hours outside of class on homework and readings for every hour in class. Leave a major responsibility to the students. Thus, it is necessary to be selective.

How does the professor decide what to select from the wealth of information and procedures which could be presented? The following ideas can be used to guide the selection of lecture material.

1 Cover key points and general themes. This serves to guide the students' reading and helps them build mental structures. These areas should be reflected in the course objectives and serve to reinforce the importance of these objectives.

2 Lecture on items that students find to be very interesting. Since lecturing is part performance, you might as well give yourself the advantage of choosing topics that students find particularly interesting.

3 Pick especially difficult topics or those that are poorly explained in the textbook. Tell the students that you will focus on these more difficult topics so that they will be able to do the homework better.

4 Discuss important material not covered elsewhere. Particularly in graduate-level courses, important new findings can be included in lectures long before they make it into the textbooks. "The lecture is the newspaper or journal of teaching; it, more than any other teaching, must be up-to-date" (McKeachie, 1986).

5 Include many examples. Students, particularly sensing students, love and need examples. Examples should include problems with numerical solutions and a modest number of short "war stories."

6 Choose material at the appropriate levels of depth and simplicity. Unfortunately, this is easier to say than to do when one has never taught the course before. Once you have taught the course, you can reduce the lecture coverage in areas where most students do well on tests and increase it in areas where students have difficulty.

Once the content has been chosen, you need to put some thought into the mode of presentation. Remember that everyone can use auditory, kinesthetic, and visual modes, and that the more modes employed, the more is retained. Unless special attention is paid to including other modes, the vast majority of lecturing will be in the auditory mode (words written on the blackboard are in the auditory mode). Yet most people prefer the visual mode. In arranging the content, include pictures, drawings, graphs, slides, computer visuals, and so forth. This may require some variation in the content and organization of the lecture.

What content areas can be left to readings and homework? Any content which experience shows students have little trouble with can be left out of lectures. If the textbook does an admirable job of covering particular areas, there is no reason to include this material in the lecture. When fine detail is required to solve problems, it is appropriate to outline the general procedure in the lecture, but leave the details to the textbook. Often, presenting a detailed example is the best way to show students how to do these problems. Whenever material is left out of a lecture, be sure that the students are explicitly told whether they are accountable for it. Clearly written objectives are helpful to ensure that students learn what they are supposed to learn.

A relatively simple organization is often best. Start with an attention-grabbing opener such as a question, a problem, a unique statement of fact, or a paradox. Then provide the students with advance organization: In other words, tell them what you are going to tell them. It is helpful if this advance organization ties into the last lecture. The main body of the lecture presents the content. In the main body you tell them the information. To finish the lecture, summarize or tell them what you told them. It is helpful to briefly mention what will be covered in the next lecture.

The bulk of the class period is spent on the main body of the lecture. One can organize the main body in a linear, logical fashion. This type of organization is appreciated by the sensing students in the audience and does not prevent the intuitive students from learning the material. A nonlinear, intuitive approach can also be effective, especially for upper-division classes, but is likely to confuse many students at lower levels. It may also be appropriate to present two or three topics simultaneously and to contrast and compare them. For example, transport phenomena can be presented in this form. Students need a hierarchical structure of knowledge, but they learn material best when they do some of the organizing. The result of this is that "a high degree of organization does not seem to contribute to student learning" (McKeachie, 1986). When students are seeing the material for the first time, use an inductive approach. Start with specific, concrete examples that are fairly simple. Use analogies if you know that the students understand the analogous theory. This can lead to much more rapid student comprehension (Meador, 1991). Then lead slowly into general principles. For students who have studied the material previously, a deductive approach can work well. Even in graduate level classes an inductive approach is appropriate if the material is new

The main body should be organized in parts which are clearly delineated. For example, in a lecture using an inductive approach, the first part could introduce the topic with a simple example, the second part could consider a more complex example, and the third part could discuss the general principles. Each part should be ten to fifteen minutes long, and certainly no longer than twenty minutes. Between parts do something else such as ask questions or have a discussion to give the students a short break and make them active. This is necessary because most students have a twenty- to thirty-minute attention span.

In planning the lecture think about the way students learn. If the scientific learning cycle (see Section 15.1) can be incorporated in some of your lectures, many students will benefit. If you consider that your lecture is part of Kolb's learning cycle (see Section 15.3), then the appropriate activities for periods when you aren't talking and appropriate homework activities will be clear.

6.3. PERFORMANCE

All lectures are performances. Poor performances lead to poor lectures regardless of the content. Master performances can lead to outstanding lectures if the content and interpersonal rapport are also masterful. The good news is that professors who are content with being "competent" do not have to "perform." Professors who want to become master teachers do need to develop skills in the performance aspects of lecturing, which are discussed by Cashin (1985), Davis and Alexander (1977), Eble (1988), Engin and Engin (1977), Lowman (1985), and McKeachie (1986), among others. Since,

Preparation + presentation = performance

we will discuss the preparation and presentation of lectures.

6.3.1. Preparation for the Performance

Actors and actresses start with a script and rehearse. Since a lecture is a play starring one actor or actress, professors also need to prepare for the performance aspects in addition to preparing the content. The main part of the script is the professor's lecture notes. These notes outline the content in a form that the professor finds useful for live presentation. The lecture notes of good lecturers vary from three or four lines on a single index card to a completely written-out speech of several pages. Experiment with different forms of lecture notes to find what works for you. Lecture notes should include specific examples, visuals, and questions to ask students.

One of the paradoxes of lecturing is that the teacher needs to be thoroughly prepared yet appear spontaneous. Underpreparation can lead to fumbling which is obvious to the students. Overpreparation can result in a rigidity that forces the professor to try to cover all topics in a prearranged order despite numerous signs from the audience that the lecture is not going well. Lectures need built-in flexibility so that the performer can adjust to the audience.

Just as playwrights put stage directions in their plays, professors need to include stage directions in their lecture notes. These include announcements and reminders to pass out handouts or to collect homework. Stage directions can also indicate pauses, where to ask questions, and breaks in the lecture for student activities. Alternative paths to provide flexibility can be included in the stage directions. Finally, stage directions can remind the professor to make any last-minute announcements (e.g., "Remember that the project progress report is due next period") at the end of the period. Stage directions are one way that the professor can help to ensure that the lecture is successful.

There is seldom enough time in a professor's schedule for a complete dress rehearsal for every lecture; however, there is time to do some rehearsing ahead of time. Obviously, reviewing and updating lecture notes shortly before the lecture are part of the rehearsal. So is a five-to ten-minute mental preparation immediately before the lecture. If the class is in another building, this preparation can be done while walking over. Review the major points and "psych" yourself up for the lecture. One sign of a professional is the ability to be enthusiastic and interesting for the lecture hour even when the topic is not a particularly interesting one.

Arrange to arrive early at the stage door (the classroom). This allows time to check out the stage. Rearrange seats, clean the blackboard, check the bulb in the overhead projector, and get ready for the class. If the room is too small, too hot, or too cold, complain to the proper authorities. Eventually something may be done to improve classroom conditions. Teaching is often a low-budget production, and the professor must also be the stagehand.

In show business there are always warm-up acts before the main act. Professors can help warm up the audience also. One useful procedure is to write a brief outline of the class in one corner of the blackboard. This will help students start to think about the class and become mentally prepared to focus on the material. The outline helps satisfy the learning principle of guiding the learner (see Section 1.4). Surprisingly, a handwritten outline is more effective than a typed outline distributed to the class, perhaps because students are more active in processing the information (McKeachie, 1986). A second useful activity is to talk to students. Many students will talk to the professor before or after class but would never dream of coming in for office hours. The professor can be proactive and seek out students instead of waiting for them to come to him or her. Just being in class early sends the message that you are interested in and excited about the class. This interest and excitement can be contagious.

6.3.2. Presentation Skills for Lectures

When a play starts, the house lights dim, the curtain opens, and the audience leans forward attentively. A formal start to a class can focus the students' attention. Professors who use an overhead projector can dim the room lights and turn on the machine. This might be a useful start even if the overhead is used only to start the class with one transparency. Another possibility is to step out of the room to get a drink of water and then make a grand entrance to start the lecture. Some professors start writing on the board a minute or so before the class starts and then signal the class it is time to start by putting the chalk down and turning toward the class. One professor we know takes off his suit coat when it is time to start (and puts it back on to signal when the class is over).

This attention to small items such as how the class starts may seem like nitpicking, but it is this attention to detail that can make the difference between a great and an average performance. Also, not all the changes need to be made simultaneously. Institute a few changes every semester and slowly become more comfortable with performing in class.

Many plays start with an attention-grabbing ploy, such as a murder or dead body. Although killing one's students is not considered good form, professors need to capture attention quickly. Some methods that other professors use include:

1 Start with an appropriate comic strip on the overhead projector.
2 Start by saying, "I want to talk about next period's test."
3 Start with an appropriate newspaper headline such as, "Engineer gives million to university to improve undergraduate teaching."
4 Show a photograph of a disaster appropriate to the class. Examples include the collapse of a bridge, a fire at a chemical plant, and a plane crash caused by the failure of a part.

If you occasionally change the type of grabber, the students will wonder what you will do next and this increases their attention.

Once you have the students' attention, you need to retain it while the lecture proceeds. Variety is the key. Change the tone, pace, volume, pitch, inflection, and expressiveness of your voice. A flat, unvarying monotone puts students to sleep, and sleeping students cannot be learning. Variety is also needed in gestures and in the format of the lecture. Even some variety in where you stand and how you interact with the students can be helpful.

A professor's voice is indispensable in lecturing. Professors who want to improve their speaking skills need to analyze their voices and then work on any problem areas (Lowman, 1985). Listen to excellent speakers such as television newscasters and try to develop a feel for expressiveness, diction, and pace. Then, take the terrifying step of recording and analyzing your speech. Since we hear our own speech in a very different way than we hear the speech of others, no one likes to hear a tape recording of their voice. Listen for particular problem areas such as repeated verbalizations, such as "uh" and "OK", or a strident tone. Repeated words can be reduced once we become aware of the problem. Strident tones can be eliminated by focusing on breathing deeper. Improper articulation is a common problem which makes it difficult for students from different sections of the country to understand a speaker. This problem may be so much a part of the professor's speech pattern that he or she does not notice it even when listening to a tape. Thus it is useful to have someone point out these problems to you in a friendly way. Articulation can be improved by practicing reading aloud (find a small child to practice on).

Another common problem of college professors is failure to project their voices. A good rule of thumb to remember is that you should be speaking to the row behind the last one in the room. But projection is more than merely speaking louder—a practice which usually just wears out the voice. True projection begins with proper diaphragmatic breathing which gives a base for the sound, and then follows with full articulation of the sounds: crisp consonants and full and liquid vowels. Like walking, speaking is too often taken for granted; but improvement in speech, just as in posture, step, and stride, can do wonders for one's personal as well as professional health. Self-help is valuable, but guaranteed improvement is best sought from a professional. If you are serious about improving your speaking voice, consult a professional voice coach (any university with a speech, audiology, or theater department has such an individual).

Beyond speaking the words, the manner in which the lecture is presented is also important. Should you read it verbatim? Use three-by-five cards? Rely on your memory? It is very difficult for people who are untrained to read a lecture effectively. And a lecture can be significantly improved if it is presented spontaneously. As a professor, you have enough command of your material so that notes or topical outlines will suffice to keep you on track. Perhaps the only thing worse than reading a lecture to students is to read the textbook to them. This is guaranteed to earn the professor poor student ratings.

Variety in mannerisms is just as important as variety in speech. Your gestures are also an important aspect of how you communicate, but they must appear natural and not be either wooden or flailing. Most importantly, they must be purposeful, such as those that indicate size, shape, emphasis, and so on; nervous jabs that are out of synch with the message are nonpurposeful and distract the audience. One very effective but underused gesture is to walk into the audience. This gets the students' attention, allows you to make contact with those in the back of a large lecture hall, and provides variety to your lecture. Since the lecture is a performance, you can preplan effective gestures like this. Also practice walking toward the back of the classroom during a class when the lecture is dragging and something needs to be done to liven it up. Once you have tried an activity a few times, you will have added something new to your repertoire.

Even the barest stage has props. Professors have a table, podium, blackboard, and overhead projector, plus whatever props they bring with them to the lecture. Props can also be used purely for dramatic appeal. Some professors bring in a glass of water and then drink the water while taking a break between two important topics. Props can also be objects brought in for educational purposes. A valve, a circuit board, a new alloy, packing for a distillation column, or different types of crushed rock can all be an informative part of the lecture. Classroom demonstrations during lecture can provide a concrete learning experience and the chance for discovery. The availability of new projection equipment has made it easier for all students to observe demonstrations, and more sophisticated equipment increases student interest (Dareing and Smith, 1991). Demonstrations do require setup time and a practice run before class. These props have a greater impact beyond their educational value alone: They also provide variety and a chance for both visual and kinesthetic learning.

The most important props in most classrooms are the blackboard and the overhead projector. Though commonplace and easily taken for granted, both need to be used effectively. Both tools can be used most effectively (1) as an external memory aid, (2) for emphasis, and (3) for visuals. When the outline is written in one corner of the board or on a transparency, it can be referred to during the lecture to show the students where they have been and where they are going. Thus the blackboard or overhead retains the information and serves as memory. The blackboard can also retain an item that you later want to compare and contrast with another item. Whatever is written on the blackboard or overhead is emphasized, and most students will attempt to copy the material. However, while doing this they may miss what you are saying, so putting too much on the blackboard or overhead is counterproductive. If you have some artistic skill, then the blackboard can serve for visual presentations. But even without such skill, you can show graphs and simple schematic diagrams on the blackboard. For more complex figures, transparencies can be made in advance, and students can be given copies of the figures.

Neither the blackboard nor the overhead projector is the best way to present large quantities of detailed information. Students may spend all their time trying to copy the material. In addition to not listening to the lecture, they invariably make mistakes in copying equations or complex diagrams. The situation is often aggravated when predrawn transparencies are shown in rapid succession. If the content requires that you cover a large number of equations or complex diagrams, hand out partially prepared lecture notes that contain the equations and diagrams and have space for student lecture notes. This greatly reduces students' errors in transmission of information and allows you to lecture somewhat faster. An alternate solution is to change the content selected for presentation. If the goal is to produce engineers who can do abstract mathematical proofs, then the lectures, homework, and tests are rightly focused on this activity. If the goal of the course is to have students become good problem solvers, then it makes more sense to spend time solving problems during the lecture.

The biggest difficulty in using a blackboard is the loss of eye contact while writing on the board. This is less of a problem with overhead projectors, but the lecturer must occasionally glance at the screen to check the message the students are seeing. Blocking the view of the students may also be a problem with both the blackboard and the overhead projector. In addition, most professors lecture too fast when using overheads. One advantage of the blackboard is that material can be left on some portion of the board so that students can go back and copy something they have missed. Overhead projectors can also retain information if the the classroom is equipped with two projectors and two screens. Once one transparency is finished, it should be transferred to the back-up projector. We suggest that new professors try both overhead projectors and blackboards. First, obtain student feedback on what can be done to improve both procedures. Then, select one method to focus on and become an expert with this technique.

Eble (1988) states that the skillful lecturers he observed, "were above all keenly aware of and responsive to their audiences." Remember that a lecture is a live performance. Watch and read the audience. Are they generally engaged with the material or is their attention wandering? If they are showing signs of boredom, what can you do to shift gears? If someone is clearly confused, try asking if you can help (see Section 6.4.2). The audience provides feedback by both their verbal and nonverbal behavior. On rare occasions the message you have from the students is that everyone is focused on you and you have the class in the palm of your hand. Enjoy the moment and try to remember what you did or what the magic content was so that you can do it again.

When something starts to go wrong, the trick is to observe and respond to the problem quickly. After many failures, we finally realized that continuing the lecture and perhaps talking louder does not work. Perhaps you have overstayed the twenty- to thirty-minute attention span of the students and it is time to go to a group activity or have a question-and-answer session. Clearly shift gears and do something which forces the students to engage the material actively. Consider doing one of the following:

- Ask for student questions.
- Switch to a socratic approach and ask the students questions.
- Ask the students to summarize the most important point in the lecture on a piece of paper.
- Give a short quiz (see Section 6.6.1).
- Do a group activity (see Chapter 7).

After about five minutes of this activity you will be able to switch back to lecturing with a renewed student attention span.

Responding properly to signs of audience problems and preventing such problems before they occur require timing. Timing is an art, but it can be learned. If you are good at telling jokes, then you have a sense of timing which can be used in your lecturing. Essentially, having good timing means knowing the appropriate time to do something. In a lecture it is sometimes appropriate to stop when a student has a question, and it is sometimes appropriate to ask the student to wait until you can come back to that student later. Sometimes the lecturer needs to speed up, sometimes to slow down, and sometimes to pause. When a student becomes a bit aggressive and hostile, sometimes it is appropriate to hash out the problem in class, and other times it is better to do it privately. All of these instances are examples of timing. Good lecturers and good actors develop a sense of timing with experience. It helps to pay attention to what works and record what doesn't work so that next time the timing can be improved.

Humor can also be part of the professor's repertoire in working with the audience. Cultivate your own sense of style. If you can successfully tell "canned jokes," then use them to start the class or break the routine. If you can't tell a joke, don't. Many professors successfully use comic strips on transparencies to start a class; however, the strip should be appropriate and in good taste. Some professors' style of humor is spur of the moment and based on things that happen in the class. Again, if you can do this successfully, it can help keep the attention of the class. If you can't, don't. Finally, avoid overkill.

A final note about performance: Some people have a flair for being dramatic. A little drama can help keep the class interested. There is an inherent drama and majesty in the ability of theory to predict and occasionally to totally miss the behavior of the real world. Build up to the conclusion and at times slip in an unexpected conclusion. A bit of challenge in the class can be fun for the students, particularly if it is nonthreatening. Ask dramatic questions or make dramatic statements. For example,

- What did X do that made him one of the most revered engineers of his era?
- Was the suicide of Professor Y justified?
- There is one pearl of wisdom in this class which will make you rich and famous if you follow it. Your challenge is to find this pearl.
- In today's class we will discuss the most misunderstood phenomena in electricity and magnetism.

A sense of timing is needed to let the drama build. Do not answer your question or explain the statement immediately. Let the students search and try to puzzle out the answer. Student learning will be much deeper if they can determine the answer for themselves, even if they beat your telling them by only a minute.

6.4. QUESTIONS

Answering and asking questions is an art in itself. Questions offer an opportunity to work on the content and develop rapport. Students asking or answering questions are active and thus are satisfying one of the learning principles discussed in Section 1.4. Questions also serve as a break in the lecture and allow some students a chance to catch up in their note taking. Finally, the instructor's availability to answer questions is one of the factors that students implicitly include in their overall ratings of instructors (see Section 16.3.2).

6.4.1. Answering Student Questions

We strongly encourage students to ask questions in class. If many students are confused, the professor can clarify the issues for them simultaneously. Thus, during the first class period we make it clear that we want students to interrupt the lecture with questions. Some professors prefer to control student questions and have students ask only at specified times. Pause fairly frequently during the lecture and ask if there are any questions. Then, give the students time to pose an intelligent question. The appropriate length of the pause requires a sense of timing.

When a student asks a question, accept it positively and then rephrase it so that the student can be sure that you understand the question and so that the rest of the class can hear it. Examples of positive reinforcement for asking questions include:

- Good question.
- That's very insightful of you, Karen.
- Bob, you're following me exactly because that's my next topic.
- Good, I was waiting for someone to ask about that.

Restating the student question can be a challenge. When students are extremely confused, they have difficulty even phrasing an intelligent question. Asking a question under these circumstances is an act of bravery (which is one reason the student should receive a positive response). Make your best guess as to what the question is, even to the point of asking the student if that form is reasonably close to what he or she wants to know.

Various responses to the question are now possible. Since students usually prefer either a brief direct answer or an involved direct answer, it's best to give direct answers most of the time. If the question opens up a new topic which will be covered in a few minutes, ask the student to wait, and if not satisfied in a few minutes to ask again. When we use this technique we try to remember to ask the student later if the question has now been answered. The student can be referred to the book; however, this works best if the answer can be found in the book during the lecture and the question is answered immediately. Otherwise, "Look it up in the book," comes across as a very negative reaction to a student's question. The question can be posed to the class to determine an answer. This works well in classes where discussion is commonplace. If the question is quite involved or the student clearly does not understand your answer, ask him or her to see you after class. This is often appropriate when the student wants to see the complete solution to a problem and time is not available to do this. Another response is to ask another question to try to lead the student to the correct response to the original question. Unfortunately, this approach tends to inhibit student questioning since it puts the student on the spot. Finally, if you do not know the answer, the safest response is, "I don't know, but I'll find out." This instructor honesty helps to increase rapport with the students.

6.4.2. Asking Students Questions

Asking questions is a rather different skill than answering questions. There are several advantages to asking questions during class (Hyman, 1982). Questions can provide as a break in the lecture which helps to keep the students active. Questions also provide feedback to the professor and students about what material is being understood. Questions provide the professor with an alternate way to emphasize particular points, clarify difficult concepts, and review material. Rhetorical questions are often useful for this purpose or for highlighting key questions. Questions can be used as examples of possible test or homework questions. They can also be used to start a discussion or to encourage student questions ("If you don't have questions for me, then I'll have some for you"). Questions can be used to help maintain discipline or keep students awake. Some professors structure their entire teaching style around questions and use a socratic style instead of lecturing.

If you often ask rhetorical questions, then some sort of signal is needed that the question is for the students. For instance, "Now I have a few questions for you." Even if you never use rhetorical questions, it's useful to let the class know that you are going to shift gears away from lecturing. "Let's take a break from the lecture and try some questions."

Students and new professors often believe that the questions asked by the professor must be spontaneous. A few are, but most are preplanned. Posing a good, clear question which requires some thought to answer but is not beyond the ability of the students requires some time and effort to prepare. Prepare ahead of time and write these questions in your lecture notes. If a good question arises spontaneously, try it and record it in your notes after class.

What are the elements of a good question? It should be relatively short, clear, and unambiguous. Only one question should be included; that is, do not run a string of questions together. If you want to ask a string of related questions, then ask one at a time and get a response before proceeding. Otherwise, you are likely to confuse the students (Hyman, 1982). The question can be at any level of Bloom's taxonomy, and if you want students to become proficient at all levels, then you must ask questions at all levels. In some cases you may want to write an equation or draw a figure on the blackboard or on a transparency to frame the question.

In engineering it is appropriate to ask questions which require a modest amount of algebraic manipulation or numerical calculation. Tell the class to take out a piece of paper and a calculator. Then write key elements of the question on the blackboard or use a preprepared overhead transparency. Students can work individually or in groups. Questions can range from very simple single-answer questions, such as unit conversions, to unusual situations where basic principles can be used to obtain an answer to open-ended questions. Wales et al. (1988) give some examples.

Usually, it is best to ask the question of the class as a group and then pause. When a question is asked of the class as a group, no one knows who will answer it and most students will try to develop an answer. If you are using the question to help keep a student awake or to control a disruptive student, then you might want to preface it with the student's name. Even students who are close to falling asleep will respond to hearing their name. After asking the question, pause. The pause is critical and for most teachers is much too short. It takes time for students to formulate an answer.

There are a variety of ways to field students' responses to questions (Hyman, 1982). If the student's answer is correct, offer praise: "Excellent," or "You're absolutely right." This gives the student strong positive feedback and tells the rest of the class that the answer is correct. If several students are straining to answer, you can call on several without responding to each individual answer. Then respond in general to all of the responses. You can also build on a student's response. "You're correct about the fluid flow. But let's consider the mass transfer in more detail...." The continued detail can consist of explanation by the professor or additional questions.

What if the answer is wrong or partly wrong? For many professors the immediate reaction is a "Yes, but...." type of response. Unfortunately, this sends a negative message to the student. It is better to be more straightforward about those aspects which are wrong. Some possibilities include:

- You're right about aspect X but wrong about Y. Let's explore Y in more detail.
- I think that you have misinterpreted my question. Let's try it again. (Use this type of response only when you really believe the student has misinterpreted the question.)
- No, I don't think that you have the right idea on this.
- Explain how you developed that answer.
- How many students think this is correct? How many think this is incorrect? Why is it correct or incorrect? (These responses should be used occasionally for both correct and incorrect answers.)

Should you call on students who volunteer to answer, or should you call on all students at some time during the class? There are advantages and disadvantages to both options. Volunteers are likely to be more articulate and are more likely to have an answer. In addition, calling only on volunteers makes the class safer for the students, since they know they won't be called on when they don't volunteer. If you call on volunteers, spread out which of the volunteers is called on. Call on a student who seldom volunteers, when he or she finally does volunteer, to help him or her participate in the class. The disadvantages of calling only on volunteers are that some students will never participate in the class and that students who decide not to volunteer may not try to solve the problem independently.

Calling on students at random or with some prearranged rotation schedule keeps all the students "at risk." The professor can force more students to participate, but the anxiety level in the class is likely to increase. In addition, the percentage of wrong answers or "I don't know" answers will increase. If class participation and the ability to answer questions and present arguments in public are important in your class, then some type of strongly encouraged participation is needed. One modification used in law schools is to allow students to put a slip of paper with their name on it on the desk if they are not prepared to discuss that day's class.

Some professors use a modified socratic approach in their lecturing. Short periods of explanation of the material are interrupted by question periods. The professor calls on particular students and makes sure that everyone is called on at least every other period or so. This is most effective in medium or small classes (less than about fifty students). The professor can develop better rapport with the students if each student is called by name. In addition, professors who become adept at reading students' nonverbal clues can choose to call on students either when they are ready or when they are not ready to respond. This procedure does help most of the students improve their ability to think and respond under pressure, which is a useful ability for an engineer. Depending on the professor's attitude and style, this approach can be either moderately or very threatening to students.

There are gender differences in asking and answering questions (Tannen, 1990). Generally, men are more comfortable speaking in public, while women are more comfortable speaking in private. Thus, the men in the class are more likely to ask and answer questions regardless of how well they know the material. They are also more likely to be willing to challenge the professor. Professors are then faced with a value question. Should they let people keep the roles they have been socialized into or should they try to change them (the men, the women, or both)?

6.5. BUILDING INTERPERSONAL RAPPORT IN LECTURES

Interpersonal rapport is the second dimension in the model of good teaching shown in Table 1-1. Although large lecture classes are not the ideal vehicle for building rapport, a professor can do many things to increase rapport with his or her students.

6.5.1. Student Contact Before, During, and After Lecture

We have already discussed the importance of coming to the lecture hall early. The few minutes before class provide an excellent opportunity to make contact with students even if you and the students have to wait in the hall while the previous class finishes. Greet students by name. For example, "Hi Susan, how are you doing today?" or "John, did you get that problem we were talking about yesterday?" Early in the semester when you don't yet know every student's name, it is not impolite to walk up to a student and ask what her or his name is. Many students will come up before or after class and ask if they can ask a question. Responding in a friendly way using the student's name sends the message that you are friendly and you know who they are—"Yes Bob, what's the problem?" (Note: In our examples we use first names. Some professors are more comfortable and think it is more professional to be more formal and use the student's last name.)

Once the lecture starts there are a variety of ways to make contact with students. The most obvious and direct way is by eye contact. In some cultures it is considered impolite to look a person in the eyes when speaking, but in ours the opposite is true. Establishing eye contact with students not only lets them know that you're aware of their presence but also makes them feel that you are speaking to them and not just at them, whether in lecturing or asking and answering questions.

If a student has come up with a thoughtful question or a clever solution to a problem, share it with the class by naming the student: "Jennifer Watkins has come up with a very interesting paradox that I thought everyone would be interested in." This shows the student that you really did pay attention and thought that her idea was important. If student presentations are part of your class, you could also ask the student to present her paradox and the resolution to the class.

Recognizing student feelings during the lecture can help increase your rapport with students. For example,

"I know several of you are angry about the test. You felt that you could have done much better if you'd had more time. I agree that the test was a bit long. I'm working on getting more time for the next test.

"This point must be confusing. Can all of you who are *NOT* confused please raise your hand. Yes, I was right, many people are confused." (Note that a large percentage of students will not raise their hands regardless of the question. Thus, the professor can make most of the class look like they are on one side of the question or the other by changing the phrasing of the question.)

Part of the trick of developing personal rapport in class is presenting a part of yourself in the lecture. We mean doing this in a professional way, not talking about personal problems. Although it does not hurt for students to see you as a real person with a family and with real problems, this realization should come from activities outside the classroom. In class, share with the students your excitement and enthusiasm for the subject, "This is really great stuff." If you had difficulty learning a topic when you studied it for the first time, share this with the students also. If you will miss a class because of a professional society meeting, share with the class the importance of the meeting and what you expect to learn there.

Another prime time to talk to students is after the class is over. Stick around for a few minutes and answer students' questions in the classroom or in the hall. If a cluster of students is waiting to talk to you, turn first to the student who rarely says anything in class. This is the student who needs the most encouragement. If there are too many students waiting to talk to you after class, consider shortening the lecture by five minutes to allow more time for informal questions.

6.5.2. Other Methods of Increasing Rapport With Students

There are a number of ways that students can be become involved in a lecture class. One method is to have a group of volunteers who meet regularly with the professor to provide feedback (McKeachie, 1986). This is useful in very large lecture classes. The student volunteers can be told to talk regularly to other students and obtain feedback. Then when the feedback is presented to the professor, the volunteers quickly learn that they can be very blunt since they are merely reporting what someone else told them. If the professor is willing to make some adjustments, this procedure can help class rapport since the students can see that their feedback makes a difference and that the professor cares. Obviously, the professor has the opportunity to get to know the volunteers well. The entire class can also be asked to do a formative evaluation early in the semester, and the professor can respond to these comments (see Section 16.1.).

The professor can increase rapport by being sensitive to nuances in relationships with students. Clearly, the professor has more power in this relationship, but by using egalitarian language in making assignments, he or she can promote student independence (Lowman, 1985). For instance, instead of ordering students to do a homework assignment, the professor might say, "Those of you who do problems 6, 7, and 9 will find that they will help you in Friday's quiz." Sharing the course objectives with the students can also make assignments seem rational. Thus, the professor might explain the reason for an assignment, "This reading will help us reach our goal of being able to...."

When possible give the students some choice. Projects can be very effective, particularly for upper-division students, because they give students a choice as to what they do (see Section 11.4). In elective classes students can also be given a choice, within limits, of what material to cover (Wankat, 1981). If the examination date is not carved in stone, students can be allowed to vote on the date. These options give them some control over their studies and increase the likelihood that the professor can become a partner in learning.

Other special procedures can be used to help build rapport, particularly in large lecture classes. These are discussed in Section 6.6. A variety of one-to-one contacts outside the classroom will help build rapport (see Chapter 10). Other aspects of the entire course can fit together so that rapport with students is enhanced (see Section 6.8).

6.6. SPECIAL LECTURE METHODS

One reason why professors continue to lecture hundreds of years after the invention of the printing press is the flexibility of the lecture format. Lectures can be modified to include almost all the learning principles. In this section we will briefly discuss three of the many possible modifications.

6.6.1. Postlecture Quiz

Students need to pay attention, and they need feedback on what they have learned. Both of these principles can be included in a lecture format by regularly giving a short postlecture quiz. One way to do this is to give a quiz on technical content during the last ten minutes of class (Peck, 1979). Since the quiz covers material that has just been presented in the lecture, open book and open notes are preferable. Usually the quiz will consist of one short-answer problem which can be solved in a few minutes. The extra time is necessary since students who have just learned the material will be inefficient problem solvers. Of course these quizzes do not replace the need for longer problems in homework assignments and for a few longer tests.

In large classes daily quiz grading can become a significant burden. This burden can be reduced because less homework will need to be graded since the quiz provides the practice that one homework problem would. As a result, one less homework problem can be assigned each day. Also, if the professor is satisfied with awarding no partial credit, a multiple-choice problem can be used. This greatly decreases the grading chore. When there are numerical answers, a method for developing multiple-choice tests which eliminates many of the drawbacks of multiple-choice tests is presented in Section 11.1.3. To start the next period the professor can spend a minute or two going over the solution to the quiz problem and use this as a springboard for discussion. The next lecture will then build on this base.

This type of approach has several advantages. Since the students know that they will be tested at the end of the period, they will pay attention to the lecture and may even read the textbook in advance. They will also ask questions during the lecture because they know they cannot wait until they get home to puzzle out a confusing point. The students also practice every class period, which requires them to be active, and then they receive feedback either immediately or in the next class period. The professor also obtains feedback immediately and knows if the students are learning the basic material. The very frequent quizzing reduces the importance of each quiz and thus reduces test anxiety. Although the professor loses about ten minutes each class period for presentation of new material, some of this is gained back since less frequent hour examinations need to be given. The large number of quiz scores makes grading at the end of the semester easier, and every student knows where he or she stands in the class.

Obviously, this procedure can be varied significantly while retaining the advantages. Some of the quizzes can be assigned as group efforts, or quizzes can be given every other period instead of every period. A short derivation or essay problem can be given instead of a numerical problem. And occasionally the quiz problem can be a review problem instead of a new problem. If the students do very poorly on one quiz, the quiz can be repeated. An alternative procedure called the "one-minute quiz" has been widely adopted in nontechnical areas. In a one-minute quiz students are asked to write one sentence answering one question or some variation of it: "What was the most important concept covered in lecture today?" This can be open book and open notes and, if desired, open neighbor. A one-minute quiz actually takes a few minutes. It has many of the advantages of a problem-solving quiz and can be used as a replacement on days when you do not have time to prepare a quiz.

6.6.2. Guest Lecturers

Guest lecturers can broaden the viewpoint of any course. If the world expert on a topic has an office next door to you, perhaps you could invite him or her to present one lecture. Engineers from industry can give an industrial flavor to presentations that most professors cannot duplicate. Such lectures from an industrial perspective can be valuable in any engineering class and not just in design courses. Many universities also have "old master" or "outstanding alumni" programs which invite interesting people back to the campus. These individuals are delighted to talk to students, and students usually appreciate the break in the routine provided by guest lecturers.

Ideally, a guest lecturer will integrate his or her presentation neatly into the course (Borns, 1989). For example, an engineer from industry could come in and talk about the design of heat exchangers at the point where the students are ready to cover this material and will benefit from the engineer's expertise. If the guest's lecture is to be integrated into the course this tightly, the professor needs to be sure that it is clear what the students can do and what is to be covered. Give the lecturer copies of the syllabus and reading assignments so that he or she can develop a lecture at the appropriate level. If any special props, such as a microcomputer, will be needed, be sure that everything is available and in working order.

Several years ago one of the students in class was the host to an old master. The student asked if the old master (the CEO of a major chemical company) could talk to the class for one period. Although this talk would not be integrated into the class lectures and course topics, we agreed. It turned out to be a rewarding experience. The students were interested and asked many intelligent questions. The old master enjoyed making his presentation and made several interesting points about an engineer's life in industry and about the importance of courses in school. Although one period of content was "lost," the students gained more than they lost. Now we take advantage of special opportunities like this once a semester, even if the guest lecturer does not talk about the course content.

Guest lecturers can be wonderful or horrible, and they need to be selected carefully. Find someone who has the special expertise that you want; then check around to find out how good a speaker the person is. Once you have found someone who does a good job, invite them back.

In general, the professor should be present when a guest lecturer speaks. This sends a message to the students that the material is important; furthermore, the professor may learn something. An exception occurs when the professor finds someone to substitute while he or she is away on a trip. Substitute lecturers usually cover the normal course material. Be sure to tell the substitute exactly where you stopped the period before, and afterward find out how much he or she covered. Delineate for the substitute exactly what material should be covered. The best substitute to use is a professor who teaches the same course in different semesters. If you arrange to trade substitutions, the workload tends to even out, and no one feels taken advantage of. Often the class TA is asked to present a lecture while the professor is out of town. This can be a good experience for the TA, but not necessarily for the class. Be sure to sit down with the TA ahead of time and go over the content you want covered in detail. If the TA regularly attends the lecture, he or she will not be rusty. After you return, discuss the TA's lecture with the TA so that the lecture can be a learning experience for that person. Another practice, once common, was to load all of the substitution work on the assistant professors in the department regardless of their teaching areas. Since this was obviously somewhat unfair, one hopes this practice is gone forever. Regardless of who does the guest lecturing, be sure to thank them in writing for their efforts. If the students comment on how much they enjoyed the lecture, be sure to mention that in your letter.

6.6.3. The Feedback Lecture

The feedback lecture is a technique developed by Osterman at Oregon State University (Osterman et al., 1985). In this approach students first receive a study guide that outlines how they should prepare for each lecture. Then during the lecture they receive a lecture outline as a further guide. The professor lectures for roughly twenty minutes. Small student groups then discuss an important question and turn in a response sheet. The professor briefly discusses the discussion question and then lectures for the last twenty minutes of the period. Homework assignments provide further practice.

This procedure formalizes the use of many learning principles. Students receive clear objectives in the study guide, and their learning is carefully guided by the study guide. The lecture outlines help them organize the material. The group activity in the middle of the lecture requires that students be active and makes the class more cooperative. Feedback occurs from other students in the group discussions and from the professor both during the group activity and after the response sheets are turned in. Some teaching of students by other students often occurs during the group discussions. The discussion questions are chosen to be particularly thought-provoking to pique the students' interest.

Such a formalized procedure obviously requires a fair amount of advance preparation, so it's unlikely the professor will arrive in class unprepared for the lecture. This method also motivates students to prepare for each class since they know that they will have to do something in each class. Students are thus less likely to procrastinate. Obviously, components of the feedback lecture and the postlecture quiz can be combined in a variety of ways.

6.7. HANDLING LARGE CLASSES

Very large classes (more than 100 to 150 students) have some special characteristics which make them different from small classes. The challenge for the professor is to make student learning as close to equal to that in a small class as possible. Our discussion of handling large classes leans heavily on the papers by Hickman (1987), Kabel (1983), and Middleton (1987), and on Johnson's (1988) book.

Large classes require more preparation, more structure, more formalized procedures and more rules than small classes. In many ways good teaching is good teaching regardless of class size, but unfortunately, large classes magnify any problem. Consider preparation. If several hundred copies of a handout are to be distributed, the professor cannot wait until half an hour before class to have the copies made. So more preplanning is required. Arrangements must also be made to distribute the several hundred copies efficiently; thus, the professor must assign the TAs and the secretary the responsibility of being in class to help with these duties. The professor's lecture preparation must also be thorough. In small classes professors quickly develop rapport with the students, and the professor can come in less than fully prepared on occasion. In large classes inadequate preparation is all too obvious and students are much less forgiving.

Large classes need to be more structured. The syllabus needs to be detailed and available on the first day of class. Examination dates *must* be listed. In effect, these dates become a contract with the students and it becomes quite difficult to change them without causing some students major problems. The textbook must be ordered early and needs to be more carefully selected since many students will rely heavily on it. Course supplements may be necessary and have to be prepared well in advance. The grading scheme needs to be formalized and set forth at the beginning of the semester. Since there will be students that the professor does not know well, grading becomes more impersonal. Rules for missed examinations and late assignments need to be stated and followed. There are likely to be more problems with uninterested students talking, reading a newspaper, or sleeping. Formal rules are required so that the students know what is acceptable behavior. An attendance policy needs to be set. Since there is a good correlation between attendance and grades, attendance should be encouraged. Even if attendance is not required, it is still useful to keep track of who attends since poor attendance often explains poor grades, and poor attendance always reduces the professor's tendency to be lenient in grading. The best solution is to assign seats and have a TA take attendance.

Unfortunately, increases in class size often do not mean an equivalent increase in the number of TAs. When this problem occurs, there must be a decrease in the number of items graded. It is possible to grade only a subset of the homework assignments and to use multiple-choice problems for parts of tests. It may also be possible to get undergraduate graders to do part of the grading (Kabel, 1983). Just keeping track of grades becomes a burden with large classes. Use an electronic spreadsheet and periodically post the current scores for every student by student number. Make it the students' responsibility to check for clerical errors.

Cheating is more of a problem in large, impersonal classes (see Chapter 12). Thus, more care is needed in administering the examinations (see Section 11.1.4). It may be necessary to have two copies of the test if the class is too large for students to sit in staggered seats. Care must be taken that all tests are collected and that there is no cheating during the chaos of hand-in time. Uniformity in grading is very important, and each problem needs to be graded by a single person. Rapid feedback is important, but more difficult to achieve in large classes. Written regrade requests are a necessity (see Section 11.2.3).

Small classes, particularly those with fewer than fifteen students, develop interactions between students and between the professor and students with very little formal effort by the professor. This is not true in large classes. A structured period for interactions needs to be provided in the professor's lecture plan, either by small group discussions or a question period. Informal meetings with the professor before and after class become more important, although not all students can be accommodated. Some type of formal procedure such as a seating chart or photographs of every student is required if the professor is to learn the students' names. Close rapport with students is essentially impossible if the professor does not learn the names of the students.

Students in large lectures can be assigned seats in blocks corresponding to their laboratory or recitation sections. Students feel less isolated since they see the same classmates more often. It is worthwhile to set aside two minutes early in the semester to have students formally introduce themselves to the students they are sitting next to. If cooperative group activities are used (see Section 7.2), the students will be working with a group of students they are familiar with. The laboratory or recitation section TA can also attend the lecture and sit with his or her students. This provides someone close by to answer questions. And it will tend to reduce disruptions since there is a person in authority close to each student.

Overall, teaching large classes is much more of a challenge than teaching small classes. Small classes can be quite a bit of fun, but teaching large classes is hard work. To do outstanding teaching in a large class is the mark of a master teacher–a rare professor.

6.8. LECTURES AS PART OF A COURSE

We should never forget that lectures are only a part of the entire course, and it is the entire course which determines how much students learn and what their attitudes will be. It is important that appropriate learning principles are satisfied for the entire course, but it is not feasible to satisfy all of them in the lecture alone. We will discuss the list of what works from Section 1.4 and consider how each item can be satisfied in a lecture-style course.

The course objectives can be covered in lectures. Probably the most effective way is to hand out a sheet of objectives for each section and discuss them in the lecture. This helps guide students. The lecture is the appropriate place to develop a structured hierarchy of material and the most appropriate place to use visual images. The textbook serves as a useful adjunct for these two tasks.

Although a modest amount of practice and feedback can be provided in a modified lecture, homework and tests serve best for providing practice and feedback. Even if homework is not graded, it is still necessary to provide feedback. This can be done by making solutions available to students.

An attitude of positive expectations will be shown by the way the professor presents the lecture and the assignments. It is also conveyed by the TAs when they mark papers and talk to students. This attitude needs to be conveyed in all aspects of the course.

Student success can be obtained first by enforcing reasonable prerequisite requirements, second, by having homework problems of graduated difficulty, and third, by providing sufficient help for students who need help. Making the class more cooperative can also help ensure success. The class can be made more cooperative by requiring students to work in groups during class and by encouraging group work for homework. If this is done, the homework grade should be a modest portion of the course grade. Group work also ensures that there is some opportunity for students to tutor other students.

Every lecturer should show enthusiasm and the joy of learning during lectures. With a little effort thought-provoking questions can be included in the lecture, in group work, and in some of the homework and test problems.

Individualizing the teaching style is difficult in a lecture, but quite a bit can be done by teaching small parts of the course in a form that will appeal to different learning styles. For example, visual and kinesthetic material can be used in addition to the usual auditory presentation. Lectures plus homework can be arranged to encourage students to go through all four steps in Kolb's learning cycle. (See Section 15.3.) Some advance organizers can be used to help global learners.

In addition to satisfying these learning principles, it is important that the students feel that the professor is accessible for questions and that the grading scheme is fair. Accessibility for questions includes answering questions both in and out of class. In-class questions can be supplemented by the professor being available both before and after class, and by having a reasonable number of office hours. Since some students are more comfortable talking to TA's, the TA's should also have office hours. Fair tests and grading can be ensured by testing on the objectives (see Section 11.1.1) and by being very careful that grading is uniform. The total course workload should be challenging but not outrageous.

If your class includes all these aspects, it will be a good class even if you are not the most polished lecturer in the world.

6.9. CHAPTER COMMENTS

We have devoted much more attention and space to lecturing than we will any other teaching method. This stems from an "if you can't beat them, join them" attitude. Most professors use some form of lecturing, so we wanted to do everything we could to ensure that lecturing is well done and satisfies learning principles. Lecturing is both the best and the worst teaching method imaginable–it all depends on the skill of the professor.

6.10. SUMMARY AND OBJECTIVES

After reading this chapter, you should be able to:

• List the advantages and disadvantages of the lecture method of teaching.
• Discuss how one selects the content and organizes a lecture. Use these principles to prepare a lecture.
• List the performance characteristics of a lecture and determine how to improve your lecture performance.
• Discuss procedures for answering and asking questions and practice improving your questioning skills.
• List what can be done to develop rapport with students during lectures.
• Discuss various modifications of lecture methods and explain how these modifications help lectures satisfy learning principles.
• Explain how large classes are more challenging than smaller classes and discuss what needs to be done differently in large classes.
• Put lectures into the context of the entire course and explain how a lecture course can fit together to optimize student learning.

HOMEWORK

1 Professor X uses the lecture method. List four problems to be aware of and briefly discuss solutions which will allow him to satisfy learning principles.

2 Consider one of the best lecturers you've ever had. Describe four qualities that made his or her lectures exceptional.

3 Pick one class period in a specific undergraduate engineering course. Select and organize the content to be covered in this one lecture.

4 Write lecture notes for the lecture in problem 3. Include stage directions and questions for the students.

5 For the lecture in problem 3, consider what methods you will use to keep the students active. Determine if one of the special lecture methods discussed in Section 6.6. would be appropriate. Explain.

6 Make a list of performance characteristics which you think are desirable in a lecture. Make a list of what you actually do. (It may be necessary to have a few students help with this list.) Develop an action plan to make the two lists approach each other.

REFERENCES

Alexander, L. T. and Davis, R. H., "Choosing instructional techniques," *Guides for the Improvement of Instruction in Higher Education, No. 11*, Michigan State University, East Lansing, MI, 1977.

Borns, R. J., "Professional guest lecturers," *Proceedings ASEE Annual Conference*, ASEE, Washington, D C, 168, 1989.

Cashin, W. E., "Improving lectures," Idea Paper No. 14, Center for Faculty Evaluation and Development, Kansas State University, Manhattan, KS, 1985.

Dareing, D. W. and Smith, K. S., "Classroom demonstrations help undergraduates relate mechanical vibration theory to engineering applications," *Proceedings ASEE Annual Conference*, ASEE, Washington, D C, 396, 1991.

Davis, R. H. and Alexander, L. T., "The Lecture Method," *Guides for the Improvement of Instruction in Higher Education, No. 5*, Michigan State University, East Lansing, MI, 1977.

Eble, K. E., *The Craft of Teaching*, 2nd ed., Jossey-Bass, San Francisco, 1988.

Engin, A. W. and Engin, A. E., "The lecture: Greater effectiveness for a familiar method," *Eng. Educ.*, 368 (Feb. 1977).

Hickman, R. S., "Effective teaching in large engineering sections," *Proceedings ASEE Annual Conference*, ASEE, Washington, DC, 211, 1987.

Hyman, R. T., "Questioning in the college classroom," Idea Paper No. 8, Center for Faculty Evaluation and Development, Kansas State University, Manhattan, KS, 1982.

Johnson, G. R., *Taking Teaching Seriously: A Faculty Handbook*, Texas A&M University Center for Teaching Excellence, College Station, TX, 1988.

Kabel, R. L., "Ideas for managing large classes," *Eng. Educ.*, 80 (Nov. 1983).

Lowman, J., *Mastering the Techniques of Teaching*, Jossey-Bass, San Francisco, 1985. (This book has extensive sections on lecture presentations.)

McKeachie, W. J., *Teaching Tips. A Guidebook for the Beginning College Teacher*, 8th Ed., D.C. Heath, Lexington, MA, 1986.

Meador, D. A., "Parallel learning curve," *Proceedings ASEE/IEEE Frontiers in Education Conference*, IEEE, New York, 136, 1991.

Middleton, C. R., "Teaching the thundering herd: Surviving in a large classroom," in Shea, M. A. (Ed.), *On Teaching*, Faculty Teaching Excellence Program, University of Colorado at Boulder, Boulder, CO, 13—24, 1987.

Osterman, D., Christensen, M., and Coffey, B., "The feedback lecture," Idea Paper No. 13, Center for Faculty Evaluation and Development, Kansas State University, Manhattan, KS, 1985.

Peck, R., "Examinations as a method of teaching," *Chem. Eng. Educ.*, 10, 76 (Spring 1979).

Tannen, D., *You Just Don't Understand*, Ballantine Books, New York, 1990.

Taveggia, T. C. and Hedley, R. A., "Teaching really matters, or does it?" *Eng. Educ.*, 546 (March 1972).

Wankat, P. C., "An elective course in separation processes," *Chem. Eng. Educ.,* 15, 208 (Fall 1981).

CHAPTER 7

NONTECHNOLOGICAL ALTERNATIVES TO LECTURE

The lecture method can be particularly effective if the goal is to have students learn objectives in the three lowest levels of Bloom's taxonomy, but it is not the best method for higher level cognitive objectives such as analysis, synthesis, evaluation, problem solving, and critical thinking. In Chapter 8 alternatives to the lecture which use technology such as TV, computers, and laser videodiscs are discussed. Chapter 9 covers methods commonly used for teaching design and laboratory courses, and Chapter 10 considers one-to-one aspects of teaching. In this chapter we look at nontechnological alternatives to lectures.

The discussion method of teaching, fairly common in the humanities, is seldom used in engineering; however, it can be a very useful supplement in lecture classes. In cooperative groups most of the learning occurs with students working together in small groups. This method has been used for the entire course or as a supplement in lecture classes. A variety of other methods such as panels or debates can be used to spark student interest and encourage student involvement. Mastery learning requires that students reach a particular level of mastery on tests but gives them repeated chances to do so. The Keller plan and the personalized system of instruction method (PSI) employ mastery learning in a format that allows a student to control the rate of progress through the course. In individual study a student studies alone or with occasional tutorial help to satisfy certain objectives. Field trips can be used as part of any course to help meet the course goals. Evidence will be presented that for certain objectives these methods can be more effective than lectures.

7.1 DISCUSSION

Perhaps because discussion is a commonly used teaching method in their disciplines, social scientists and professors of education have studied it extensively (Cashin and McKnight,

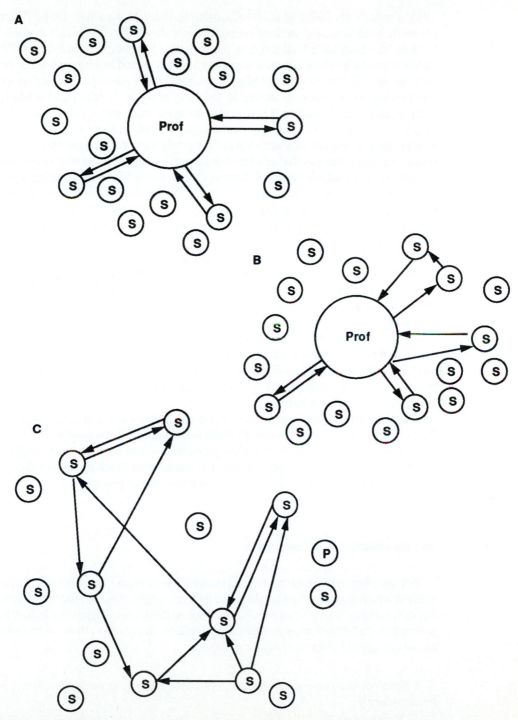

FIGURE 7-1 INTERACTION STYLES. A. Questions. B. Instructor-lead discussions. C. Student-centered Discussions.

1986; Davis et al., 1977; Eble, 1988; Lowman, 1985; McKeachie, 1986). There is ample scientific evidence cited in these sources which shows that discussion is not an efficient method for transmitting facts and data, particularly when compared to lectures. For teaching the three lowest levels of Bloom's taxonomy, discussion and lecture students do equally well on tests, and the lecture students learn the material more quickly. However, discussion and questioning are superior to lecture in teaching analysis, synthesis, evaluation, problem solving, and critical thinking. There is also some evidence that students remember material they learn through discussion or questions longer. Therefore, discussion should be considered as a teaching method in engineering when the professor wants to work on these higher-order processes. Since many of the benefits of discussion can be obtained with rather short periods, the engineering professor does not have to change her or his entire teaching method.

In order to participate in an intelligent discussion, students have to know something about the topic. Thus, we like to use discussion as a break in a class which is basically a lecture class. Another use of discussion is the cooperative group learning method which is included in both this section and in Section 7.2.

Discussions and questions (see Section 6.4) aim to involve students in the material and to interact with others. The main difference between question sessions and discussion is in the style of interaction. The interaction in a question period is clearly between the professor and individual students (see Figure 7-1a). The professor is definitely in charge, whether asking or answering questions. This is not an exchange among equals. In the student-centered discussion shown schematically in Figure 7-1c, all participants are roughly equal. The professor may participate but does not lead, and in small "buzz" groups the professor is often working with another group. The instructor-led discussion shown in Figure 7-1b is intermediate between these two methods. The instructor is clearly in charge but encourages significant interaction between students.

The instructor's control is greatest in the question format and least in student-centered discussions, so there is less that can go wrong in the question format and this procedure appears to be more efficient. However, student gains in problem solving and critical thinking are highest in well-functioning student-centered discussions (McKeachie, 1986). In addition, student changes in attitude are highest in student-centered groups.

7.1.1. Advantages and Disadvantages of Discussion

All teaching methods have advantages and disadvantages. The professor needs to select a method which is appropriate for the material to be taught, fits the students to be taught, and fits the professor's style. Since the competition to discussion is often the lecture method, the advantages and disadvantages of discussion will be compared to those of lecturing. Among the advantages of discussion are the following:

1 Students learn how to do analysis, synthesis, evaluation, problem solving, and critical thinking better.

2 There appears to be better retention of material.

3 Discussion is an effective method for changing student attitudes (affective objectives).

4 Intellectual development (see Section 14.2) is greater.

5 Students are more active and become more involved.

6 In engineering, discussion is a novel method which gains the students' attention. It also breaks up the routine of the lecture.

7 Discussion can improve students' group interaction and communication skills.

8 In student-centered discussion students can be leaders and can teach other students.

9 Discussion is more likely to lead to commitment to a field (McKeachie, 1983).

10 Discussion does not have to be a "big deal" and can be included in a class which basically follows the lecture format.

There are disadvantages to discussion:

1 "Developing the ability to conduct effective discussion is even more difficult than learning to lecture effectively . . ." (Eble, 1988).

2 The process can be time-consuming and the rate of transfer of information is low.

3 Students do not show improved learning of knowledge, comprehension, and application objectives.

4 It may be difficult to obtain student participation, particularly in engineering.

5 Students must know something before an intelligent discussion is possible.

6 The instructor has less control and may be uncomfortable with the method.

7 Entire group discussions are not possible with more than about twenty students and work best with ten students or fewer (Davis et al., 1977). This problem can be surmounted by using small student-led groups.

8 The discussion approach may be less acceptable to students, particularly engineering students who want to learn from an expert.

9 Meaningful discussions may be difficult with immature students.

10 Engineering students often think that group interaction and communication skills should be taught in another class, not in engineering classes (Hayes et al., 1985).

Engineering classes often include objectives which can be appropriately taught by using discussion methods. For example, evaluation and comparison of competing designs, evaluation of unproven scientific theories or data such as "cold fusion," and determining the best way to allocate scarce resources are all appropriate topics for discussion. If one of the course goals is to help students define and explore problems to develop a variety of solutions, then brainstorming (see Section 5.6), which can be considered a type of discussion method, is appropriate. Small cooperative groups, which are appropriate for other aspects of problem solving (see Section 7.2), include a significant amount of discussion. Ethical dilemmas seldom have clear-cut answers and so are appropriate for discussion classes. Discussing ethics in class is also one approach to changing attitudes and possibly producing ethical engineers. Although the ethics dilemma does not have a correct solution, there are many incorrect solutions and students can learn to recognize them. If communication, interpersonal, or leadership skills are on your agenda (and many engineers in industry think they should be), then discussion methods are one of the appropriate teaching methods. If you want the students to be more active, to develop their higher-order processing skills, and to pay attention during the lecture, then question and discussion periods of five to ten minutes can be inserted in the middle of lectures.

7.1.2. Conducting Discussions

Conducting discussions is an art, for good discussions don't just happen; paradoxically, they must be structured to occur spontaneously. And this is why conducting excellent discussions is difficult. Discussion experts (Cashin and McKnight, 1986; Davis et al., 1977; Eble, 1988; Lowman, 1985; McKeachie, 1986) have a variety of suggestions concerning what to do to improve discussions. First, the professor must prepare for the class. Since discussion classes can wander through a broad range of material, the professor needs broad knowledge in the area. In lecture classes the professor can be one lecture ahead of the students, but in discussion classes this is not generally true. At the beginning of the class period the professor needs an agenda for the discussion session, even for only five minutes of discussion. It does not work to tell the students, "Let's discuss . . . ," followed by silence. Until they have been trained, engineering students won't know what you want. If you plan on using discussion techniques anytime during the semester, start early. Students expect the entire course to be similar to the first two weeks, so you need to have some discussion during the first two weeks, even if just as brief breaks in a lecture class.

Engineering students are generally very task oriented, so give the students a task. For example, "We have discussed the engineering and political factors which affect the siting of new airports. You have all studied the siting of a new airport for Chicago. Let's see if we can come up with a consensus of where to put the new airport." Note that the purpose of the discussion is really not to find a solution. The purpose is to expose the students to the process of reaching a solution. In the give and take of a good discussion on this topic, the students should learn something about the interaction between engineering and politics, about communication, and about the process of obtaining a consensus. This topic would also be useful for a panel discussion (see Section 7.3.1) or a debate (see Section 7.3.2). Since the process will be different in these three techniques, they complement each other.

Engineering students tend to enjoy problem-solving discussions. However, the discussions tend to become fragmented since students present comments at different stages of the problem-solving strategy. As the professor, you can exercise some control. Break the problem into parts and clearly tell the students which part to discuss. If a large class is broken up into small groups, the instructions to the small groups can clearly state: "Now that we have defined the problem, let's explore alternatives by brainstorming. You have three minutes." The leader ensures that everything stays on track either in the big group or in the small groups. Note the time constraint. If the purpose is to learn the process, a long period is not required since the lack of a complete solution is not a major problem. There is another advantage to the time constraint. If a group gets off track, it isn't allowed to go very far astray before being called back on track.

Particularly in large groups, the first contribution is the hardest. Be patient. If silence doesn't work (and at least two or three minutes may be needed), there are some alternatives. Ask a student to share a comment that he or she has made earlier. Since this is essentially a prepared comment which already has instructor approval, it is less threatening than having to volunteer something new. Make a provocative statement yourself. Challenge but don't threaten the students. Prepare ahead of time by planting a comment with a student or a TA. We suggest this be done only very early in the semester since this procedure can backfire. Say something encouraging like, "I'd really like to hear someone's opinion," and then try more silence.

If the difficulty in getting a discussion started keeps you from using discussions, you can always revert to questioning, but before doing so try dividing the class into small groups for discussions. Many students find it much easier to talk in small groups, particularly when the instructor is not sitting at the front waiting to pounce on the first remark. This is especially true of women students (Tannen, 1991). Many students participate when they feel the responsibility to do so, and the instructor can include this responsibility in the charge given to small groups. "I want each group to be sure that every student has the opportunity to speak." Finally, regardless of what the instructor does, some of the small groups will work. As the noise level in the room increases, other groups will start to talk.

Once students start talking, the instructor needs to be encouraging and accepting both verbally and nonverbally. The comments in Section 6.4 are all appropriate for discussion, except that it is not necessary to respond to each comment in turn. Let several students talk and let students respond to each other. In lectures the instructor talks—discussion is the students' chance.

What does a professor do? Davis et al. (1977) suggest several techniques which can be useful for working with the entire class once the discussion starts.

1 Post ideas on the board and verify the ideas. This allows you to use correct jargon. It also gives you something to do other than talk.

2 Serve as a gatekeeper to keep students on the topic. It's easier to keep students on the topic if the assignment is a clear task which has been broken into parts. You can exert some control by calling on students who raise their hands. Most students have been socialized through many years of school to talk only when called on. If you want to call on students who raise their hands, it is a good idea to enforce this rule so that everyone is treated equally.

3 When the discussion falters, request examples or illustrations.

4 Encourage and recognize contributions. The act of writing a contribution down is recognition. Also recognize contributions verbally: "Good!"

5 Test the consensus. Is the class ready to move on to the next part of the problem?

6 Summarize the discussion. To summarize the student discussion well you must listen. This is another reason to talk less.

The professor's role with small groups is discussed in Section 7.2.

What problems might arise? One is disagreement and conflict between students. Conflict can be resolved in a very positive way if the class is structured correctly (Johnson and Johnson, 1979). Early in the semester the instructor needs to set the climate that problems are to be solved together. That is, the discussion climate is to be cooperative and not competitive. There doesn't have to be a winner. There should be a firm rule that topics, not personalities, are to be argued. Students need to learn that they can agree to disagree and still have a high opinion of the other person. When conflict occurs, you should ensure that everyone has the same accurate information and then help the students to recognize similarities and differences. When this is done, it may become clear that the conflict is either over semantics or over one fairly small point. The result will often be a near-consensus. Another way to help students come to a consensus is to use the principles of debate and have them switch sides and argue for the other side. This works because students often see that the other side also has some valid arguments. If the conflict becomes heated, you need to deal with feelings either during or after class. Purposefully introducing controversy into a class is the topic of Section 7.2.3.

Nonparticipants are another problem. The discussion may drag if there are too many of them. Even if there are only a few, they may not be involved in the class and probably are not benefiting fully from the discussion. Quiet students may be quite involved in the material, but the other students are not benefiting from their input and the quiet students are not improving their communication skills. Pay special attention to these students to get their ideas and opinions. If a nonparticipant ever raises his or her hand, call on that student. If they are interrupted, come back to them. If they look ready to speak, encourage them verbally and nonverbally. These students may not speak because they are slow at formulating responses. Passing out one or two discussion questions as a homework assignment the period before may get these students past this barrier. They are more likely to participate in a small group, particularly if you specifically request that everyone participate. Women are less likely to participate in discussions than men, especially in a class where ideas are attacked (Tannen, 1991). They are likely to speak out more when asked for personal anecdotes or about what was useful to them in a reading or an assignment.

The overparticipant or monopolizer is another problem. In an instructor-led discussion the instructor can say that someone else's ideas need to be heard. This ploy often works since many students do not want to monopolize the discussion. If the monopolizing continues, you can call the student aside. We suggest trying a positive approach such as expressing concern about the person's need to work on listening skills. If the monopolizing behavior continues, a sterner admonition may be required. Once the group is comfortable doing discussions, the instructor may not have to step in to control the monopolizer. The other group members will tell the person to shut up, and they will often be stronger than an instructor should be.

Often the problems of nonparticipation and monopolization can be solved simultaneously by asking that each student speak at most twice during the class (Palmer, 1983). This forces the impulsive to slow down and weigh what they want to say. The resulting silences give the quieter students permission to talk.

Discussion should be considered part of the entire course. When the advantages of discussion are compared to the list of learning principles in Section 1.4, discussion by itself can satisfy some but not all of these principles. Discussions allow students a chance to be active, practice certain tasks, and provide feedback if they are willing to participate. The professor can easily communicate high expectations for the students and challenge them with thought-provoking questions and discussion problems. The class can be made cooperative, and in small discussion groups students can teach other students. Finally, the professor can certainly radiate enthusiasm. What discussion does not do efficiently is guide the learner, develop a structured hierarchy of material, and provide visual images. In addition, the practice, feedback, and challenges are only of one type and do not include detailed numerical calculations. Lectures, homework, and tests in addition to discussion can satisfy all the learning principles.

7.2. COOPERATIVE GROUP LEARNING

The topic of this section is cooperative group learning, not cooperative (or co-op) education which consists of alternating periods of work and of education on campus. In cooperative group learning students work together to understand the material, do homework, complete projects, and prepare for tests. Research has shown that a cooperative learning environment is conducive to learning higher-order cognitive tasks such as analysis, synthesis, evaluation, and problem solving (Johnson et al., 1991; Smith, 1986, 1989). Group work has long been common in engineering education in laboratory and design courses (see Chapter 9), and the Whimbey pair method for teaching problem solving discussed in Section 5.5 can also be considered a cooperative group method. What is new in this section is the use of groups for content-oriented classes which would normally be taught by lecture. We will start by considering informal learning groups, extend our comments to formal learning groups, and finish with a discussion on structured controversy.

7.2.1. Informal Cooperative Learning Groups

Informal cooperative learning groups are "spur-of-the-moment" groups formed for a particular short-term task and then dissolved. Their direct ancestor is the "buzz" group which has been commonly used for discussion for many years. Informal groups can be quickly formed in the middle of a lecture and students can be assigned a task such as solving a problem, answering a complicated question, or developing a question for the professor. These groups serve as a way to encourage students to be active in a large lecture class, provide for discussion, serve as a break when the students' attention starts to falter, and provide a more cooperative atmosphere in the class. In addition, these small groups have a modest number of students teaching students and provide students with an opportunity to practice teamwork. Inclusion of a short break from lecture with an informal group helps to individualize the class for the extroverts and field-sensitive individuals.

Informal cooperative groups also allow you to start experimentation with cooperative learning. Including these groups within a lecture class is not difficult and takes no more preparation time than the lecture. Since the groups are informal, assignment into groups can also be informal. We have found that the easiest thing to do at the start of the semester is to have students cluster in groups of about four based on choosing students who are sitting close to each other. This can be done in a normal lecture hall, although lecture halls are not ideal for discussions. The first time the class breaks up into small groups the professor must be very directive. A solitary student should be assigned to a group even if the student has to move. Later in the semester you may want to experiment with different groups. There is an advantage in having students move and work with students they do not know. Since small group dynamics are different in same-sex groups, you may also want to experiment with groups of the same gender (Tannen, 1991).

Once the groups are formed, tell the students to briefly introduce themselves to each other. You may want to assign a leader and a reporter, or let the group act informally. If no assignments are made and you notice that the group is not working, you can assign a discussion leader to get things going.

As the professor you must structure the small group experience and provide an agenda. Give a clear problem statement and a deliverable. Although the groups are formed on the spur of the moment, your agenda must be preplanned. As noted in Section 7.1, asking students to discuss a topic is not sufficient. The task for small groups should fit the following (Hamelink et al., 1989):

1 Have several possible solutions.
2 Be intrinsically interesting.
3 Be challenging but doable.
4 Require a variety of skills.
5 Allow all group members to contribute.

Hamelink et al. (1989) note that "if the task has one right answer or involves simple memorization then competitive education methods are far superior."

Most engineering students are pragmatic and want to do something, so there must be a deliverable. For example, if the problem is to come up with a list of five possible solutions, the deliverable is this list. If the problem is to come up with a consensus about some question, then the deliverable is the consensus. These deliverables should be presented to the entire class. If a reporter has been assigned, that student can make the presentation. Otherwise, let the group choose who will report, or call on a group member at random. Small groups should be told in advance how this reporting will be done. When the groups present, you can post the solutions on the blackboard or the overhead projector. Also note in advance that the first group to present has a major advantage since everything they report will be new. The last group may repeat items which have already been presented.

The problem statement should be very clear. Be sure to delineate what the deliverable is, either orally or with written instructions. If different groups have different instructions, then written instructions will probably be less confusing. Tell the groups roughly how much time they have. Then, say something like, "Let's get started. I want to hear some noise."

During the group discussion the professor and the TA can circulate among groups. If a group is working well together, there is no need to interrupt. Groups which have trouble getting started need a little help. A group with only introverts may have trouble. In the future you can mix the groups up and avoid the exact grouping which caused the trouble. At this time you might want to assign a discussion leader and a recorder to get things started. (One nice thing about informal grouping is that problem groups last for only about 10 minutes, and the next time the class can start over with new groups.) The instructor should also watch the time. Although it is not necessary for students to finish the task (Felder, 1990), we find being assigned a task with no chance of finishing to be frustrating. Thus, we like to watch the groups and to close the discussions when about half of the groups are essentially finished with the task. The entire process, including the reports to the whole class, can be completed in 10 to 15 minutes. Thus, informal groups can be conveniently inserted within a lecture.

These informal groups can satisfy many of the learning principles discussed in Section 1.4, and they also provide for some individualization in teaching style. They can satisfy most of the five elements necessary for cooperative group success (see Section 7.2.2). Cooperative groups help make the class seem more friendly and thus help the professor establish rapport with the students. Finally, informal groups are simple to implement and thus are a good approach with which to start.

7.2.2. Formal Cooperative Learning Groups

Formal cooperative learning groups are formed to teach each other and to work on longer-term tasks than are informal groups, even lasting for the entire semester. These groups often produce a project which is graded as a team effort. Since these groups are longer-term and grading is involved, a bit more thought might be put into forming and structuring them. Students who have worked in informal groups will have a good start in working in formal groups.

Getting started with cooperative learning groups can appear daunting at first since most professors have not experienced this teaching method. However, you do not have to convert the entire course to group work. Informal groups can be interspersed into the lecture, and one project can be done with a formal group. Then as you become more familiar with the strengths and weaknesses of this approach, you can convert more or less of the class to cooperative groups. Step-by-step procedures for getting started are outlined below (Smith, 1986):

1 As the instructor, you need to have clear objectives and a plan. The clearer the objectives, the easier it will be to get the groups started and functioning well.

2 Assign the students to groups. Smith (1986, 1989) suggests the use of random groups, while Goldstein (1982) recommends placing one good and one poor student (based on grade point average) into each group before randomly assigning the other students. Johnson et al. (1991) suggest even more instructor control, with high- medium-, and low-achieving students in each group. Both good and poor students can benefit from working together. The good students will teach the poorer students, and both will benefit. If the groups are to do significant discussion, there is also an advantage to having groups which are all women or where women are not in the minority (Tannen, 1991). All-women groups give women a chance to practice leadership. However, at least during part of the semester men and women need to be together in groups since they need to learn to work together; men, in particular, need to learn to work with a female team leader. Other criteria which can be used in selecting groups are discussed in Section 9.1.2. Depending on their purpose, groups should have from two to six students. Topping (1992) suggests starting with dyads since the teacher has more control. The class should meet in a room with circular or square tables, so that the groups can sit facing each other.

3 Carefully explain the task of the group. Early in the semester be very explicit about the task and the job of the group. It is desirable to promote interdependence. That is, one student cannot get a good grade when the group fails to perform its task satisfactorily. Thus, the grading procedure needs to be explained carefully. Some students will resist being graded on the results of the entire team. The rationale for this is that most industrial jobs are too big to be done by individuals and teamwork is a necessity. The team must function together to get the job done correctly and on time. Students might as well get used to the concept of teamwork now. If projects are chosen to be large enough that one student cannot complete them in the time available, there will be less complaining. Teamwork and cooperation should be emphasized in this explanation.

Grading on tests should also be carefully explained. One option is to give group take-home tests with each group receiving a single grade. The students can also be tested in pairs where time is available to confer with one's partner (Buchanan, 1991). If individual tests are given, it is important not to grade on a curve since grading on a curve fosters competiveness, not cooperation. Either mastery or a fixed scale should be used for grading individual tests. The professor can also assign bonus points to group members if everyone in the group gets above a cut-off score. (Johnson et al., 1991).

4 Monitor groups to ensure that everyone is working together and intervene if there are problems. You may need to know something about group dynamics to help groups if there are problems. Also, you may want to impose some structure on the groups such as requiring that everyone contribute once before anyone can contribute a second time. Or the recorder can be asked to keep a running account of the number of times different students speak. You and a TA can circulate and serve as resource persons when the groups are unsure about something. If there are technical problems, caution the students to check something or give a mini lecture to explain a complicated point.

5 Provide closure to the group session. Ask the students in each group to prepare a summary of their results for that day. If appropriate, ask for an outline of their future plans. Provide homework or additional assignments for the group.

6 Evaluate the achievements of each group and of the individuals in each group. Discuss with each group how well they are collaborating. Give them advice on how to improve. Students who have been pitted against their fellow students for years cannot be expected suddenly to blossom as cooperators without some practice and guidance. Be sure that class grading does not reinsert competitive behavior into the class. For example, individual tests can be mastery (see Section 7.4.1) or can be graded on an absolute scale. Group grading strategies are discussed in Section 9.1.

Now that you no longer spend the bulk of the time lecturing, what do you do? First, set clear objectives and provide learning materials such as a clear textbook, articles, and a study guide. As noted in Chapter 6, this plus a test is sufficient to ensure that students will learn the lower-level cognitive objectives (Taveggia and Hedley, 1972). You may also want to give different students different material to master. Then the contributions of all group members are essential for the group to have the complete picture.

Next, develop the activities the students will do in class and out. These are projects and open-ended problems with a clear deliverable. Problems must be challenging yet solvable with the basic principles, be realistic and attention-grabbing, and have multiple solutions. Particularly, early in the semester problems should be clearly defined. Later in the semester definitions can be quite vague.

Third, set up the groups and get them started. A good start will convince many otherwise skeptical students that they can learn efficiently in a cooperative group. It is important that the first problems not be trivial or closed-ended because at least the better students can do these more efficiently on their own.

During the functioning of the groups, the professor and the TA are both resource persons and troubleshooters. The professor can help when a group is struggling technically. This is important early in the semester when many groups want reassurance that their path is correct. Some groups will click, and some won't. The professor needs to help groups which aren't functioning well. There are several things that you can try. Remind them that the evaluation is a group evaluation and then let the group muddle through. It may be helpful to provide more structure to a group by assigning a group leader for this set of problems, or to focus on what the students are doing and remind them to do one problem solving step at a time. (This is the same procedure as that used in Section 7.1.2 for discussions.) You may also want to watch the interaction patterns for a while and then discuss group dynamics with the group. Finally, the groups may need to be shuffled. During this process of working with the groups, monitor the contributions of all group members.

The professor also serves as a time keeper and moves the group onward through a series of tasks. Students who are not experienced in working in groups often need to be guided through the process. The professor must also be sure that there is time for the group reports to the entire class, and that there is time for group processing at the end of the period.

An alternative group problem-solving procedure is a group-based socratic approach (Felder, 1990). Groups are given a problem to work on in class. Then a series of questions are used to guide the students toward the solution procedure. Students are given short periods (two to three minutes) to work on each question. This is followed by a brief discussion, with the instructor providing the answer if the groups have not had time to finish. The groups are then asked the next question in the sequence required to solve the problem. This procedure gives the professor considerable control and ensures that every student will be active and no student will become totally lost. However, it does reduce group interactions and group responsibility. This type of strongly directed group process is probably beneficial for freshman and sophomore classes where considerable direction is still desirable.

One advantage of cooperative groups is that the professor focuses on what the students are doing, not what the professor is doing (Astin, 1985). Since the students are the ones who must learn, this focus is appropriate. The group procedure also encourages most students to be active.

Five elements of group success, which should be remembered when groups are set up and operated (Smith, 1989; Johnson and Johnson, 1989; Johnson et al., 1991), have been identified.

1 Positive interdependence means that students believe that for one to succeed they must all succeed. The professor can promote positive interdependence by appropriate grading procedures, by making sure that that the group depends on the resources of all the students, or by requiring that a division of labor be used to complete the task. Early in the semester positive interdependence can be promoted by giving the group only one set of instructions.

2 Face-to-face promotive interaction means that students work together discussing, explaining, teaching, and solving problems. This face-to-face interaction promotes learning since it helps support the students' efforts to learn and motivates them.

3 Individual accountability and **personal responsibility** must be stressed so that an individual cannot "hitchhike" on the work of others without contributing. The professor can monitor attendance and contributions, call on students at random for presentations, and give individual examinations.

4 Social skills to work together are needed. Students need help in learning how to lead, teach, reach consensus, resolve conflicts, and communicate. For example, an engineering professor can encourage groups to check that everyone understands. Engineers in industry are expected to do these things, and students who learn how while in school will have an advantage on their first job.

5 Group processing is a necessary maintenance activity to keep a group working smoothly. What have members done to support the functioning of the group? What can they do in the future? Group processing can be checked by requiring each group to submit a summary of their processing. Johnson et al. (1991, pp. 3–10) help explain group processing by quoting Willi Unsveld, a mountain climber. "Take care of each other. Share your energies with the group. No one must feel alone, cut off, for that is when you do not make it."

There are gender differences in how people react in groups (Tannen, 1990). In general, women have been socialized to develop group rapport and to seek interaction. Thus, many female students are experienced in social skills and group processing. Male students, on the other hand, have been socialized to seek independence and not the interdependence necessary for proper group functioning. Thus, initial resistance and attempted sabotage of group work is much more likely to come from male students.

The results that have been achieved with cooperative groups include superior learning of higher-level cognitive processes and superior problem solving (Hamelink et al., 1989; Johnson, et al., 1991; Smith, 1986). In addition, cooperative groups report the formation of positive relationships and increased social support with the development of professional self-esteem (Johnson et al., 1991; Johnson and Johnson, 1989; Smith, 1989). Students in cooperative learning environments liked the subject more and wanted to learn more about it (Johnson et al., 1991). Cooperative learning also increases retention of students in college (Johnson et al., 1991; Tinto, 1987). In minority programs cooperative groups have led to greatly increased retention and a large increase in facilitators going on to graduate school (Hudspeth et al., 1989; Shelton and Hudspeth, 1989). Many students (and professors) are searching for an educational community (Palmer, 1983). Cooperative group education can deliver this sense of community.

7.2.3. Structured Controversy

Structured controversy is a special type of cooperative group in which students confront an emotional issue in a structured format and strive for a consensus (Smith, 1984). This procedure is useful for issues which combine technology and public policy. Appropriate issues for a structured controversy include the siting of roads, landfills, nuclear facilities, and government research centers; regulations for air pollution and control of acid rain; proposals to outlaw greenhouse gases such as Freon; and the legality of company rules which prevent women of child bearing age from working at certain jobs.

The professor first develops packets of materials with all the facts and with opinions both for and against. The packet in favor of one side has all the positive arguments and facts. The con packet has all the negative arguments and facts. A complete picture can be seen only by combining both packets. For many controversies there are organizations which have essentially already prepared either the pro or the con package. Normally, the built-in biases of materials from advocacy groups is a problem, but not in the structured controversy procedure.

The class is divided into groups of four students with one pair of students being assigned on the pro side and one pair of students on the con side. Each pair receives the appropriate packet and is told to study it thoroughly and to prepare a position statement. This preparation can be done as homework if the pairs can meet together. In the four-person groups each pair first presents its position. The other pair is told not to refute the presentation (this is not a debate), but to listen and ask for clarification. Then the issues are discussed. The other pair then presents its position while the first pair listens and asks for clarification. Then there is a group discussion where all four group members try to achieve a consensus position. The consensus positions are then reported to the large group, and an attempt is made to achieve an overall consensus position.

Before starting a structured controversy, the professor needs to state the discussion rules clearly. These rules are the same as those for handling controversy in discussion (see Section 7.1.2) (Johnson and Johnson, 1979). Ideas, not personalities, are argued. The focus is on attaining the best group decision or consensus, *not* on winning. Listing, restating, understanding and integrating all facts—this is forced by the structure of the groups since no side has all the facts. All sides must be understood, and evidence used to determine logical fallacies in the positions. Finally, everyone must participate.

It is useful to give the students specific rules for reaching consensus. Palmer (1983) lists the following:

1 *Do not* argue to achieve your rankings or solution.

2 *Do not* change your mind just to avoid conflict. Be suspicious of too rapid agreement.

3 *Do not* use coin flips or majority votes. These do not represent consensus.

4 When there is a stalemate, search for a compromise position which is acceptable to all parties. However, do not award a member for finally agreeing by giving in later.

5 Look at differences of opinion as healthy and natural. These differences of opinion help the group arrive at a better final decision.

6 Use consensus procedures with groups where the members are comfortable with each other.

With a procedure like this it is the process and not the answer which is important. Thus, after the group discussion the professor should clearly set out the procedure and the rules which make reaching a consensus possible. Experience in activities such as this should make engineers much more effective communicators when working with the public on controversial issues.

7.3. OTHER GROUP METHODS FOR INVOLVING STUDENTS

In this section we will briefly explore three other group methods which can be used to involve students: panels, modified debates, and "quiz shows." These methods are useful as breaks in a lecture course and often serve as marker events for students. Thus, they are useful additions to the teacher's bag of tricks. However, we would not recommend using them for the entire semester.

7.3.1. Panels

The use of a panel consisting of three or four experts is a good way to start a question-and-answer period about a topic which has more than one correct answer. Professional seminars often use panels on topics such as job hunting, interviewing, what the first year in industry is like, what industry wants from young engineers, obtaining research funding, and achieving tenure. Panels can also be used for controversial technical topics, particularly those where technology and policy interact.

First choose the topic and decide on the date for the panel; then pick the panel and obtain the panel members' agreement to participate. Each panelist should prepare a very short (three or four minute) presentation which can serve as a springboard for questions and discussion, and on the day before the session remind the panelists of the meeting and the topic. Tell the class ahead of time about the panel meeting and assign readings which will prepare them for.

During the panel discussion, of which you are the moderator, introduce each of the panelists and ask for a brief statement. When the time has expired, gently ask for a summary and introduce the next panelist. When the last panelist has finished, ask the class for questions. If there are none, start the period with a question. Once the questions start you can control the session by calling on students. As many students as possible should be involved. It may be appropriate to ask a specific panelist to answer a question because occasionally one panelist will tend to monopolize the conversation.

An interesting alternative to this procedure is to assign students to the panel. The students are assigned the task of becoming an "expert" on a particular topic before the panel discussion. Serving as a panelist can be an alternate assignment to giving an oral presentation, serving on a debate team, or being on a quiz team. If the students are unfamiliar with panel discussions, they will need a clear set of directions, preferably written. The panelists will certainly become very involved in the discussion, and if a good topic is chosen, so will the class. The result will often be a much smoother class than having four separate oral presentations which are not directly connected.

7.3.2. Modified Debates

A variety of forms of modified debates are useful whenever there are two or more sides to a question. In our teaching class we debate the question, What is the best teaching method? One can also debate topics concerning resource allocation such as the site of a new airport or how much government money should go to super large science. One can structure a debate around competing designs or controversial technology. Reynolds (1976) found simulated historical debates useful in a class on the history of technology.

In a classical debate there are two teams with two members each. One side takes the pro side of an issue, and the other takes the con side. The debate pattern is affirmative-negative-rebuttal-rebuttal. Good debaters are taught to prepare for both the pro and con sides of the question. The argument requires inference based on reasoning. Evidence consisting of facts and the opinions of authorities is used to bolster the argument. In classical debates, there is little room for personal opinion and no room for personal attacks. Debaters are taught to attack the logic and doubtful facts of the opposition. Each team in a classical debate tries to win; it is not an exercise in cooperative consensus building.

Debate is an excellent way to involve students in the material, work on communication skills, and require group effort. Unfortunately, in most classes the classical debate approach involves too few students. More students can be involved by increasing the size of the debate teams and by having more than two teams. For example, in our teaching class one team champions lectures, another PSI, and another cooperative groups. In a debate on siting a new airport each team champions a different site.

Many ways of running a modified debate are possible. We have found three groups with three members each to be convenient. Students are assigned to groups in advance and are given the topic to support. (An interesting alternate for advanced students is to have them prepare for all positions without knowing in advance which side they will be for.) The groups are told that each student will talk for four or five minutes. The first speaker from each group is told to take an affirmative position and present only positive statements. After each group has presented its affirmative positions, a second speaker from each group takes a mixed affirmative and negative position. The last speaker rebuts any damaging statements from the first two rounds and summarizes the team's position. The teams decide who goes first, second, and third and what will be presented.

The professor must first assign teams. We try to balance the teams as much as possible. The professor must choose the topic and pick the sides. The rules of the debate are spelled out, and the idea of an argument backed by evidence is explored with an example. Reynolds (1976) also requires that debaters prepare a position paper in advance which is turned in immediately after the debate.

One of the nonparticipants can serve as the debate moderator. Others can serve as judges; this makes the entire class active participants. It helps to give the judges a rating sheet so that judging is somewhat uniform. We use a rating sheet with five 5-point scales: analysis, evidence, argument, refutation, and delivery. The rating sheets are collected, and the team with the most points is declared the winner.

In our classes debates have always proven to be marker events. The students prepare hard and try to win. The competitive nature of the debate is a strong motivator for many students even though the results have little effect on the their grades. A debate is also another opportunity to practice communication skills, to improve analysis and evaluation, and to work together in teams.

7.3.3. "Quiz Shows"

Another break in the usual routine is to use one class period as a quiz show following the format of Trivial Pursuit, Jeopardy, or College Bowl. This can be done with either individuals or groups. Students are told to become experts on the class material. The participants can be selected in advance, or at random on the day of the quiz show. As in most competitive activities, this procedure works best if the teams or contestants are evenly matched. The professor can act as moderator and ask the questions. The contestant who presses a buzzer or rings a bell first gets to answer the question first. Points are awarded for correct answers and subtracted for mistakes. The winner or winning team is the one with the highest score at the end of the show.

This format works best for knowledge-level questions since they have the most straight-forward one-line answers. The professor needs to generate the questions and answers ahead of time. A panel of judges can be selected from the noncontestant students to decide if answers are correct. Another noncontestant can judge which student was first at pushing the buzzer. Since this type of quiz show is intense, twenty to twenty-five minutes of the period is probably sufficient.

We have never had the opportunity to use this procedure in class (we have used Trivial Pursuit in a student fund raiser) but think it would be a good break in a class where the students have to learn a large number of facts. Because of the competitive nature of a quiz show, many students will prepare diligently to try to win.

7.4. MASTERY AND SELF-PACED INSTRUCTION

Requiring students to achieve mastery in each topic is an appealing idea. However, this concept is more complex than it first appears. Once the concept has been explored, two instructional methods utilizing mastery will be discussed: self-paced (the Keller plan) and instructor-paced mastery courses. This is a logical but not chronological sequence. (The development could have logically occurred in the order presented but did not. In engineering education the Keller plan became quite popular before the key element, mastery, was isolated.) This section is important since these teaching methods are the only methods which show a clear advantage in the amount students learn. The extensive review by Taveggia and Hedley (1972), which found no difference in learning based on content examinations, did not include mastery-type classes.

7.4.1. Mastery

Mastery is a very simple, yet powerful, idea: Make students understand material well before allowing them to move forward. For hierarchical material this concept makes a great deal of sense. For any material, retention is better and relearning is easier when material has been mastered. In addition, success is motivating and the opportunity to master a subject often convinces students that they can learn.

If mastery is to be required, the material must first be divided into units or modules and objectives must be developed for each unit. Then the students must be tested for mastery of the objectives. Students who have not mastered the material need prompt feedback and probably some type of aid in learning the material. Repeated tests may be required. Thus, in a mastery course all students could theoretically earn A's, but the time required would vary significantly. In courses graded on a curve, grades correlate with ability, while in a mastery class the time required correlates with ability (Bloom, 1968; Stice, 1979). The need for repeated tests requires some modification in class schedules. Two different ways to do this are discussed in Sections 7.4.2 and 7.4.3.

What does mastery mean? For simple, lower-level cognitive objectives an unequivocal definition is easy. For example, the student can spell 100 words perfectly, or the student can quote the Gettysburg Address, or the student can repeat the definition of technical words without error. Since 100 percent is not required to achieve an A when straight-scale grading is used, mastery can be defined as 90 or 80 percent accuracy. Once the number (80, 90, or 100 percent) has been agreed on it, is easy to determine if the student has mastered the material.

But how does one determine mastery for higher-level cognitive objectives? In engineering most problems involve either application or analysis. Even for relatively simple technical concepts an infinite number of problems can be generated. How does the professor decide if the student has mastered the material? This question has been argued strongly by critics (e.g., Gessler, 1974). We think these arguments miss the practical point. Any professor who routinely awards partial credit for problems can separate student tests into mastery, near-mastery, questionable, and not-mastered piles. The near-mastery pile includes the tests of students who clearly understand the theory and how to apply it but have made a mistake in algebra or arithmetic. These students should probably be allowed to move forward. Students whose examinations are placed in the questionable pile can be talked to individually to see if they understand the concepts. Alternately, they can be told to study more and take another test—the only penalty is time, not a grade. Our conclusion, based on nine years of experience with mastery tests, is that there is no practical difficulty in using mastery learning for application and analysis problems in engineering.

Synthesis problems may present a practical difficulty. However, grading synthesis problems or grading for creativity presents a practical difficulty with any grading scheme. Our pragmatic solution has been to include a few synthesis problems where appropriate and then to score them very leniently. Mastery is probably not an appropriate grading scheme for design courses which include a significant amount of synthesis.

A second major question is, Can all students master the material if given sufficient time? The answer is probably no, but the percentage who can is much higher than the percentage that do with other teaching methods. Bloom (1968) found that 80 to 90 percent of the students in a mastery class could achieve test scores which would have given them an A in a lecture class (where 20 percent earned A's). In many engineering classes concrete-operational thinkers (see Section 14.1.1) will be unable to master the material. There are also students who could master the material but are unwilling to work hard enough or decide they do not want to be engineers. The vast majority can and do master the material. As a rough rule, Bloom (1968) thought that 90 percent of students can benefit from mastery learning, 5 percent will stumble, and 5 percent will master the material with any teaching technique.

If we adopt the concept of mastery learning, what is good instruction? Instruction which helps the student efficiently master the objectives is good instruction. This means that instruction must be individualized. Bloom (1968) states that the optimum would be a talented, dedicated tutor for each student. Before dismissing this as utopian, note that throughout grade school and high school many middle-class students have exactly this situation—their parents tutor them. The Keller plan can come close to reaching this ideal (see Section 7.4.2).

There are a host of practical questions. How big should the modules be? What are the important objectives? (This question should be asked in every course regardless of the teaching method used.) How does one arrange the schedule to allow for test retakes and extra learning time? If almost everyone masters the material, how does the professor grade? What method is used for presenting content? How do students receive feedback? How do students receive help if they do not understand a concept? These practical issues are discussed in Sections 7.4.2 and 7.4.3.

The results from many different types of mastery courses show that based on tests students learn more than they do with other teaching methods (Bloom, 1968; Hereford, 1979; Kulik et al., 1979; Stice, 1979). The previously cited extensive comparison of teaching methods (Taveggia and Hedley, 1972) found no differences between teaching methods in the amount students learned, but did not include mastery courses in the comparisons. In addition to learning more, students in mastery courses like the subject, are motivated to learn, and have an improved self-concept.

All teaching methods have disadvantages. These will be discussed when the detailed course types are considered in Sections 7.4.2 and 7.4.3.

7.4.2. Self-Paced Courses (Keller Plan or Personalized System of Instruction)

Self-paced courses handle the scheduling problem by letting students decide what pace they want. They are allowed to take mastery tests whenever they wish and thus can move through the course at their own pace. Several variants of the self-paced or personalized system of instruction (PSI) have been adopted in engineering. It is useful to consider the basic plan which was first developed by Keller (1968) in a psychology course.

What the student first sees in a Keller plan course are a course outline and a set of instructions. The student then gets a study guide and studies alone or in groups. When ready, he or she reports to a proctor and takes a test. The proctor grades the test with the student present. If the test is in the uncertain category, the proctor asks the student a few questions. If the student passes, the proctor marks the student as passed and gives him or her the next study guide. If mastery has not been achieved, the student studies some more before returning to take a different test on the same topic. The student continues to take tests on the area until the topic is mastered. After each test he or she automatically receives some tutoring as the proctor points out the mistakes and explains why the answers are wrong. After all required units are completed, there may be optional units and/or a final examination. A Keller plan course has the following six recognizable characteristics:

1 The course is self-paced. In the pure form no pressure is put on students to complete units at a given time. Many professors have found that for practical reasons students need to be encouraged to complete modules at some minimum rate.

2 The course is modularized, there are clear objectives for each module, and learning materials such as a study guide and a textbook are available. Clear objectives and the availability of learning materials are the necessary and sufficient requirements so that students learn as much as with other teaching methods.

3 Mastery. Mastery appears to be the key reason why students in PSI courses learn more than in nonmastery courses.

4 Undergraduate proctors as tutors to grade mastery tests and provide immediate feedback and help to students. The use of undergraduate proctors is extremely helpful and is appreciated by the students taking the course. Proctors can approach the ideal of providing individual tutoring for each student. In addition, the proctors learn a good deal and often become motivated to go on to graduate school. However, proctors do not seem to be essential for success as long as there is reasonably rapid feedback and help is available.

5 Lectures and demonstrations are used for motivation but not for transfer of basic information. This is clearly not necessary for the success of students using the method, and in instructor-paced classes lectures can be used for information transfer (see Section 7.4.3). Lectures may be necessary for the success of the professor since it is widely believed that "teaching and talking go hand in hand" (Keller, 1985).

6 Written and oral communication are used for testing. It is clear that a mastery class can be successful with only written communication on tests, and we see no reason why only oral communication could not be used.

Based on over twenty-five years of experience and experimentation since Keller first tried self-paced courses, it appears that there are three successful ingredients:

1 The course must be modularized with clear objectives and available learning materials.
2 Mastery must be required, but the exact level set (e.g., 80, 90, or 100 percent) is not critical.
3 Prompt feedback is necessary.

Regardless of who does the grading and provides the feedback, one result of a mastery course is that poorer students are forced to obtain more practice and receive more help than better students. This is the reverse of what often happens in nonmastery courses. The other details used by Keller are not critical for success (of course, if self-pacing is not used, the course is not a Keller plan course but can still be a mastery course).

Many variations in grading have been used in PSI courses. Keller (1968) based about 75 percent of the grade on the number of mastery quizzes which were successfully passed and 25 percent of the grade on the final. There is no penalty for taking a quiz and failing it. Some professors have required that students complete all required sections and then have awarded an A when this was done. The course grade distribution was either an A or an F/incomplete. This procedure has been extensively criticized. Some professors award a C when the basic modules have been completed and allow students to work for a higher grade with optional modules, an optional final, or other optional learning activities such as computer programs. This is a type of *contract grading* where the student contracts to do a specified quantity of work to earn a grade. The professor can also base the entire grade on the final examination which the student takes after completing the required modules. Grades in mastery plan courses are usually higher than in nonmastery courses. Mastery courses have been criticized for this; however, since the students are learning more, why shouldn't they earn higher grades?

No longer a lecturer, the professor becomes a facilitator of learning and chooses the content to cover, develops the objectives, selects learning material such as articles and textbooks, and writes the study guides. The professor must write the mastery tests and decide what constitutes mastery. He or she supervises the proctors or TAs and checks the grading. In many schools proctors are hired, though in small classes the professor may do the grading. The professor helps to motivate students and helps with the tutoring, particularly when the student has difficult questions. The professor is responsible for selecting the grading scheme and for assigning the final grades.

Billy Koen of the University of Texas first introduced the method into engineering education in a nuclear engineering course in 1969. A very wide variety of engineering courses have been taught by variations of the PSI method in every engineering discipline (e.g., Baasel, 1978; Conger, 1971; Craver, 1974; Grayson and Biedenbach, 1974; Harrisberger, 1971;

Flammer, 1971; Koen, 1970; and Koen et al., 1985). Because of a variety of time, money and administrative constraints, engineering professors have often modified the standard Keller plan. Pressure is often applied to students to keep them progressing in the course. Most professors do not present the motivation lectures or demonstrations. A TA or the professor may substitute for the undergraduate proctors. Tests may be only in written form with no opportunity for oral explanations. Since these changes keep the three key components intact, these courses are usually successful.

As noted in the previous section, students learn more in PSI courses than in nonmastery courses, and students do better on common final examinations (Keller, 1968; Kulik et al., 1979; Stice, 1979). Stice (1979) found that 75 percent of students preferred PSI to lecture courses. Small classes received particularly high ratings (this is not a surprise; see Section 16.3.3), and ratings were high in all classes (Hereford, 1979).

There are some problems with self-paced courses. The first time a professor teaches a PSI course the time commitment is roughly twice that for a nonmastery course (Craver, 1974; Hereford, 1979; Stice, 1979). This experience has prevented some professors from continuing with PSI. The good news is that subsequent offerings take about as much professorial time as lecture classes. Proctor costs are real, and PSI courses may be a bit more expensive than other classes. Hereford (1979) found that proctor costs ranged from $16.11 to $50.80 per student with a mean of $33.60. Because of inflation these costs have undoubtedly increased. However, there are major benefits of using carefully selected undergraduate proctors, and if they can be afforded they are a plus.

One advantage of PSI courses is that students are not competing with each other for a grade. Thus, they can be encouraged to cooperate. However, in most PSI variations no formal effort is made to arrange for cooperation, and some students work through the course in total isolation. These students talk only to proctors, and if the student masters the material this contact may be minimal. This shortcoming can be overcome without compromising the PSI procedure by developing cooperative groups and encouraging students to work together.

Procrastination can be a major problem because it can lead to excessive drops, incompletes, and lower grades. Drops increase because students realize that they are far behind and feel that they cannot catch up. Incompletes increase if students are allowed to receive an incomplete if they don't finish on time. This can be controlled by allowing incompletes only if the student meets the university's requirements for an incomplete which usually means illness, involuntary military service, or death in the family. Grades often decrease since the grade is based on the number of units the student has finished. In addition, procrastination spreads out the tests students take. This is a burden for the graders since they must be expert in a wider range of material and must have more tests available. Procrastination is worse with freshmen and seniors and is much worse with instructors who are inexperienced in using PSI (Hereford, 1979). Clearly, there are things the instructor can do to reduce procrastination. Students can be told the rules on incompletes, and they can be given both an average rate of progress and a minimum rate of progress. The professor or proctors can call and confront students who fall behind. All of these are successful in reducing procrastination, but they do compromise the concept of self-pacing. Extreme measures to control procrastination lead to an instructor-paced course.

7.4.3. Instructor-Paced Mastery Courses

A variety of instructor-paced mastery courses have been devised (Block, 1971, 1974; Bloom, 1968; Stice, 1979; Wankat, 1973). In the original development of this procedure (Block, 1971, 1974; Bloom, 1968) the instructor used whatever group teaching procedure that he or she wanted. The students took regularly scheduled formative examinations which were scored but not graded. The instructor marked the tests as mastery or not mastery. For each problem missed the student received information about alternate learning resources to learn the material. This diagnosis of problems is the key step in this procedure. The learning resources could consist of specific passages in other textbooks, articles, programmed texts, audiovisual material, workbooks, and so forth. The use of an alternate to the first way the student has studied helps to individualize the instruction for each student. Students were expected to study and learn the corrective material on their own time. Since the formative tests were not graded and did not affect the student's grade in the course, students were encouraged to cooperate with each other and with the professor to learn the material. In other words, the class and the professor became a team that tackled the real enemy—the content to be learned. All the students proceeded through the course unit by unit at the same rate. Students who had not mastered a previous unit were also simultaneously studying the unit they had not mastered. At the end of the semester the class was given a final examination which was scored and graded. The course grade depended entirely upon the final. Bloom (1968) found that 80 percent of the students received A's on the same final that 20 percent of the students in a nonmastery course had received A's on. When the formative examination results were compared to the previous year as a measure of progress, 90 percent of the students received A's. In this case the instructor spent extra time on those topics with which students were having additional problems.

In an absolute sense mastery was not required in these applications as it is in PSI courses. The frequent formative evaluations and diagnostic feedback were apparently sufficient for the students to learn more than in a usual class. The course was also modularized and had clear learning objectives. Feedback to the students was highly emphasized and was individualized to help each student learn. Unlike the situation in PSI, the instructor did "teach" in addition to structuring the course. As in PSI, students were not competing with each other. This was true even on the final since the grade necessary for an A was predetermined by what students in a lecture class had achieved.

The success of this type of course calls into question the need to make students achieve exact mastery on every test, and also makes moot the argument about what mastery is. However, based on our experience, a few students slip through who do not know the material well, and they do poorly on the final. This can be prevented with a instructor-paced mastery class which requires students to pass each formative test.

Our experience has been in developing and using such an instructor-paced mastery course (Wankat, 1973). The course was developed as an elective course for seniors and graduate students. To avoid procrastination problems, which can be severe with seniors, students were forced to move with the instructor. Each week the first mastery quiz on the old material was given on Tuesday, a lecture on new material on Thursday, and a repeat quiz on the old material on Saturday. The results of the first mastery quiz were posted on Wednesday. Students who did not master the material were required to come on Saturday and had to turn in homework before taking the repeat quiz. On Saturday the professor and the TA graded the quiz while the student watched; the mistakes were explained so that the student did not repeat them. Because of budget constraints proctors were unavailable and the staffing was the same as for a lecture course. If students did not pass the first repeat quiz, they had to return the next Saturday. Because of university scheduling, the quizzes on Tuesdays were timed, but the Saturday quizzes were not. With this arrangement some students fell quite far behind. The insertion of a two-week computer design module with no new Tuesday quizzes in the middle of the semester allowed them to catch up. Students who mastered the twelve required modules received a C. They could improve their grades by exercising one of three options: writing a computer program, mastering an optional module in a maximum of three attempts, or mastering the final in one attempt. Many graduating seniors worked for a C or a B and did not try to earn a higher grade.

The instructor informally compared the results to previous years and found that the students learned more. In most years when the course was taught there were no D's, no F's, and no incompletes. There were slightly more A's, many more B's, and fewer C's than when the course was taught as a lecture. Student ratings were very favorable. However, students who earned C's thought that they had put in more work and learned more than required for a C in other courses. Interestingly, the mastery course took an unfavorable schedule (Saturday morning classes) and turned it into an advantage. Many students studied diligently to avoid coming to the Saturday class. The instructor's time requirements were very similar to those reported for PSI classes.

7.5. INDEPENDENT STUDY CLASSES: INCREASING CURRICULUM FLEXIBILITY

In its simplest form, an independent study class consists of a study guide, a textbook, and a final examination. A student follows the study guide, reads the textbook, works any appropriate problems, and takes the final examination when ready. The student's grade is determined entirely by the final examination. If the study guide includes detailed objectives and the textbook is well written, any student with enough self-discipline to work through the material should do well on test questions at the lower levels of Bloom's taxonomy. Obviously, this approach will not work well for fostering higher-level cognitive skills, communication skills, and teamwork. Although uncommon in engineering, such independent study courses are fairly common in the humanities and social sciences. Independent study courses have the advantage of ultimate flexibility in scheduling. It is not necessary that the student complete the course in one semester, and either more or less time can be used.

Many variants of the basic independent study course are possible. Lectures can be carried on radio or TV, and taped lectures may also be available. Then students have the option of listening to the broadcasts in addition to reading the text. This additional mode of information transfer may be useful to some students. The broadcast schedule can also serve to provide some structure and as an indication of how fast they must progress in the course.

Another variant of independent study is to have a tutor available to answer questions and check homework problems. Otherwise, the student's pace and learning are independent and the course grade depends on test results. We have used this procedure to satisfy a very important prerequisite requirement in the chemical engineering curriculum. No test was given, and no course credit was earned; however, students were allowed to take the prerequisite course as a corequisite. Because of the structure of prerequisites in chemical engineering, this procedure allowed transfer students to graduate in two instead of three years. Over about a ten-year period we have had very good success with this use of independent study in a select group of motivated students. Since these students were seeing the material in the required course for the second time when they took it for credit, it is perhaps not surprising that they tended to do well. The only quality control applied was the requirement that the tutor be a chemical engineer and sign a letter stating that the student had covered the required book chapters; the tutor also had to list the homework problems that were worked.

Various other independent study options could be useful in providing flexibility in otherwise inflexible curricula. In addition to allowing students to take a prerequisite course as a corequisite, independent study could be used to allow students to continue taking engineering classes after flunking a required course. This would be particularly useful at schools where courses are offered just once a year and would reduce some of the pressure on students and professors. The independent study course would again satisfy the prerequisite requirements only—the student would have to retake the course for credit when it was reoffered. Since the reoffering would, in effect, be the third time, many students would be able to pass an otherwise impossible course. Independent study options would also be of interest to select students during the summer or when on co-op assignments.

The professor's task in these independent study options is first to decide what the essential material is and then develop the key learning objectives for this material. Next he or she must determine the required sections of the book and some representative homework problems. Finally, if the option will be used for a credit course, the professor must select the test(s) that will be used to grade the student.

Independent project and thesis courses are fairly common in engineering. They often involve fairly close work with a professor or graduate student and may involve student teams (see Sections 10.4 and 11.4).

7.6. FIELD TRIPS AND VISITS

Field trips and visits to local facilities are an underutilized teaching method in most engineering courses. Seeing real equipment or manufacturing operations provides students with a concrete, visual, and often kinesthetic learning experience. Such first-hand experience can make abstract equations seem much more real, and the trips can be motivating to many students. These trips can also serve as marker events. (We remember field trips that were taken 25 years ago, while we rarely remember individual lecture classes.)

Unfortunately, many engineering professors believe the myth that a field trip has to be an all-day affair which requires much time to set up. Such longer trips are often necessary to see particular types of engineering operations. However, local trips to facilities on campus or at the university's research park can often be completed in one class period, or even part of a period, and can provide a useful supplement for many courses. For example, many freshmen or sophomores will benefit from a "field trip" to the senior laboratory down the hall. This can be done in the last ten or fifteen minutes of a class. A class studying power production can visit the university's power plant, which is also of interest to a class studying cooling towers. Classes in structures or foundations can visit the sites of new buildings or bridges. Environmental engineers can visit the local wastewater treatment plant. Industrial engineers can obviously benefit from visiting any manufacturing facility, but less obviously can also learn from seeing the university's printing and mailing rooms or from a visit to a local travel agent. Practical information on steam transmission can be found in the basement of many campus buildings. Many research laboratories have specialized equipment which will at least give students an idea of what something looks like.

Field trips and visits offer many advantages: They are often a welcome break in the routine, are visually and kinesthetically rewarding, are often marker events, and provide the concrete experience of seeing real equipment and engineering operations, which can be motivating, with "real" engineers explaining the equipment or operation. Disadvantages include the loss of time for covering content and the loss of some control of what happens. In addition, appropriate trips require work to set up and this must be done well in advance, long distance trips are very time-consuming and arrangements must be made for students to miss one day of classes, trips away from campus cost money and often the professor has to find an "angel" to cover the cost, and some students do not take the trip or visit seriously since it is not covered on the test.

Our experience has been that ten-to fifteen-minute visits are very useful motivators for sophomores. Longer field trips are useful for seniors who have not had industrial experience, but the scheduling can be difficult. Optional trips arranged by a student organization are a useful alternative, and students often have an easier time raising money. In our class on teaching methods visits to local audio-tutorial laboratories, computer teaching presentations, and laser videodisc demonstrations have been among the highlights of the semester.

7.7. CHAPTER COMMENTS

This chapter has been a smorgasbord of different methods which can be used either as part of basically a lecture class, as a break in a class, or instead of lecturing. All these methods try to involve the student with the content and work to make him or her active. The methods included here certainly do not exhaust the possibilities. With some creativity engineering professors can develop new variations to involve their students.

You may decide that you have no interest in using cooperative group or mastery techniques; however, both of these methods clarify the need for clear learning objectives. Cooperative group learning has emphasized that professors should focus more on what the students do and less on what the professor does. Mastery learning shows clearly that a criterion-referenced grading scheme can be used, and that professors do not have to grade on a curve. All these truths can be adopted in other teaching methods.

Introducing change in the classroom can be difficult. Professors cling to lecturing because it gives them power, minimizes risk, and is socially acceptable within their department. Many students also prefer known teaching methods because the known is more secure. If a new teaching method has been used previously, student acceptance can be increased by showing the improved grade distribution obtained with the new method as compared to the old method (Tschumi, 1991). Elective courses are a good place to experiment since professors are scruntized less and the students are all volunteers.

7.8. SUMMARY AND OBJECTIVES

After reading this chapter, you should be able to:

• Outline the use of and discuss the advantages and disadvantages of the following teaching methods.

Discussion	Mastery and PSI
Cooperative group learning	Individual study courses
Panels, debates, and quiz shows	Field trips

• Incorporate appropriate methods into an engineering class taught by lecture.
• Develop an engineering course taught by a method where lecturing is clearly a supplemental teaching method instead of being the major teaching method.

HOMEWORK

1 Choose a specific undergraduate engineering course which is normally taught using the lecture method. Determine how you can incorporate two of the teaching methods listed in the first objective in Section 7.8 into the lecture course. Explain what you would accomplish by doing this. Develop your script for one day using one of the methods, and for another day using another method.
2 Choose the same engineering course selected in problem 1. Determine how to teach it using a nonlecture method. Prepare a detailed script for two days of class.

REFERENCES

Astin, A. W., *Achieving Educational Excellence*, Jossey-Bass, San Francisco, 1985.
Baasel, W. D., "Why PSI? How to stop demotivating students," *Chem. Eng. Educ.*, 78 (Spring 1978).
Block, J. H. (Ed.), *Mastery Learning: Theory and Practice*, Holt, Rinehart and Winston, New York, 1971.

Block, J. H. (Ed.), *Schools, Society and Mastery Learning*, Holt, Rinehart and Winston, New York, 1974.

Bloom, B. S., "Learning for instruction," Evaluation Comment, University of California, Los Angeles (2) (May 1968). Also RELCV topical paper and reprint no. 1.

Buchanan, W., "An experiment in pairs testing with electrical engineering technology students," *Proceedings ASEE Annual Conference*, ASEE, Washington, DC , 1764, 1991.

Cashin, W. E. and McKnight, P. C., "Improving discussion," Idea Paper No. 15, Center for Faculty Evaluation and Development, Kansas State University, Manhattan, KS, 1986.

Conger, W. L., "Mass transfer operations—A self-paced course," *Chem. Eng. Educ.*, 5(3), 122 (1971).

Craver, W. L., "A realistic appraisal of first efforts at self-paced instruction," *Eng. Educ.*, 448 (March 1974).

Davis, R. H., Fry, J. P., and Alexander, L. T., *"The discussion method," Guides for the Improvement of Instruction in Higher Education, No. 6*, Michigan State University, East Lansing, MI, 1977.

Eble, K. E., *The Craft of Teaching*, 2nd ed., Jossey-Bass, San Francisco, 1988.

Felder, R. M., "Stoichiometry without tears," *Chem. Eng. Educ.*, 188 (Fall 1990).

Flammer, G. H., "Learning as the constant and time as the variable, *Eng. Educ.*, 511 (March 1971).

Gessler, J., "SPI: Good-bye education?" *Eng. Educ.*, 252 (Dec. 1974).

Goldstein, H., "Learning through cooperative groups," *Eng. Educ.*, 171 (Nov. 1982).

Grayson, L. P. and Biedenbach, J. M. (Eds.), *Individualized Instruction in Engineering Education*, ASEE, Washington, DC 1974.

Hamelink, J., Groper, M., and Olson, L. T., "Cooperation not competition," *Proceedings ASEE/IEEE Frontiers in Education Conference*, IEEE, New York, 177, 1989.

Harrisberger, L., "Self-paced individually prescribed instruction," *Eng. Educ.*, 508 (March 1971).

Hayes, R. M., Friel, L. L., and Hess, L. L., "Learning groups in statics instruction," *Proceedings ASEE/IEEE Frontiers in Education Conference*, IEEE, New York, 152, 1985.

Hereford, S. M., "The Keller plan within a conventional academic environment: An empirical 'meta-analytic' study," *Eng. Educ.*, 250 (Dec. 1979).

Hudspeth, M. C., Shelton, M. T., and Ruiz, H., "The impact of cooperative learning in engineering at California State Polytechnic University, Ponoma," *Proceedings ASEE/IEEE Frontiers in Education Conference*, IEEE, New York, 12, 1989.

Johnson, D. W. and Johnson, R. T., "Conflict in the classroom: Controversy and learning," *Rev. Educ. Res.* 49(1), 51 (Winter 1979).

Johnson, D. W. and Johnson, R. T., *Cooperation and Competition: Theory and Research*, Interaction Books, Edina, MN, 1989.

Johnson, D. W., Johnson, R. T., and Smith, K. A., *Active Learning: Cooperation in the College Classroom*, Interaction Book, Edina, MN, 1991.

Keller, F. S., "Good-bye, teacher...," *J. Appl. Behav. Anal.*, 1, 79 (1968).

Keller, F. S., "Testimony of an educational reformer," *Eng. Educ.*, 144 (Dec. 1985)

Koen, B. V., "Self-paced instruction for engineering students," *Eng. Educ.*, 60(7), 735 (1970).

Koen, B. V., Himmelbau, D., Jensen, P., and Roth, C., "The Keller plan: A successful experiment in engineering education," *Eng. Educ.*, 280 (Feb. 1985).

Kulik, J. A., Kulik, C. C., and Cohen, P. A., "A meta-analysis of outcome studies of Keller's personalized system of instruction," *Am. Psychol.* 34(41), 307 (1979).

Lowman, J., *Mastering the Techniques of Teaching*, Jossey-Bass, San Francisco, 1985.

McKeachie, W. J., "Student anxiety, learning and achievement," *Eng. Educ.*, 724 (April 1983).

McKeachie, W. J., *Teaching Tips. A Guidebook for the Beginning College Teacher*, 8th ed., D.C. Heath, Lexington, MA, 1986.

Palmer, P. J., *To Know As We are Known: A Spirituality of Education*, Harper-Collins, San Francisco, 1983.

Reynolds, T. S., "Using simulated debates to teach history of engineering advances," *Eng. Educ.*, 184 (Nov. 1976).

Shelton, M. T. and Hudspeth, M. C., "Cooperative learning in engineering through academic excellence workshops at California State Polytechnic University, Ponoma," *Proceedings ASEE/IEEE Frontiers in Education Conference*, IEEE, New York, 35, 1989.

Smith, K. A., "Structured controversies," *Eng. Educ.*, 306 (Feb. 1984).

Smith, K. A., "Cooperative learning groups," in Schomberg, S. F. (Ed.), *Strategies for Active Teaching and Learning in University Classrooms*, Continuing Education and Extension, University of Minnesota, Minneapolis, MN, 1986.

Smith, K. A., "The craft of teaching. Cooperative learning: An active learning strategy," *Proceedings ASEE/IEEE Frontiers in Education Conference*, IEEE, New York, 188, 1989.

Stice, J. E., "PSI and Bloom's mastery model: A review and comparison," *Eng. Educ.*, 175 (Nov. 1979).

Tannen, D., *You Just Don't Understand*, Ballatine Books, New York, 1990.

Tannen, D., "Teacher's classroom strategies should recognize that men and women use language differently," *Chronicle Higher Educ.*, B1 (June 19, 1991).

Taveggia, T. C. and Hedley, R. A., "Teaching really matters, or does it?," *Eng. Educ.*, 546 (March 1972).

Tinto, V., *Leaving College: Rethinking the Causes and Cures for Student Attrition*, University of Chicago Press, Chicago, 1987.

Topping, K., "Cooperative learning and peer tutoring: An overview," The Psychologist, 5(4), 151 (April 1992).

Tschumi, P., "Using an active learning strategy in introduction to digital systems," *Proceedings ASEE Annual Conference*, ASEE, Washington, DC, 1987, 1991.

Wankat, P. C., "A modified personalized instruction—lecture course," *Proceedings ASEE/IEEE Frontiers in Education Conference*, IEEE, New York, 144, 1973.

CHAPTER 8

TEACHING WITH TECHNOLOGY

This chapter also focuses on the how-to of teaching, except that here technological means are used to deliver the instruction. The delivery media include television and video, computer and laser videodisc, and audiotutorial. Many different teaching *methods* such as lecture, interactive tutoring, discussion, and drill can be used with different delivery media. Television and video are discussed first because these media are often used with the traditional lecture method of Chapter 6. In universities, educational television has been used to deliver lectures to remote sites or at different times. Television and video are also useful as backups for live lectures and for providing feedback to students. A computer can be used as a tool to reduce the repetitive nature of calculations (see Sections 8.2.1 and 8.2.2 on spreadsheets and equation solvers and simulation programs), while most of the teaching uses traditional teaching methods and a live delivery medium. A computer can also replace the traditional live delivery through computer-aided instruction (Section 8.2.3) or interactive laser videodisc (Section 8.2.4). The audiotutorial technique involves a combination of technology, laboratory, and other teaching methods.

In this chapter it is necessary to draw a distinction between the teaching method and the delivery medium (see Figure 8-1). A teaching method (lecture, discussion, drill, etc.) is chosen and then paired with a delivery medium (live interaction, live TV, videotape, noninteractive computer, etc.) to reach the learner. The general flow sheet is shown in Figure 8-1a, and specific applications are shown in Figures 8-1b to 8-1g. In Chapters 6, 7, 9, and 10 the delivery medium is usually live interaction. In this chapter various technological media are used to deliver the instruction.

Over the years, the introduction of new technology for education has generated initial high excitement, but that has been followed by disillusionment, although eventually most technologies find a niche in the educational system. Throughout this chapter we will consider what delivery of instruction by technological media can do better than the nontechnological

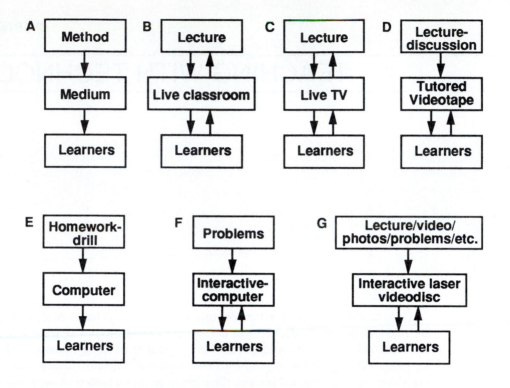

FIGURE 8-1 INTERACTION OF TEACHING METHODS AND DELIVERY SYSTEMS.
A. General flowsheet. B. "Normal" lecture (Chapter 6). C. Live TV (8.1.1) D. Tutored videotape
(8.1.2). E. Non-interactive CAI drill (8.2.2). F. Interactive CAI (8.2.2). G. Interactive laser
videodisc (8.2.3).

delivery alternatives such as lecture, discussion, cooperative groups, and PSI. Gibbons et al.
(1977) present the following list of guidelines for the successful use of technology in
education:

1 Plan use for a specific audience.
2 Define objectives which are relevant to the audience.
3 Pick a technological medium and a teaching method which are appropriate to the topic.
4 Pick educators interested in using the technology.
5 Plan for personal interaction, particularly among students.
6 Monitor the course and change materials and methods as appropriate.

Of course this list can be applied to any teaching method if the words "teaching method"
replace "technology." If use of the technological medium does not have an advantage as
compared to nontechnological delivery, the combination of technological delivery medium

and teaching method will probably not survive after the innovator has moved on to other activities.

8.1. TELEVISION AND VIDEO

We will discuss television and video as delivery media for the education of engineering students (Section 8.1.1), describe a particular form of instruction with video called tutored videotape instruction (Section 8.1.2), discuss the steps the professor should take to improve television teaching (Section 8.1.3), and finally, briefly consider the use of television as feedback for students (Section 8.1.4).

8.1.1. Instructional Delivery by Television and Video

What can delivery of instruction by television or video do better than other means of delivering instruction? First, television and video make it possible to provide instruction at remote sites. This ability has been extensively used for continuing education and graduate programs for engineers employed in industry away from universities. Second, television can be used to break a huge class into much smaller sections. Third, videos provide flexibility in that they can be observed at any time. And fourth, a professor can use them to make "electronic" field trips to observe technology.

"Distance education," or the use of television and video to deliver instruction at remote locations, has become important in both continuing education and graduate education as well as in many fields in addition to engineering. It has even spawned *The American Journal of Distance Education* (Penn State University, 205 Rackley Building, University Park, PA 16802). Both live television delivered by satellite and video are used, although the applications are somewhat different. Most universities that have used television have used it for graduate-level courses. Practicing engineers can continue their education to a master's degree with minimal disruption of their careers and of their family life. Seigel and Davis (1991) also found increasing acceptance of television for undergraduate engineering courses, and thirty-nine of the schools surveyed allowed this option. On a national scale, the National Technological University (NTU) collaborates with many universities to present a wide variety of live broadcasts most of which are of interest to engineers and scientists. These vary from single two- to three-hour broadcasts for continuing education, to three-credit courses, to ninety-hour certificate programs, to masters programs in engineering. For more information write to the National Technological University, 700 Centre Ave., Ft. Collins, CO 80526. The Public Broadcasting System (PBS) has televised college-level courses since 1981. Almost 2 million students have earned credit toward undergraduate degrees at a variety of universities (Anonymous, 1991). Unlike the courses at the NTU, the PBS courses are of much more general interest, and PBS is not the credit-granting institution.

Live educational television in engineering has usually employed a teaching method where a professor lectures in a television studio. Often there is a live audience of students taking the course for credit at the university and at a number of remote sites. A typical studio has a camera for the professor, an overhead camera for the notes the professor writes on a tablet, and a camera for the audience. Students in the studio audience can ask questions of the professor, and their questions are picked up by microphones so that the remote sites can also hear them.

The remote sites usually have some form of two-way communication with the professor. The most common form is two-way audio over telephone lines with speaker phones. This is certainly the cheapest form of two-way communication, and in most instances it is adequate. In engineering some form of visual feedback is also very useful since it is difficult to discuss equations or drawings with audio alone. An electronic blackboard such as the AT&T Gemini 100 is one option (e.g., Gupta, 1981; Walker and Donaldson, 1989), but it was not widely adopted by schools and is now obsolete. Audiographics has proven to be a more acceptable technology for interactive delivery of both audio and graphics to remote sites (Chute and Elfrank, 1990). Audiographics uses networked personal computers to transmit text and graphics. Speakerphones can be used for two-way voice transmittal while television provides visual and audio transmittal, but audiographics can be used without the television link. A third option which is useful for equations is electronic mail, but this has also not been widely used since many remote sites are not on networks. Fax could be used, but the transmission delay might be a problem. Perhaps the best, but most expensive, solution is two-way video which is used by Washington State University (Howard and Peters, 1986). This example illustrates a common problem in education: An adequate technology is used because it is significantly cheaper than a technology with fewer limitations.

The television delivery system must be of high quality (Canelos and Mollo, 1986). This means that a professional-quality studio must be available. It is not feasible to do high-quality television or video without professional-quality equipment. The downlinks from the satellite must also be of professional quality for live television. These requirements mean that television courses are expensive. Live television becomes cost-effective when a large number of sites share the same broadcast. For a single remote site it is probably cheaper to have the professor travel to the site or to use video instead of live television. The quality of the professor's presentation is also critical, and this is discussed in Section 8.1.3.

The major instructional difficulties with live television are the lack of contact between the students and the professor, and the cost and difficulty in doing anything other than "straight" lecturing. Television can be an impersonal environment for learning. The term "distance" applies to psychological distance as well as geographic distance. Field-sensitive individuals in particular will have more difficulty adjusting to a television course. (See Section 15.2.1) Since the majority of engineers are field-independent, this will be less of a problem in engineering than in other fields. Still, the professor needs to do whatever is feasible to create a sense of contact and to help build rapport. Visits to the remote sites during the semester can help tremendously. Professors can also have phone office hours every week, although many students will not take advantage of this opportunity. Discussion and questions are more difficult in a live television course even with the students in the studio. Thus, the professor must increase the effort made at soliciting and answering questions. Television encourages student passivity, which is not productive for learning.

Television excels at showing visuals; unfortunately, most engineering programs do not take advantage of this characteristic. For example, a course in structures could include video of the site before, during, and after construction of a building or bridge. A course in robotics could show an actual assembly line in operation before and after the installation of robots. With planning and organization appropriate visuals can be included. For example, the professor can take a hand-held video camera to a construction site or into a plant. With some modest editing the result can be used as part of the television broadcast for both local and remote sites. Full utilization of television requires some creativity on the part of the professor.

Television and videotape classes on campus have the advantage of additional flexibility for students. If a student cannot schedule the class, he or she can always watch the videotape at a more convenient time. We had this experience in a live television class. Halfway through the semester a student was unable to attend the lectures, but he was able to keep up with the class by watching a videotape of the live broadcast. In addition, he presented an oral report to the class on videotape. Since finals were scheduled separately, he was able to take the final with the rest of the class. Although not widely used, videotapes could also be helpful to students with limited mobility who might prefer to watch a video in their homes rather than come to the campus every day.

Videotapes can also be useful supplements to other classes. A videotape can show how a laboratory experiment should be done (e.g., Kostek, 1991). Then instead of each group being shown how to do the experiment when it is their turn, they can be handed the videotape. We have used this procedure with good results even though the videotapes were homemade. Since the students were very motivated to learn from the tape and since they watched it in groups of three, the homemade character of the tape was not a problem. Videotapes of the apparatus or of various pieces of equipment can also be made to save time in the students' getting-acquainted process before the experiment begins. They can also be very useful for electronic field trips. In a biochemical engineering class, for example, students can use a videotape to observe the operation of specialized and often delicate equipment (Austin et al., 1990). Once produced, the videotape can also be used at schools which do not have the equipment but want to present an up-to-date course. Druzgalski (1988) notes a similar application where biomedical equipment and techniques can easily be videotaped and shown to classes which might not otherwise be able to see the procedures. Squires et al. (1991) used company-produced videotapes of plant tours to show students a chemical plant without the time and expense of a field trip. The advantage of involving the companies was that once they decided to support the videotape they paid for a professional company to produce it. These tapes can then be used at many schools to justify the production costs. Other applications of videotape supplements to other teaching methods await the ingenuity and energy of individual professors.

How well do students learn from television or video courses? Based on a variety of studies, the answer is that there are no significant differences between student learning from either television or video and from more traditional courses (Canelos and Mollo, 1986; Chute and Elfrank, 1990; Gibbons et al., 1977; Moore, 1990; Scidmore and Bernstein, 1986; Walker and Donaldson, 1989; Wergin et al., 1986). Although some studies have found that on-campus students do better, others have found that off-campus students do as well or slightly better. The net result is that the medium used is not critical. Much more important are the quality of the delivery and the message.

8.1.2. Tutored Videotape Instruction

The other system which has been used extensively for the delivery of classes to remote sites is tutored videotape instruction (TVI) which is illustrated in Figure 8-1d. This method was originally developed at Stanford University (Gibbons et al., 1977). With this technique a video is produced on campus by essentially the same procedures as live television. It seems to be most effective if the video is made of a live class. The video is then shipped to the remote sites where it can be shown at a convenient time. When the video is shown at a remote site, a local engineer who is qualified to help teach the material serves as a tutor. The video is shown for roughly five to ten minutes and then halted for questions and discussion. The next segment of the video is then shown followed by a question-and-discussion period. This procedure is repeated until the video is finished. The tutor may discuss example problems at any time. If there is a time constraint on class length, the video should be about thirty minutes long so that there is time for the questions and discussion. If the tutor is unable to answer any questions, the professor can be called on the telephone at prearranged times. This is apparently rarely necessary. The professor prepares homework and examinations, sends them to the tutor, and then supervises the grading after the tutor returns them.

This procedure is more flexible than live television and has more live contact except that the contact is with the tutor instead of with the professor. Groups should have from three to ten students with the optimum size appearing to be from three to eight students. This procedure can then act as a cooperative learning group (see Section 7.2.2) and will have the advantages of cooperative groups.

The selection of tutors is important. Gibbons et al. (1977) suggest that tutors at remote sites should be:

1 Practicing engineers at the site.
2 Have a personal interest in reviewing the subject but not be so expert that they will be bored by the tapes.
3 Have a desire to help teach the course.
4 Be sensitive to the needs of the students and able to draw them into discussion. Tutors with a discussion style are more effective than tutors who want to answer all the students' questions.

Tutored videotape instruction has also been used to advantage on campus (Gibbons et al., 1977; Robinson and Canelos, 1989; Scidmore and Bernstein, 1986). TVI allows the school to offer a course even when the professor is not available because of sabbatical or other commitments. Graduate students are happy to serve as tutors and probably find the assignment more enjoyable than being a grader. For on-campus applications of the method Scidmore and Bernstein (1986) found "as much, if not more, success with undergraduate tutors as with graduate student tutors. Undergraduate tutors have frequently just completed the course and are closer to the students' problems than a graduate student." TVI has also been used in undergraduate classes to break supersized classes down into much more manageable sections. With a tutor assigned to each section the students have the benefits of contact and of seeing the professor lecture on the material. Since small classes are always appreciated by students, this application of TVI should receive particularly high student ratings if the class size is kept within the suggested range of three to ten students per section.

One possible abuse that does not occur with live television is failure to update the videotapes. Once prepared, tapes often continue to be used even though they may have become outdated. TVI can also be abused if the professor who produces the tape abandons the class or if sections are allowed to grow too large in order to keep tutor costs down.

The TVI method appears to be a very effective instructional technique. Gibbons et al. (1977) found that TVI students performed better than students in live lecture classes, who performed better than students in live TV classes, who performed better than students in video classes without a tutor; but the results were not statistically significant because of the small numbers of students in the sample. There was also evidence that the poorer students benefited most from the TVI teaching technique. Gibbons et al. (1977) hypothesized that the small class size and the ability to interrupt the lecture frequently for discussion were more important factors in the success of the method than the use of videotape. (That is, the method was more important than the medium.) Scidmore and Bernstein (1986) compared on-campus TVI students to on-campus students in lecture courses. For three years of use in sixteen sections spread out over three different electrical engineering courses, the TVI students consistently averaged better on a comprehensive final examination than did the lecture students.

8.1.3. Instructional Hints for Television and Video

As with all techniques and classes, it is the instructor who controls the quality of instruction. Obviously, with television and video there is the added requirement that the production must be well done. Production details are discussed by Canelos and Mollo (1986) and Yoxtheimer (1986). However, even great production facilities cannot compensate for poor instruction. The following hints are from Canelos and Mollo (1986), Garrod (1988), Yoxtheimer (1986), and our personal experience.

1 Be prepared and well organized. Since television magnifies problems, you must be prepared and organized.

2 Arrive early at the studio. Extra time is required for setting up the cameras, and the producer will become very agitated if, as the starting time approaches, the "star" of the show is absent.

3 If possible use an overhead camera for visuals instead of a blackboard. It is difficult to obtain in-focus pictures of the entire blackboard.

4 Make sure the presentation is of high quality. Material which is prepared ahead of time must be neat and carefully proofread. If you write notes as the lecture is presented, have the ideas prepared ahead of time. Write few words and few equations. Write neatly. Orient the material horizontally since television uses dimensions with a height of three and a width of four. If large quantities of written material are required, use prepared material which has been handed out to all students in advance. Be sure that the handouts are also of high quality.

5 Use the principles of good teaching and good lecturing. Aim for variety in the presentation. A head talking in a monotone is even more boring than a boring presentation in person. Break the lecture into small parts with time for questions, discussion, and group activities.

6 Work to obtain group participation. Learn the names of the students both in the studio and at the remote sites. Allow extra time for questions from the remote sites. Repeat all questions since the microphones may not pick up student questions.

7 For feedback watch the tape. If necessary, adjust your teaching style. If the tapes will be used, edit or reshoot unsatisfactory portions. Prepare a practice tape before the semester starts and discuss your performance with a television expert.

8 In a TVI course develop written instructions for the tutors for live, in-class activities. Do not assume that they can develop these by themselves. Encourage the tutors to stop the tape frequently for discussion and other activities. Meet with and get to know them since the professors and the tutors form a team. Monitor the tutoring throughout the term.

9 Have copies of the tapes and the written materials available at the library or learning center.

8.1.4. Videotape Feedback for Students

Videotape is the premier technology for showing students how others see them. The gift which Robert Burns prayed for is now here, and it is videotape. If oral communication or interpersonal teamwork is required, videotape feedback to students is invaluable. Fortunately, such use can be relatively inexpensive. It is not necessary to have a studio, and it may be preferable not to have one. Many students are afraid to appear before a camera, and videotaping in a normal classroom is less threatening. The equipment needed includes a camera, a tripod, a VCR, and a TV monitor. In the classroom only the camera and the tripod are necessary, and it is probably better not to have the other equipment in sight. Although it is convenient to have a TA or undergraduate assistant serve as camera operator, this is not absolutely necessary. The camera can be prefocused on the tripod and then be turned on before the students start.

Procedures for videotaping oral reports are discussed by Wankat et al. (1977). First, get the students accustomed to the camera. This can be done by asking every student to make a very short, ungraded oral presentation in front of the camera. Although they may learn something from watching these short videos, the main purpose is to reduce anxiety when they make their regular presentations.

The regular presentations should follow the normal format for oral presentations in class. The reports should be timed. Students should be encouraged to use visual aids such as an overhead projector. These visuals will probably not show up on the tape, but this is unimportant since the purpose of the tape is feedback, not communication to others. Have the class ask questions and continue to videotape the speaker while he or she responds.

No one likes the sound of their voice on a tape, and many people do not like the way they appear on camera. Since the student is likely to be embarrassed, show the tape privately. Most students will be very severe critics of their presentation when they see the tape. If a student becomes upset while watching the tape, be sure to give some positive feedback and point out what worked. The camera is very blunt in showing problems, and usually there is no need to point out what is obvious to the student. Give the student a few pointers on what to do to improve, but do not overload him or her with too much advice. However, after more than ten years of taping student presentations, both undergraduate and graduate, the more common response we have seen is a positive one. Students discover that all the nervousness they feel while speaking does not show up on the tape; it's all internal—their knocking knees aren't visible for all to see. Also, they generally concede that the talk went better than they thought it would, or that they didn't sound as bad as they thought they would. For every student that is appalled at seeing himself or herself, many more enjoy watching themselves. Our favorite reaction was that of one student who sat back as he watched himself and in all seriousness proclaimed, "Damn, I'm good looking!"

Additionally, videotaping oral reports greatly improves the instructor's ability to give feedback. While watching the tape, instructors regularly see mannerisms and nuances they did not see the first time. The student also receives much more individual attention, which is important for improving presentation skills. Once students see their presentations, they seldom complain about the grade they receive on the oral report. Finally, the old adage of a picture being worth a thousand words holds very true with videotapes of oral presentations. You can tell a student over and over that he or she says "um" too often, but the impact of watching oneself "um" and "er" through fifteen minutes of material is much more powerful and immediate. The reality becomes painfully obvious.

Although much less common in engineering classes, videotapes can also be very helpful for interpersonal training. A camera is a very effective device for showing students their behavior and the reactions to their behavior in groups. In counseling programs it is common to use a room equipped with one-way mirrors so that the presence of an observer and of the camera does not disturb the group. Even without a one-way mirror, videotaping can be valuable for providing feedback. Once they get started, most groups tend to forget that the camera is there. The camera operator should attempt to be unobtrusive and should never give directions to the group. The quality of the camera work is not very important; it is much more important to capture the group in action. One problem with groups is that members behave differently at different times. It may be necessary to videotape several hours of group interaction to obtain the entire range of any individual's repertoire of group responses.

If the group proceeds well, the entire group can watch the videotape for feedback. Be sure to stop the tape at appropriate places for a discussion of what has happened. If one member of the group was obstructive, it is probably appropriate to show the tape to that person privately. Otherwise, there may be a tendency for the group to beat up on that person now that he or she cannot deny the behavior. Discuss with the student what can be done to improve her or his skills in groups.

8.2. COMPUTERS IN ENGINEERING EDUCATION

There has been a computer revolution in engineering education, but to date it has been much less far-reaching than many prophets predicted. Computers and calculators have greatly increased the ability of students (and practicing engineers) to perform calculations. Since computers and calculators allow professors and students to do a much better job at calculation, they have been widely adopted in engineering education. As a result, professors have changed the nature of the problems presented, and they have changed many of the mathematical techniques taught. This has been an important change in the way engineering is taught (and practiced). However, we have not seen significant adoption of computers for the delivery of instruction.

In this section we will first explore the use of computers as tools in the classroom. The commonly used generic computer tools are spreadsheets, equation solvers, and symbolic algebra programs. Simulation programs tend to be much less generic but will be discussed with the other tools. Then we will discuss computer-aided instruction which uses a computer to deliver instruction. Finally, interactive laser videodisc instructional methods will be considered.

Before any computer application is adopted, the professor needs to determine whether five prerequisites for instructional use of computers have been met. The first three are from Trollip (1987/88).

1 Accessibility. Both the hardware and the software must be readily accessible to both students and faculty.

2 High-quality software. The software must do something that the students want it to do, it must have clear and unambiguous screen displays, the interaction between user and machine must be easy, the software must be easy to use, the software must be relatively fast, and above all, the software must be robust.

3 Faculty interest. The faculty must have sufficient interest and energy to follow through with the project. The amount of interest and energy required depends on the project. For adopting generic tools such as spreadsheets, the amount is modest, but for writing computer-aided instruction packages it can be staggering.

4 Advantage. A computer must be able to do something better than the student can do it working without the computer. If there is no perceived advantage, then students will not use the computer and the faculty will drop the experiment.

5 Student computer background. Students must be taught how to use both the hardware and the particular software. If this has not been done in a prerequisite course, then they must be taught in the current course. Particularly for weaker students, learning about unfamiliar hardware and software in a discipline-oriented course can lead to cognitive overload and a poorer performance (Whitney and Urquhart, 1990).

8.2.1. Computer Tools

Engineering professors have recently discovered the use of generic software such as spreadsheets and equation solvers for the solution of engineering problems. As the available packages have become more powerful, robust, and user-friendly, it has become clear that they represent an extremely useful middle ground between hand solutions and computer programming. Some students will do almost anything to avoid programming, but the generic packages are user-friendly enough that, with a little training, almost all students can be induced to use them. Thus, in many applications computer tools are a significant advance over both hand calculations and programming. Because of this advantage, computer tools, particularly spreadsheets, have been widely adopted.

Students need to learn how to use the various software tools. Probably the ideal arrangement is to teach engineering students how to use the software as freshmen and then use it in all subsequent engineering courses. If students have not learned a particular software tool before it is introduced in class, most of them will not use it unless they receive help. Keedy (1988a) suggests the development of core manuals for software using the "20-80 rule." That is, identify approximately twenty concepts and the associated keystrokes which represent 80 percent of the power of the package—and everything the students need to do. When students first learn the package, they don't need to know the most efficient way to do something; instead they need to know the easiest way to learn and remember. Once the 20-80 items have been identified, write a short core manual which explains how to use these selected features. Interested students will learn other operations on their own or from other students once they know how to use the basics of the software.

Applications of spreadsheets in engineering courses have exploded since 1987, and they have been used in all engineering disciplines: for example, aeronautical engineering (Stiles et al., 1989), chemical engineering (Misovich and Biasca, 1991; Rosen and Adams, 1987), civil engineering (Anderson et al., 1988; Mortimer, 1987), electrical engineering (Lofy, 1988), freshman engineering (Genalo and Dewey, 1988; Keedy, 1988b), and petroleum engineering (Cress, 1988, 1989). Chapra and Canale (1988) show how spreadsheets can be used to implement a variety of numerical methods, and Burman (1989) illustrates the conditioning of climatic data with spreadsheets. This list of references is only the tip of the iceberg, and anyone who wants to can easily find many more.

As long as a spreadsheet has appropriate graphing and scientific function features and is fast enough, the choice of spreadsheet is almost immaterial (Lofy, 1988). In addition, students who learn how to use one type of spreadsheet can easily learn to use a different spreadsheet on their own. Thus, there is no need to worry about them seeing a different spreadsheet when they graduate.

The advantages of spreadsheets are discussed by Cress (1989), Misovich and Biasca (1991), and Mortimer (1987), among others. Spreadsheets are easy to learn; one two-hour laboratory is sufficient to learn the basics. Spreadsheets remove much of the tedium from doing calculations and allow the professor to assign more meaningful problems. "What if?"

experiments are easy, and students can explore the effect of changing parameters, thereby gaining a feel for the magnitudes of parameters in problems. And in certain circumstances they can see what effects can be ignored. Spreadsheets are also easily adapted to discovery learning methods. Instead of being told, students can discover the effect of variable changes for themselves.

Spreadsheets are in many ways easier to use than programming. They are structured and encourage students to structure their calculations, even for hand calculations. A spreadsheet can easily show tabular solutions. It is easy to debug since syntax errors are shown immediately, and the instant display of numerical results makes it easier to spot obvious mistakes. Input and output are easy since any cell can be displayed or changed at any time. The inclusion of graphics capabilities means that students can easily prepare presentation-quality graphics and can search for trends visually instead of looking at a mass of numbers. In addition, spreadsheets are easily documented since each cell can be labeled.

Students invariably prefer spreadsheets to programming. In addition, it doesn't appear to make any difference if they learn programming or spreadsheets first (Genalo and Dewey, 1988). Students are also able to generalize the use of spreadsheets to other classes and will use them in follow-up courses.

Spreadsheets are not without problems. They are slow, large-scale branching is difficult, and it is difficult to use variable names (Mortimer, 1987). If students are unfamiliar with spreadsheets or do not use them for a significant period of time, their introduction along with engineering material may decrease the learning of material (Merino, 1989). This could well be due to oversaturation with new material. If spreadsheets are introduced early in a course and used throughout the course, this should not be a problem. For very large problems the use of spreadsheets becomes cumbersome if not impossible. For these problems discipline-specific programs are often preferable. For the solution of large systems of equations and for many numerical methods, equation-solving software is preferable.

In many engineering classes spreadsheets allow students to get to real engineering problems faster, and permit them to focus on thinking since the program does the routine calculations. We are strongly in favor of the integration of spreadsheets into the engineering curriculum at all levels.

Much of what has been said about spreadsheets also applies to equation-solving programs. There are many examples of the use of these programs in engineering education: FORMULA/ONE (Baxter, 1988), MathCAD (Head and Fry, 1989; LaBlanc, 1991; Rogers et al., 1989), MATLAB, and TK!Solver (Fowler, 1985; Harbach and Wiggins, 1985; Keedy, 1988b). With an equation solver the user lists equations and the program automatically generates a list of variables. The user gives values for the known variables and asks for a solution. The program then finds a direct solution or iterates to find a solution after the user supplies initial values.

Equation solvers perform the input and output routines for the user, including graphing routines, and they choose the algorithm, although the user may be able to override this choice. They are quicker to set up than programming. The user can do "what if?" calculations and can thus learn by discovery. The programs can be used for optimization by trial and error and thus are useful for design problems. The simple features of an equation solver can be learned in one or two laboratory sessions, but some of the more advanced features take considerably more time before the user becomes proficient.

The programs require that students know how to write the equations. However, they do not need to know how to solve the equations, and in the worst case the program can become a black box. Generally, the programs have little logic capability and cannot do branching. Each program has limitations; for instance, TK!Solver cannot solve differential equations either symbolically or numerically. MathCAD can solve differential equations numerically, but the manual is difficult to follow (Head and Fry, 1989). Unfortunately, equation-solving programs do not appear to be as generic as spreadsheets, and experience with one program does not necessarily translate to facility with another.

Since equation-solving packages have more power than spreadsheets, we recommend their use in upper-division engineering courses. Some coordination within a department is appropriate to ensure that professors are using the same package. Spreadsheets should be taught first since they are more generic, are more visual, are easier to learn, and are applicable to the problems taught in lower-division courses. Additionally, students who learn the power of spreadsheets are more likely to believe that the time invested in learning to use an equation solver will be well spent.

Symbolic algebra programs such as MAPLE, MACSYMA, Mathematica, Derive, and Theorist are starting to have an impact on the teaching of calculus throughout the United States and Canada (Tucker, 1990; Goodman, 1991). A significant number of math teachers think that these programs can remove some of the repetitive calculations and drudgery from learning calculus and that the net result will be better learning of the theory of calculus. However, very preliminary results suggest that it may be the better students who benefit from computer use and that the poorer students may actually do worse than in standard classes (Watkins, 1991). Although the majority of calculus students are still taught by traditional methods, the movement toward using symbolic algebra programs appears to be gaining momentum. This brings up several questions for engineering education. Should engineering professors be exploring the use of symbolic algebra programs in engineering education? What can symbolic algebra programs do in engineering education? How do we accommodate students who have been taught calculus with a symbolic algebra program and who want to continue to use the program?

A student who wants to use advanced software such as a symbolic algebra program should be encouraged to do so, but at the same time, he or she needs to realize that the program may not be available on tests. For relatively complicated problems, symbolic algebra programs appear to be useful in engineering analysis (Prudy, 1990) and engineering design (Lee and Heppler, 1990). Students using these programs were able to work much more complicated problems in greater depth than students who did not use the programs. The number of errors was reduced since the program did the routine manipulations and could check whether the solution was algebraically correct. Symbolic algebra programs can plot functional relationships. At least some of them can generate FORTRAN code from equations written in the symbolic language, which obviously saves time and reduces errors. However, symbolic algebra programs can make mistakes, do not mesh with each other, and demand large amounts of memory (Heppenheimer, 1991).

Unfortunately, a significant amount of time is required to become proficient with symbolic algebra programs (Prudy, 1990). Students who have used a program in a calculus class are already past this barrier. If other students are to use them, they will benefit from instruction either in class or in an optional short course. This need for instruction brings up the old question of whether engineering professors should teach computer applications (or communication or whatever) instead of engineering. If a professor believes that there will be a net gain in the amount of engineering that students do in class, then the time spent learning a symbolic algebra program is well spent. Naturally, many engineering professors will also need to learn how to use the programs if they are to employ them in class.

Commercial application software such as CAD programs, ADAMS, ASPEN, NASTRAN, SPICE and pSPICE, and specialized simulation programs also have a place as tools for teaching engineering. These programs are extremely powerful, specialized, and realistic since they are written for practicing engineers. Unfortunately, they are usually not particularly user-friendly and are expensive to license. Commercial programs are often used in design classes (see Section 9.1) since they allow students to attack realistic problems. However, a university needs a large commitment to support computing (Eisley, 1989), and professors need to be committed to teaching students how to use the programs.

Specialized simulation programs written for a particular problem can be very useful since they allow students to "experiment" with otherwise inaccessible equipment (Squires et al., 1991) or to gain experience which would not normally be available until they are employed in industry (Kabel and Dwyer, 1989). Unfortunately, the commitment in time and money needed to produce large simulation programs robust enough for student use is huge. The only justification (other than a labor of love) for such a development is sharing programs throughout the country. Unfortunately, there are still barriers such as equipment incompatibility, a "not invented here" syndrome, and a lack of distribution networks which make such sharing difficult. A promising start for the disemination of specialized programs has been made with centers such as the NSF/IEEE Center for Computer Applications in Electromagnetic Education (Iskander, 1991).

Students using computer tools can suffer from the black box syndrome. As the program becomes more complex, it becomes increasingly likely that the student will not understand or perhaps even care what it is doing. When this occurs, the possibility of "garbage in, garbage out" becomes increasingly likely, and the student may not be able to detect errors. We believe that students should do simple hand calculations and then repeat the problem with a computer. This helps them understand what the computer is doing, gives them confidence, and shows them how the computer can save time. Whelchel (1991) professes the opposite view and states that technology has become too complicated for students to understand all the techniques; thus, he suggests trusting the software.

8.2.2. Computer-Aided Instruction

In computer-aided instruction (CAI) a computer is used to teach the material to the student. Thus, the computer becomes more than a calculational tool; it supplements or replaces the traditional forms of instruction. In the past CAI was hardware-limited, but at many schools this is no longer true. Software has now become the limiting step, and we will focus on what the software does and on the problems of software development. It is not possible here to describe exactly what a well-written CAI program can do. Readers interested in exploring this teaching method will need to explore the capabilities of a well-written CAI program on their own.

Simons (1989) identifies three major modes for CAI. In the drill-and-practice mode a student is presented with a question or problem, the student responds, and the computer provides feedback on the response. This mode can serve as a supplement to traditional instruction. In many ways drill and practice is similar to textbook homework assignments followed by feedback from a TA or a grader. The advantages of a computer are that the feedback is instantaneous and private. And since the student is already using the computer, he or she is more likely to use computer tools to solve the problem. Although problem statements must be clear and unambiguous, the real art in developing a drill-and-practice program is writing the interactive feedback. The program must follow Figures 1f and 1e. A good program will help the student see where the error is and to avoid similar errors in the future. The feedback must be highly individualized for what a particular student does, and the environment must be highly interactive.

The disadvantages of drill and practice are similar to the disadvantages of other CAI modes. The student must get past the barrier of using a computer, which for weaker students may be a major impediment. Students can practice only when the computer is available, whereas a textbook can be used practically anywhere (e.g., while waiting at a doctor's office or sitting on a bus). Finally, developing good programs (discussed in detail later) is a major task.

The tutorial mode is a more complex, higher-level program than drill and practice. A tutorial contains instructional material and may be a replacement for traditional delivery methods such as lecturing and textbooks. In addition to content material, the tutorial should contain example problems and figures, include questions and problems, and have richer feedback than typical drill-and-practice program. Tutorials can guide a student to different lesson parts depending on his or her response. Since many students find a completely externally controlled tutorial frustrating, most tutorials now also allow the user to control movement through paging or a menu. The tutorial can guide the student through problem solving with prompts and then gradually reduce the number of prompts until he or she is solving difficult problems without help.

The third mode listed by Simons (1989) is simulation (discussed in Section 8.2.1 as a computer tool). We classify a simulation program as a tool if the program is written for general use. If it is explicitly written for instruction, we classify it as CAI. In engineering, the advantage of commercial simulation programs is that they have a potentially broader market, and more money will probably be spent on their development. In addition, these programs are clearly realistic since they are used by practicing engineers. A CAI simulation program is more likely to consider decision making explicitly and to have feedback if the student has difficulties. The simulation should have many options and decisions for the student so that he or she can practice the functions of an engineer.

Does CAI work? Yes, but only if the students use the programs. This is not a trivial statement since it not unusual for many students to refuse to use CAI programs (Canelos and Carney, 1986). Of course, this is not unique to computers. Some students refuse to read a textbook or attend a lecture, but the problem does appear to be worse with computers.

Do students who use CAI learn better than by traditional methods? It depends on the program. Many instances of improvement are reported (Canelos and Carney, 1986; Turner,

1988), but there have also been reports of no improvement when compared to traditional methods (Turner, 1988). If a computer program is simplistic and just does what a textbook can do (e.g., as shown in Figure 8-1e), then there is no gain in using CAI. If the computer makes a diagnosis of where the student's difficulties lie and refers the student to the appropriate information, then improvement is observed. Combining immediate feedback from a drill-and-practice program with diagnostics appears to produce increased learning. Complex simulation programs, particularly those involving dynamic operations, can utilize the full visual power of computers, and significant increases in learning can be observed (Turner, 1988). Thus, it is the quality of the message and not the use of the computer that is important.

One major difficulty with CAI is the amount of time required to author a CAI program. In his pessimistic article, Trollip (1987/88) states, "Whether or not assistance is sought, it comes as a nasty surprise to most who start in the field of instructional computing just how difficult it is to produce useful, good material, and how long it takes to do it." Most engineering professors do not have all the skills necessary to develop CAI programs, and a team must be assembled. The necessity of working with professionals outside engineering may be hard on an engineering professor's ego but is necessary to achieve a quality product (Onaral, 1990). Nelson et al. (1985) organized a twelve-member team to write CAI for a one-semester statics course, a multiyear project. This CAI was to support an existing textbook, and only twenty-five hours of CAI was being prepared. Much of the effort was expended to be sure that the CAI programs would give the user as much control as he or she normally has with a textbook! Of course, when completed, the CAI program will have the advantage of interaction and immediate diagnostic feedback. The use of new authoring languages such as the cT language can reduce the effort (Kuznetsov, 1990), but writing a CAI program remains a formidable undertaking.

This effort should be compared to writing an engineering textbook where a single author can do the job in about the same time period. Writing a textbook makes sense only if the book can be marketed and used by other professors. Then these professors and their students will benefit from the effort expended by the author. Fortunately, a highly sophisticated and effective marketing and sales system exists for textbooks. Although most textbook authors do not feel that their universities place enough value on writing textbooks, this activity is recognized for promotions and tenure.

The same arguments apply to CAI software, but unfortunately the marketing and sales system is not yet well developed for courseware. However, recent developments at both universities and commercial textbook publishers indicate that a marketing and sales system is being developed rapidly (Onaral, 1990). There are several additional obstacles to the wide distribution of CAI courseware:

1 Computer incompatibility. With the very rapid changes in hardware, software needs continual updating. This is difficult when the market is small.

2 Professorial indifference or in some cases hostility to CAI. Even if a professor wants to use CAI, there is the "not invented here" syndrome.

3 Cost. Not only must the very high development cost be recovered, but this has to be done from a smaller base than for a textbook. In addition, students will revolt if they have to pay for both a textbook and CAI software.

4 Rewards. Many universities give little if any credit toward promotion and tenure for the development of instructional software (Trollip, 1987/88). With the national publicity that the EDUCOM/NCRIPTAL software awards have received, this situation may be changing.

5 Identification with television. CAI is often identified with television and may be seen to encourage students to abandon books and traditional scholastic values (O'Neal and Vasu, 1991). This can feed professorial hostility.

Because of these problems, we think that the outlook for CAI is limited to courses that have large enrollments across the country. In engineering education, large enrollment classes include calculus, chemistry, physics, computer programming, and certain lower-division engineering classes. The engineering classes with large enrollments include circuits, thermodynamics, statics, dynamics, and fluids. Courseware for many of these courses has been developed under the $70 million Project Athena at MIT. Courseware is also available in electromagnetics (Iskander, 1991). The hope is that some of this courseware can be economically used at other institutions.

8.2.3. Interactive Laser Videodisc

Interactive laser videodisc (ILV) is a new technology which allows for the combination of a number of technologies. The capabilities of ILV are discussed by Flammer and Flammer (1986) and Meyer (1991). One side of a videodisc can store up to 54,000 still frames or twenty hours of audio combined with still frames. Each still frame can hold about 400 kilobytes; thus, the storage capacity for computer information is immense. Each side can store thirty minutes of live action. This means that live action is by far the most storage-intensive use of the disc. The videodisc has random access of any frame in two to three seconds. Access of adjacent frames is faster. Any frame, even from live action, can be frozen and looked at as long as desired. The access can be controlled manually or by computer. ILV systems can incorporate a number of media:

- Photographs and text material.
- Overhead transparencies.
- Slides.
- Motion pictures.
- Videotape.
- Computer text and graphics.

One major advantage of new ILV systems is that a full screen of computer text and/or graphics can be recorded directly onto a videodisc frame.

An intelligent videodisc coupled to a powerful computer has all the capabilities of CAI systems with the added availability of multimedia presentation (see Figure 8-1g). Thus, ILV adds a new technological capability to what can be done with CAI. Because of the ability to record directly from a computer to the ILV frame, existing CAI programs can probably be converted to ILV without a huge investment of time and money.

With write-once discs and videodisc recorders the costs of producing laser videodiscs has plummeted. McInerney and Kyker (1987) produced their own discs for $3000 and $5000, which is probably the current absolute minimum. These costs do not include the cost of the authors' time. This is thus not an inexpensive technology, even when done as cheaply as possible. However, a single disc can probably hold an entire course so long as no live action is included. In addition, once the master disc has been made, additional discs are not very expensive. Because of the random access nature of the discs, updating them is relatively easy even though they are write-once. A number of frames can be left blank, and they can be written onto when the program is updated. The computer access code can be changed so that obsolete frames are no longer accessed and the new frames are.

What are the advantages of ILV? First, there is the huge storage capacity which might allow the storage of an entire course on one disc. Second, ILV consolidates a variety of media into a single package. The problem of switching from one medium to another is solved (Meyer and Zoltowski, 1989; Meyer, 1991); this is particularly useful if short segments from a variety of media are to be used. Third, this technology allows the developer to bring in any visuals that are desired. Since most people prefer to learn visually (see Section 15.2.2), ILV should be an effective learning method. Several uses for ILV in engineering education have been proposed or tried.

1 Remote locations. ILV is a video technology which can be used in the same way as video. This can be in either a tutored or an untutored environment. One advantage of ILV over video is that shipping costs are significantly lower since the whole course can be sent in one small package. ILV has been used for continuing education in composites (Gillespie, 1989).

2 Individual self-study and PSI courses. ILV combines CAI with a host of visuals and could serve in self-study courses. In a PSI course a computer can test and grade a student and keep track of what tests the student has taken as well as the student's progress.

3 Student tutorials. ILV can be used in the same way as CAI, but again with enhanced visual capabilities (Flammer and Flammer, 1986). The remote access capability of ILV makes it preferable to video since students can quickly find the one segment they want to use (Meyer, 1991).

4 In-class use. The consolidation capabilities of ILV allow an instructor to show multimedia without switching problems. Dynamic simulations from a computer can be shown in faster than real time because the final results of simulations can be shown.

5 Laboratory simulations. The combined computer simulation and visualization power of ILV allows the development of very realistic laboratory and plant simulations. Students can "experiment" with situations which would be too expensive or too dangerous in real life. A nuclear power plant simulation, for example, could show a melt-down situation if the operator makes too many mistakes.

6 Outreach programs. McInerney and Kyker (1987) made a disc for high school teachers to interest students in physics.

ILV has a variety of different applications which involve different capabilities of the equipment. If a professor has already developed a multimedia presentation, then conversion to ILV would be relatively simple and could be done for a few thousand dollars. If a CAI program has been developed, conversion to ILV would again not be too expensive. Costs become very high when development is started from a zero base; thus, our comments about CAI are appropriate for ILV. ILV development makes sense only if the resulting disc can be shared among a large number of users. This requires a distribution network for either on- or off-campus use. One added expense is the equipment on which the discs will be used after they have been prepared. This cost is higher than for either CAI or video. Gillespie (1989) gives a list for IBM equipment.

Since ILV is a very new technology, it is too early to tell what impact it will have on engineering education. Advances in Hypertext software have reduced the advantages ILV once had compared to CAI. Our best guess is that ILV will carve a unique but small niche where it is clearly the best delivery system.

8.3. AUDIOTUTORIAL METHOD

The audiotutorial (AT) method is an educational system which uses technology and instructors to satisfy a number of the learning principles outlined in Section 1.4. First developed in 1961 by Sam Postlethwait in the biology department at Purdue University (Postlethwait, 1980, 1984), the method has had a considerable impact on the teaching of biology (Creager and Murray, 1971). AT has evolved significantly since 1961, and many different variants are practiced. We will describe it as currently practiced in two biology courses at Purdue University.

The course is divided into modules which require one week to complete. It is an instructor-paced course, but during each week students may select when to go to the AT facility. The facility is open Monday through Friday and on Sunday evening. Students are given a detailed course syllabus and a description of how the course works. For each week's module the student receives:

- Clear objectives.
- A reading assignment in the textbook.
- A journal article to read.
- Supplemental notes and study questions.
- A schedule of extra credit activities.

When students go to the AT facility, they first sign in with a TA. They are then assigned a carrel which contains simple instructions as to the order in which to proceed through the assignment. Lectures have been recorded on audiotape, and the students listen to them. The lecturer frequently tells the student to turn the tape off and do activities such as answering questions or working on problems. In addition, students are directed to pick up and examine various objects in the carrel. They then must answer questions about these objects. They are also directed to a few laboratory experimental stations to conduct simple experiments such as looking at samples through microscopes, and so forth. TAs move through the room to help students who are having difficulty. A few of the modules use videotapes when there is a major advantage to having video in addition to audio. Obviously, with a large number of carrels (forty-seven), providing audio recorders is cheaper than video equipment. When ready, the student can participate in graded activities.

1 Every day there are small group teach-about-biology (TAB) sessions during which students make a short presentation on an item (specimen, demonstration materials, experimental results, etc.) which was seen during the student's self-study. The students do not know ahead of time what the item will be. The student is to teach the other nine students in the group about this item, and the TA grades each presentation.

2 The student has two opportunities during the week to take a "C-level" quiz, and the highest score is recorded. Mastery is the goal but is not always achieved in these quizzes. Successful completion of the C-level quizzes gives the student a C in the course.

3 During the semester the student takes two "AB-level" tests which are given in large lecture halls for one hour.

4 The student can take an optional final during finals period.

5 Several laboratory sessions must be completed and written up during the course of the semester.

6 The student can earn extra points by attending a general assembly for a lecture, film, demonstration, and so forth, and then writing a summary about the session. The summary is written in the last twenty minutes of the period. Extra points can also be obtained by writing summaries of journal articles.

7 The student can obtain bonus points by taking an optional quiz or by doing optional dissections.

The points are added up at the end of the semester, and a straight 90-80-70-60 scale is used for grading. Approximately half the students enjoy the course very much, work hard, and receive A's or B's. Procrastination lowers the students' grades, and approximately one-quarter of them receive D's or F's.

This course requires some responsibility on the part of the students since they must set aside time to go in. The deadlines have proven to be critical to the success of this course since first-year students still need structure. Students are aware of their progress, and they know the grade they have earned at every point in the semester.

In an AT course the professor plans and modularizes the entire course. This requires a mastery of the content. The professor records the tapes and revises most of them every year, prepares the objectives, and writes the supplemental material. When the textbook changes, most of the supplemental material must be extensively revised. He or she writes quizzes and tests, supervises the grading, and must be a manager if a large number of TAs are employed.

The professor arranges for the general assembly sessions, even giving a few lectures, and finally, sets the tone for the entire course.

This system has built in many of the learning principles discussed in Section 1.4. Students are forced to be active. There is frequent feedback, and students can ask for help whenever they need it. They must teach others in the TAB sessions. Since a straight grading scale is used, the class is not competitive. Students are encouraged to cooperate and help others. A variety of visuals are used, and the class is multimedia (audio, video, real objects, tutoring, presentations, etc.). The student can to a large extent control his or her pace through the material and is given time to practice. Clear learning objectives and material for learning these objectives are provided, and the student is guided through this material. The audio lectures structure the material for the student. Each student is challenged in the course, yet each student can be successful. Finally, Postlethwait was extremely enthusiastic and clearly expected that every student would do well.

Applications of the AT approach in engineering were reviewed by Lindenlaub (1974). The method has been used for entire courses normally considered "lecture" courses. Several laboratories have been taught as AT courses. In addition, the AT method has been used to supplement both regular and laboratory courses. Use of the AT method appears to have peaked, even though many of the principles could clearly be adapted to other technologies. The problem is cost. Compared to a lecture course with one professor lecturing to anywhere from 400 to 500 students with a handful of TAs for office hours and grading, the AT class is expensive. Not only is additional equipment necessary, but a room must also be dedicated to this one class and more TAs are needed. The students in an AT class on average put in more time on the class and learn more. Retention for other classes and for graduation is higher. Unfortunately, most universities do not factor student learning into their system of determining the economics of a class.

We have included the AT method in this chapter because it can serve as a model of how technology can be incorporated into a system which satisfies essentially all the learning principles.

8.4. CHAPTER COMMENTS

It is extremely difficult to give the flavor of teaching with a technology in a book using print as the medium. If you are interested in any of these techniques, obtain samples and be a student for an hour or two using the sample to learn a topic. These demonstrations will give you a feel for whether you want to proceed in exploring use of the technology.

This chapter was also somewhat difficult to write since we have not been personally involved in developing CAI, laser videodisc, or AT courses. We have read extensively and seen extended demonstrations of these methods, but this is not a substitute for the first-hand experience we have had with all the other teaching methods discussed in this book.

8.5. SUMMARY AND OBJECTIVES

After reading this chapter, you should be able to:

• Describe and discuss the advantages and disadvantages of the following teaching methods:
  • Live television.
  • Tutored videotape instruction.
  • Videotape feedback for students.
  • Computer-aided instruction.
  • Interactive laser videodisc.
  • Audiotutorial.
• Discuss the advantages and disadvantages of using generic software packages in an engineering course.

HOMEWORK

1 Pick one of the teaching methods listed in the first objective of Section 8.5. Visit a facility where this technique is in use and act as a student for one class period to experience the method.

2 For the teaching method chosen for problem 1, outline in detail how the method could be used either to teach or to supplement a specific engineering course.

3 Outline how you would use generic software in a specific engineering course. If appropriate, consider how the students would learn to use the software.

4 One of the arguments against extensive use of simulations in engineering education is that students will use the software as a black box and will not look at the output critically. Delineate this argument. Read Whelchel's (1991) paper and discuss his arguments. Then develop methods to prevent this from being a major problem.

REFERENCES

Anderson, M. L., Buttry, K. E., McCullough, E. S., and Wetzel, R. A., "Lotus 1-2-3 applications in civil engineering education," *Proceedings ASEE Annual Conference*, ASEE, Washington, DC, 1516, 1988.

Anonymous, "The Public Broadcasting System's Adult Learning Service," *Chronicle Higher Educ.*, A26 (Sept. 4, 1991).

Austin, G. D., Beronio, P. B., Jr., and Tsao, G. T., "Biochemical engineering education through videotapes," *Chem. Eng. Educ.*, 176 (Fall 1990).

Baxter, H. R., "Teaching electronics using an equation solver," *Proceedings ASEE Annual Conference*,

ASEE, Washington, DC, 1857, 1988.

Burman, R., "Conditioning climatic data for classes using a spreadsheet," *Proceedings ASEE Annual Conference*, ASEE, Washington, DC, 900, 1989.

Canelos, J., and Carney, B. W., "How computer-based instruction affects learning," *Eng. Educ.,* 298 (Feb. 1986).

Canelos, J., and Mollo, R. A., "Using instructional television in engineering education: It can be workable, but watch out for the pitfalls," *Proceedings ASEE/IEEE Frontiers in Education Conference*, IEEE, New York, 338, 1986.

Chapra, S. C., anDCanale, R. P., *Numerical Methods for Engineers*, McGraw-Hill, New York, 1988.

Chute, A. G., and Elfrank, J. D., "Teletraining: Needs, solutions and benefits," *International Teleconferencing Association 1990 Yearbook*, McLean, VA, 1990.

Creager, J. G., and Murray, D. L., *The Use of Modules in College Biology Teaching*, Publ. 31, Commission on Undergraduate Education in the Biological Sciences American Institute of Biological Sciences, Washington, DC (March 1971).

Cress, C., "Introducing spreadsheet modeling across the engineering science curriculum," *Proceedings ASEE Annual Conference,* ASEE, Washington, DC, 909, 1989.

Cress, D., "Integrating spreadsheet modeling into a fluid mechanics course," *Proceedings ASEE Annual Conference*, ASEE, Washington, DC, 2141, 1988.

Druzgalski, C., "Videotape—A next step after computer graphics in biomedical engineering instruction," *Proceedings ASEE Annual Conference*, ASEE, Washington, DC, 1684, 1988.

Dutton, J. C., "A comparison of live and videotaped presentations of a graduate ME course," *Eng. Educ.*, 343 (Jan. 1988).

Eisley, J. G., "Using commercial applications software in instruction," *Proceedings ASEE Annual Conference, ASEE*, Washington, DC, 404, 1989.

Flammer, B. L., and Flammer, G. H., "The intelligent (interactive) videodisc and its promise for engineering education," *Proceedings ASEE Annual Conference,* ASEE, Washington, DC, 821, 1986.

Fowler, W. T., "TK!Solver, a microcomputer-based mini-expert system with a 'student expert'," *Proceedings ASEE/IEEE Frontiers in Education Conference*, IEEE, New York, 12, 1985.

Garrod, S. A., "How to adapt your teaching style for teaching on TV," *Proceedings ASEE Annual Conference*, ASEE, Washington, DC, 2204, 1988.

Genalo, L. J., and Dewey, B. R., "What if: Spreadsheets for freshmen," *Proceedings ASEE Annual Conference*, ASEE, Washington, DC, 1138, 1988.

Gibbons, J. F., Kinchelve, W. R., and Down, K. S., "Tutored videotape instruction: A new use of electronics media in education," *Science*, 1985, 1139 (18 March 1977).

Gillespie, J. W., Jr., "Impact of interactive videodisc courseware on composites education," *Proceedings ASEE Annual Conference*, ASEE, Washington, DC, 745, 1989.

Goodman, B., "Calculus education. Toward a pump, not a filter," *Mosaic,* 22, no.2, 12 (Summer 1991).

Gupta, M. S., "Remote teaching by electronic blackboard," *Eng. Educ.*, 163 (Nov. 1981).

Harbach, J. A., and Wiggins, E. G., "TK!Solver in senior level design courses," *Proceedings ASEE/IEEE Frontiers in Education Conference*, IEEE, New York, 2, 1985.

Head, J. H., and Fry, D. W., "Solving intermediate-level classical mechanics problems using MathCAD," *Proceedings ASEE Annual Conference*, ASEE, Washington, DC, 857, 1989.

Heppenheimer, T. A., "Computer algebra: Speed is not all," Mosaic, 22(4), 28 (Winter 1991).

Howard, E., and Peters, D., "New dimensions in education by two-way television networks: Graduate engineering and management programs," *Proceedings ASEE Annual Conference*, ASEE, Washington, DC, 866, 1986.

Iskander, M. F., "Computer applications in electromagnetic education," *Proceedings ASEE/IEEE Frontiers in Education Conference*, IEEE, New York, 208, 1991.

Kabel, R. L., and Dwyer, C. A., "Scaleup, software development, and collaboration," *Academic Comput.*, 3(8), 14 (1989).

Keedy, H. F., "If they don't need it, don't give it to them," *Proceedings ASEE Annual Conference*, ASEE, Washington, DC, 1916, 1988a.

Keedy, H. F., "Introducing engineering software tools to freshman engineering students," *Proceedings ASEE Annual Conference*, ASEE, Washington, DC, 1142, 1988b.

Kostek, T. E., "Video recording laboratory experiments for open laboratory environments," *Proceedings ASEE/IEEE Frontiers in Education Conference*, IEEE, New York, 475, 1991.

Kuznetson, H., "Interactive computer courseware for teaching and testing statics and strength of materials," *Proceedings ASEE Annual Conference*, ASEE, Washington, DC, 233, 1990.

LaSalle, A. J., "Meeting a need: Televideo course in computing," *Proceedings ASEE/IEEE Frontiers in Education Conference*, IEEE, New York, 412, 1986.

LeBlanc, S. E., "The use of MathCAD and Theorist in the ChE classroom," *Proceedings ASEE Annual Meeting*, ASEE, Washington, DC, 287, 1991.

Lee, T., and Heppler, G. R., "Algebra systems for enhancing the learning environment for control systems," *Proceedings ASEE Annual Conference*, ASEE, Washington, DC, 113, 1990.

Lindenlaub, J. C., "Audio tutorial instruction at Purdue University," in Grayson, L.P. and Biedenbach, J.M. (Eds.), *Individualized Instruction in Engineering Education,* ASEE, Washington, DC, 103, 1974.

Lofy, F. J., "Design of electronics circuits using spreadsheet techniques," *Proceedings ASEE Annual Conference,* ASEE, Washington, DC, 1567, 1988.

McInerney, M., and Kyker, G. C., Jr., "Interactive videodiscs at the Rose-Hulman Institute of Technology," *Proceedings ASEE Annual Conference*, ASEE, Washington, DC, 1965, 1987.

Merino, D. N., "Effectiveness of computer based problem solving in teaching engineering economics," *Proceedings ASEE Annual Conference*, ASEE, Washington, DC, 293, 1989.

Meyer, D. C., and Zoltowski, C. B., "Use of the laser videodisc medium as an educational delivery tool," *Proceedings ASEE Annual Conference*, ASEE, Washington, DC, 174, 1989.

Meyer, D. G., "On the delivery of visualizations for teaching engineering design using the laser videodisc medium," *Proceedings ASEE Annual Conference*, ASEE, Washington, DC, 591, 1991.

Miscovich, M. J., and Biasca, K. L., "The power of spreadsheets in a mass and energy balances courses, *Chem. Eng. Educ.*, 46 (Winter 1970).

Moore, M., "An overview of distance education research," Indiana Conference on Technologies for Distance Education, Purdue University, West Lafayette, IN, 1990.

Mortimer, K., "Spreadsheets ain't just for bean counters," *Proceedings ASEE Annual Conference*, ASEE, Washington, DC, 1529, 1987.

Nelson, C. C., Hughes, B. J., and Virgo, R., "CAI applications in statics," *Proceedings ASEE/IEEE Frontiers in Education Conference*, IEEE, New York, 284, 1985.

Onaral, B., "Engineering courseware: A developer's perspective," *Acad. Comput.*, 4(4), 12 (1990).

O'Neal, J. B., Jr., and Vasu, E. S., "The limitations of instructional technology," *Proceedings ASEE/IEEE Frontiers in Education Conference*, IEEE, New York, 496, 1991.

Postlethwait, S. N., "Improvement of science teaching," *BioScience,* 30(9), 601 (1980).

Postlethwait, S. N., "Using science and technology to teach science and technology," *Eng. Educ.*, 204 (Jan. 1984).

Prudy, D. J., "Symbolic algebra in mechanical engineering," *Proceedings, ASEE Annual Conference*, ASEE, Washington, DC, 717, 1990.

Robinson, J. W., and Canelos, J. J., "Innovative format using television and teaching assistants in a heavily-enrolled electrical engineering service course," *Proceedings ASEE/IEEE Frontiers in Education Conference*, IEEE, New York, 317, 1989.

Rogers, D. A., Helt, J. L., Lee, D. J., and Aslakson, J. T., "Using "MathCAD" in introductory antenna analysis," *Proceedings ASEE Annual Conference*, ASEE, Washington, DC, 903, 1989.

Rosen, E. M., and Adams, R. N., "A review of spreadsheet usage in chemical engineering calculations," *Comput. Chem. Eng.,* 11(6), 723 (1987).

Scidmore, A. K., and Bernstein, T., "TVI: The Wisconsin experience," *Proceedings ASEE/IEEE Frontiers in Education Conference*, IEEE, New York, 345, 1986.

Seigel, A. E., and Davis, C., "Televising undergraduate courses: A survey," *Eng. Educ.*, 482 (July/Aug. 1991).

Simons, J. M., "Computer-assisted instruction programs as a productivity tool in engineering education," *Proceedings ASEE/IEEE Frontiers in Education Conference*, IEEE, New York, 274, 1989.

Squires, R. G., Reklaitis, G. V., Yeh, N. C., Mosby, J. F., Karmini, I. A., and Anderson, P. K., "Purdue-industry computer simulation modules," *Chem. Eng. Educ.,* 98 (Spring 1991).

Stiles, R. J., Russell, J. H., Smith, M. L., and Winn, R. C., "Teaching aeronautics with spreadsheets," *Proceedings ASEE Annual Conference*, ASEE, Washington, DC, 407, 1989.

Trollip, S. R., "Wrestling with instructional computing," *Acad. Comput.,* 2, 26 (Dec. 1987/Jan. 1988).

Tucker, T. (Ed.), "Priming the calculus pump: Innovations and resources," MAA Notes, No. 17, Mathematical Association of America, Washington, DC, 1990.

Turner, J. A., "Books vs computers: The message, not the medium, is said to be important for student," *Chronicle Higher Educ.,* A22 (Oct. 5, 1988).

Walker, M. B., and Donaldson, J. F., "Continuing engineering education by electronic blackboard and videotape: A comparison of on-campus and off-campus student performance," *IEEE Trans. Ed., 32*, 443 (1989).

Wankat, P. C., Houze, R. N., and Barile, R. G., "TV taping of laboratory oral reports," *Proceedings ASEE/IEEE Frontiers in Education Conference*, IEEE, New York, 37, 1977.

Watkins, B. T., "The electronic classroom," *Chronicle Higher Educ.*, A26 (Sept. 4, 1991).

Wergin, J. F., Boland, D., and Haas, T. W., "Televising graduate engineering courses: Results of an instructional experiment," *Eng. Educ.*, 109 (Nov. 1986).

Whelchel, R. J., "Engineering as a generalist profession: Implications for engineering education," *Proceedings ASEE/IEEE Frontiers in Education Conference*, IEEE New York, 151, 1991.

Whitney, R. E., and Urquhart, N. S., "Microcomputers in the mathematical sciences: Effect on courses, students, and instructors," *Acad. Comput.,* 4(6), 14 (1990).

Yoxtheimer, T. L., "Developing video courseware: A starting point," *Proceedings ASEE/IEEE Frontiers in Education Conference*, IEEE, New York, 333, 1986.

CHAPTER 9

DESIGN AND LABORATORY

"Engineering without labs [and design] is a different discipline. If we cut out labs [and design] we might as well rename our degrees Applied Mathematics" (Eastlake, 1986).

We agree with this modified quotation that design and laboratory sessions are at the heart of an engineering education. There is nothing wrong with a degree in applied mathematics, but that is not the degree that students and companies think they are getting. Design and laboratory classes are also important in accreditation (see Section 4.6) and in the ASEE Quality in Engineering Education Project (ASEE, 1986). Despite almost general agreement on the importance of design and laboratory work, there is a tendency to cut these programs since they are expensive, messy, hard to teach, time-consuming, and not connected to the university's other mission—research.

Since this book cannot delve into the technical details which become important in teaching both design and laboratory courses, the discussion will necessarily be more abstract. We will consider the purposes of design and laboratory work and then consider particular methods for teaching these courses.

9.1. DESIGN

Many engineers contend that designing is the heart of engineering. All the mathematics, physics, chemistry, and engineering science courses are background for what makes engineering different from applied mathematics or the physical sciences. Yet there is no universally accepted working definition of what design is, and the Accreditation Board for Engineering and Technology continually struggles despite frequent criticism to be sure that sufficient design is included in the curriculum (Jones, 1991). An inadequate design curriculum is often noted as a deficiency during ABET visits.

In this section we will first discuss appropriate goals for the design part of courses and then explore methods for teaching design and including design throughout the curriculum. Finally, we will examine four different methods for teaching design—design projects, case studies, guided design, and design clinics. What we will not attempt to do is to list activities or projects which are appropriate for teaching design in different areas of engineering.

9.1.1. Design Goals

Instead of trying to define design, ABET (1989) describes activities and processes which may be included in design (see Section 4.6).   Design

- Produces a system, component, or process to meet a specific need.
- Is an iterative process which utilizes decision making with economics and employs mathematical, scientific, and engineering principles.
- Includes some of the following: setting objectives, analysis, synthesis, evaluation, construction, testing, and communication of results.
- Has student problems which are often open-ended, require use of design methodology and creative problem solving, require formulation of the problem statement and an economic comparison of alternate solutions, and may require detailed system details.

From this description of design a wide variety of possible goals for the design part of a course can be generated:

Problem definition and redefinition. Students will learn to define and redefine problem statements as they work their way iteratively through open-ended problems.

Synthesis and creativity. Students will be able to synthesize new designs using the principles of creative problem solving (see Section 5.6).

Troubleshooting. Students will be able to take an existing design that does not work up to specifications and make it work. Since troubleshooting is quite different than designing a new device or process, students need a chance to practice (Middlebrook, 1991; Woods, 1980, 1983).

Use of engineering, mathematics, and science principles. Students will be able to integrate a variety of engineering, mathematics, and scientific principles into the solution of design problems.

Computer tools. Students will use computer tools such as spreadsheets, general mathematical packages (MAPLE, Mathematica, MATHLAB, etc.), and engineering discipline-specific simulation packages (ASPEN, PRIMAVERA, SPICE, etc.) to do detailed routine calculations. "A course which does not use a professional software is preparing our students for a type of work which does not exist any more" (Paris, 1991).

Decision making. Engineers must be willing to take the responsibility of making decisions knowing that something could go wrong since perfection can never be attained (Florman, 1987).

Generic design procedures. Students will learn and use generic design procedures. Examples of these procedures are given by Warfield (1989) and Magleby et al. (1991).

Economic evaluation. Students will evaluate solutions based on economic and other criteria to determine the best solution among several alternatives.

Completion of a deliverable. It may be possible to have students carry out all steps of a design including the construction and testing of a deliverable. When this is possible, it is extremely motivating for them to see their design built and used.

Industrial or real-life experience. Design projects are normally much more realistic than engineering science problems which are concocted to illustrate a single principle. An additional goal may be to have students solve actual industrial problems.

Oral and written communication. Students are expected to develop professional skills to communicate their results.

Planning and managerial skills. Students can learn how to plan effectively and direct fairly complicated projects.

Interpersonal skills. While working in groups students can learn interpersonal skills, become adept at teamwork, and start developing leadership skills. Teamwork has become increasingly important as technology becomes more complex (Florman, 1987).

Confidence. Students can develop confidence in their ability to function as engineers. This may be the primary objective of the course (Overholser et al., 1975).

This is a long but certainly not all-inclusive list. Dekker (1989) and Harrisberger (1986a, b) discuss other possible goals. No single course can satisfy all these goals, although some professors make a valiant effort. However, an entire curriculum can be designed so that these and other objectives are satisfied. The professor's task is to select appropriate goals for the design portion of a course. These goals must be appropriate for the student level and the time allotted to design. For example, completely open-ended, unstructured problems with no guidance are not suitable for first-year students but may be very appropriate for seniors. Once the goals have been determined, teaching methods can be selected (see Sections 9.1.2 through 9.1.6).

9.1.2. Teaching Design

The ABET criteria state, "Some portion of this [the design] requirement must be satisfied by at least one course which is primarily design, preferably at the senior level, and draws upon previous coursework in the relevant discipline" (ABET, 1989). This sentence is often interpreted as requiring or suggesting a capstone course in design. Beyond this, ABET does not specify how the sixteen credits of engineering design should be distributed. It is explicitly stated that courses can be split between engineering design and engineering science. We feel that the strict dichotomy between engineering science and engineering design is a false one. Engineering design should appear throughout the curriculum. Culver et al. (1990) discuss two general ways that this can be done, Green (1991) gives details for an electrical engineering curriculum, and Juricic and Barr (1991) give specifics in mechanical engineering.

Spreading design throughout the curriculum allows the faculty to develop a design experience where students start working open-ended 'problems as freshmen or sophomores. These first projects are presented with a significant amount of guidance using a procedure such as guided design (see Section 9.1.5). Procedures for teaching freshmen design are discussed by Evans and Bowers (1988) and Evans et al. (1990). Design ideas can be included in traditionally nondesign classes (e.g., Henderson, 1989; Miller et al., 1989; Peterson, 1991; Riffe and Henderson, 1990) in both the second and third years. The traditional design classes and design laboratories would be retained in the senior year. However, students would be better prepared for these classes, and professors would see fewer students who are totally unprepared for open-ended design problems.

Introducing some ideas of engineering economics into the curriculum during the first or second year allows a professor to include relatively simple design problems and economic optimizations which can only be talked about instead of being done if the students have not studied economics (Sullivan and Thuesen, 1991). Talking about something is a totally ineffective teaching method; students must do what they are to learn. In our experience, students find some of the economics (costs, cost indices, payout periods) easy, while other parts (such as discounted cash flow) are more challenging. It helps if the textbook talks about economic factors, but unfortunately many engineering textbooks are written in an economic vacuum. Since students see the design problems with economics as real engineering, these problems are motivating as long as the professor does not overburden the problem with detailed calculations. Computer tools such as spreadsheets, mathematical packages, and simulation programs are appropriate here to remove the burden of routine calculations.

How do you add design problems to an already overloaded curriculum and overloaded courses? Cover less in the lectures and expect students to learn some of the material on their own. If the design problem includes this material, students will learn it, and they will learn how to learn on their own. As the professor, you can help by having clear objectives, making sure resource material is readily available, and believing that students can master the material on their own. The design problem can be included as a small project, a case study, or as a guided design project.

An important part of design is creativity and synthesis. Since most traditional curricula cover only application and analysis during the first three years, it should be no surprise that many students have difficulty with creativity and synthesis in senior design. Including creative exercises and synthesis problems throughout the undergraduate program should make most students more creative designers. The methods for teaching problem solving and for fostering creativity discussed in Chapter 5 are appropriate for the design component of classes. Specific methods for teaching creative design classes are discussed by Cundy et al. (1987) and Jansson (1987).

There are many significant difficulties in teaching design at any level. The first difficulty confronting the professor is the development of good design problems. Every engineering professor has one or two good design problems stored in her or his head. Design problems cannot be recycled and reused since students quickly convert the course into an exercise in rewriting files. New problems are needed every year. Thus, the first year is not the problem; it is the second, third, and following years. There are sources of problems which can be tapped by professors but are unlikely to be tapped by students. Published case studies such as the

ASEE case studies (see Section 9.1.4) and the American Institute of Chemical Engineers Contest Problem are useful. Industrial interaction can produce interesting problems with the added benefit that the problems are "real" (Emanuel and Worthington, 1987; Sloan, 1982). Ring (1982) suggests that cities can be used as a source of problems. Other possible sources of projects are bicycle design (Klein, 1991) and designs for handicapped people (Hudson and Hudson, 1991). Finally, we would like to suggest that a group of professors from different schools collaborate on developing design problems. Each year a professor from a different school could develop a problem and all the schools would use the problem and grade their own students. The labor of preparing new problems could thus be reduced significantly.

Since design problems are usually team efforts in industry, it is appropriate that they be team efforts in school. How does the professor select the teams? Some professors allow students to pick their own teams. This does not follow industrial practice and tends to result in teams which are very uneven in ability. Emanuel and Worthington (1987) suggest that the professor assign groups and include the following selection criteria:

• Mix leaders and followers within a team.
• Distribute abilities and experience among teams.
• Place one person with initiative on each team.
• Mix foreign students among teams to force communication in English.
• Do not put roommates on the same team.
• Mix men and women in teams.
• Use teams with three members.
• Be sure at least one team member lives close to the campus as this facilitates copying, computer use, and so forth.
• If travel to a company is required, be sure at least one student in each group has a car.

The MBTI (see Chapter 13) has been used for team selection (Emanuel and Worthington, 1989; Sloan, 1982). However, dysfunctional teams can result if team members try to act in accordance with their Myer-Briggs type instead of as is appropriate for the situation (Emanuel and Worthington, 1989). After trying both selection procedures, Emanuel and Worthington (1989) stopped using the MBTI for selection but instead used it to help the groups function better during the semester.

Groups malfunction during the semester for a variety of reasons. Perhaps the most common problem arises when one student does not do a fair share of the work. If the class is to be a learning experience in teamwork, the professor should not ignore these problems. The MBTI can be used as a diagnostic tool to help explain the problems; however, students must not be allowed to use their type as an excuse for behavior. Instead, students should be told that the types show both strengths and weaknesses, and that the weaknesses need to be worked on if they cause group problems. Even without the MBTI the professor should encourage students to discuss group problems, and then he or she should meet with the group to try to find resolutions. Design groups can be considered as a type of cooperative group, and many of the comments in Section 7.2 are appropriate for instruction and management of these groups.

Grading of groups can also be a problem. Since the group is producing a group report, it is appropriate to give the students a group grade. However, students often feel that this is unfair if one student has not done a fair share of the work. This problem can be resolved in several ways. The professor can talk to individual students and then to the group, and if appropriate assign a lower grade to the shirking student (Baasel, 1982). Second, the students can be given a group grade and the group can assign points to each student. Most groups will assign each student an equal number of points, but groups where one student has obviously shirked responsibility will differentiate. However, Eck and Wilhelm (1979) point out that these groups often engage in significant conflict over grade distribution and that some type of arbitration scheme may be necessary. Third, every student can be required to turn in an individual progress report every week (Stern, 1989). From these the instructor can usually make a rational decision on how to partition the final grade (and incidentally can usually predict which groups will turn in good reports). Fourth, all students can be assigned the same grade despite the claims of unfairness. Finally, a formula can be designed for partitioning the grade so that a variety of inputs are included. (Emanuel and Worthington, 1989).

What about important technical content which was either skipped in prerequisite classes or which students did not learn? A significant portion of many design courses covers economics; however, we do not think this time should also be used as a catch-all to reteach other material. Provide the students with resources (perhaps their own textbooks) and have them learn the material on their own. Engineering students can be surprisingly efficient learners when they see the need to learn material in order to complete a design assignment.

A final significant problem in design classes is time—both student and instructor time. Students need to learn how to develop a work plan and how to schedule a design project. In addition, some help in improving efficiency is appropriate. If design is included throughout the curriculum, then efficiency, time management, and scheduling can be discussed every semester. This repetition is helpful in learning how to apply these ideas. The problem with instructor time is that many universities undervalue design classes and overload professors who are teaching these classes. Design classes are time-consuming because of the need to develop problems, consult with student teams, and grade lengthy reports. Providing sufficient resources for design requires an administrative solution, which should include sufficient rewards for professors teaching these courses (Jones, 1991).

9.1.3. Design Projects

The most common way to teach design is with design projects. Students, usually in groups, are given a design problem and told to do the design. Since engineers learn design by designing, this is certainly an appropriate procedure. In addition, people remember the things that they do. We can remember our senior design (and laboratory) projects after twenty-five years, but we don't remember details of any of the lectures. The projects must be open-ended to be considered design. Multiple solutions of a well-defined problem are optimization, not design (Dekker, 1989).

The amount of guidance students need depends upon their maturity. Freshmen need significant guidance, and a guided design procedure (see Section 9.1.5) should be considered. Seniors need the opportunity to solve significant design problems with little guidance. It is helpful to most students if they have the opportunity to work up to a totally unguided design project by working on increasingly difficult designs with decreasing guidance during their junior and senior years. Design projects can be classified in many ways. Dekker (1989) suggests classifying them on the basis of various dichotomies.

Fun versus serious
Academic versus real world
Paper versus hardware
Creative versus structured
Individual versus group
Disciplinary versus interdisciplinary
Small versus large

Fun projects can include brainstorming a float design for a parade, creating a "Rube Goldberg" design, and so forth. Hardware projects, which mix design and laboratory skills, can be extremely motivating because students can see what they have designed. Dekker (1989) suggests that designing will be creative if students design something which is unknown. For example, have them design a "dollar bill picker-upper." Creative designs can be encouraged by showing students one design for accomplishing a task and then asking them to develop a competing design. This can be made more realistic by giving them the patent and telling them to develop a design which does not infringe on the patent. This procedure also brings up the subject of patents and patent law in a meaningful way. Since many companies use interdisciplinary design teams, an interdisciplinary project is a useful experience. Pierson (1987) discusses administering interdisciplinary design projects. A series of small projects allows for variety, different leaders, multiple grading opportunities, and a gradation in the degree of guidance. Large projects allow for more realistic problems, can be more open-ended, and require much more detailed planning and scheduling. It is useful if students see some of each of these different types of projects during their period in school.

It is obviously desirable to use projects that are real and have input from practicing engineers from industry or government. A variety of ways of obtaining this input are discussed by Bishop and Huey (1988), Griggs and Turano (1990), Harrisberger (1986b), Manning et al. (1988), Pierson (1987), and Sloan (1982), among others. Setting up the appropriate industrial or government contacts can be very time-consuming for a professor. Once the contacts have been made, he or she needs to arrange for sponsors. Companies are much more serious about design projects if they pay for the direct costs (not including labor) of the projects. Requiring payment helps to ensure their continued interest. Projects must be screened since some may be too easy or too difficult for the time allotted. A company engineer needs to be involved in the evaluation of projects, but if many different companies are sponsoring projects, the professor needs to control grades to ensure uniformity in grading. Other additional problems include the need for student travel to and from the client and the occasional lack of cooperation when a company refuses to release necessary information.

Balancing the difficulties in working with companies on design projects are the benefits. The opportunity to work with practicing engineers can give students contacts for future jobs and references. There is no question in their minds that the project is real and relevant. Professionalism is obviously important, and students are more likely to behave as professionals. Finally, successful performance can give students confidence that they can be successful engineers.

Regardless of the type of project, both oral and written reports should be required to stress communication. Weekly progress reports are useful to help prevent procrastination and to pinpoint problem groups. If a company is sponsoring the project, a presentation to the company is in order, but only after a full-scale dress rehearsal in front of the faculty.

9.1.4. Case Studies

A case study is a detailed description of how a professional approaches a problem. A number of engineering case studies are available from the former ASEE Engineering Case Library which is now the Center for Case Studies in Engineering at Rose-Hulman Institute of Technology. In addition, many descriptions of solving tough engineering problems are available in books (e.g., Herring, 1989; Kidder, 1981) and in the trade literature. Patents can also serve as case studies (Whittemore, 1981). Smith and Kardos (1987) and Durbin (1991) give other references. Video-based case studies are being developed (Sullivan and Thuesen, 1991).

Case studies can be used in a variety of ways and are useful in both design and nondesign classes. Professors can assign case studies to students for reports and as the basis for discussion (Henderson et al., 1983). Many case studies consider ethical questions and serve as an excellent basis for discussion. They also help to introduce the engineering profession and serve to motivate some students. This use of case studies is very appropriate in introductory engineering classes to help show students that the material being studied is relevant.

Case studies can be used in design classes, although they are not a substitute for project work since they are less open-ended (students will consider the case study the solution). Instead, case studies should complement projects. They are particularly useful in showing the human aspect of engineering. And they can show the importance of nontechnological factors, such as marketing, in the success of products.

Case studies can also be the basis for instructor-generated projects. In this instance the instructor does not show the students the case study, but obtains project ideas, data, a scenario, and so forth, from the case study. Professors can also have students work through the case study step by step and use it for feedback. This procedure is discussed in the next section.

Case studies are extensively used in law and business schools. Myers (1991) discusses the history of case studies and provides references for applications outside engineering. He notes that Harvard Business School introduced a seminar for professors to teach them to teach with case studies.

9.1.5. Guided Design

Guided design is a structured way of having students work through case studies. This procedure is particularly appropriate for introducing students to open-ended design problems since there is considerable guidance and feedback throughout. The guided design procedure was developed by Charles Wales and his coworkers (e.g., Bailie and Wales, 1975; Stager and Wales, 1972; Wales and Stager, 1973, 1977; Wales et al., 1974a,b; Wales and Nardi, 1982). Guided design was first developed for engineering classes and has since spread to a variety of other disciplines.

The guided design procedure is well summarized by Wales and Nardi (1982). The professor uses printed handouts to guide teams of five to six students through an open-ended problem in "slow motion." After the groups are formed, the guided design procedure starts with a printed handout which explains the problem situation and the student roles. The student groups then define the problem statement and set goals. This is done by a cooperative group discussion (see Section 7.2). After five to twenty minutes the groups receive a printed sheet which tells what the professional engineer did. It is important to stress at this point that this feedback sheet does not represent the solution but shows what one professional did. This point can be made quite clearly if some of the feedback sheets contain actions which are not particularly clever or are ethically dubious. The student groups then discuss the printed feedback sheet and compare their responses to that of the professional engineer. Since it is the design process which is being taught and not a particular answer, the professor must be careful in evaluation.

The guided design procedure then advances step by step through a specific problem-solving or design procedure. For example, the problem-solving strategy in Section 5.3 can be used. Wales and Nardi (1982) recommend the following ten steps:

1 Outline situation
2 Define goals
3 Gather information
4 Suggest possible solutions
5 Establish constraints
6 Choose solution path
7 Analyze factors needed for solution
8 Synthesize solution
9 Evaluate solution
10 Make recommendations.

For each step the students first complete the step and then receive and discuss the feedback. Guided design projects can take from two hours to several weeks. At the end of the guided design students can be required to communicate their results orally and in writing. While the guided design proceeds in class, the students can be assigned readings and homework for outside class. The groups can be encouraged to meet outside class as cooperative learning groups.

Although guided design was first developed as a procedure where all information transfer was in printed form (books and handouts), it can easily be adapted to the laboratory portion of a lecture class (Eck and Wilhelm, 1979). For students who are unfamiliar with working in cooperative groups, it is useful to use the first laboratory period for exercises in interpersonal communication, such as paraphrasing, self-disclosure, maintenance contributions to the group, and an ethics exercise with student observation (Eck and Wilhelm, 1979). After every group exercise it is useful to do some group processing to help students improve their skills.

What does the professor do in guided design? The professor must prepare or select the case studies and put them into a guided design format. Some prepared projects are available (Wales et al., 1974; Wales and Stager, 1973; Eck and Wilhelm, 1979). If a prepared guided design project is not available, then the professor can convert old design projects, convert case studies, or develop new projects in the guided design format. Potential users need to be aware that developing a good guided design project from scratch is very time-consuming.

Wales and Nardi (1982) discuss the development of new guided design projects. First the project must be outlined and divided into labeled steps following the problem-solving or design strategy of choice. A story line and realistic roles must be developed for the students. This step is important since it establishes a need for the project and helps to motivate the students, who must be active in each step. Since learning the process is the important goal, students should gather information only once and have an opportunity to practice the decision-making steps. Students should be asked to make important decisions. Asking them to make trivial decisions reduces the credibility of the entire project.

The form of the written feedback is important. This feedback models what an experienced engineer does to solve the problem. Be sure to write that the engineer would have done the following, not that *you* should have done the following. Since the problems are open-ended, the feedback can serve only as a model of possible actions. The students may tend to resist this at first, and the professor must be careful not to reinforce their beliefs that this is the correct solution. Be sure that the feedback responds to the questions that the students were asked in the instruction.

In a guided design class students must learn the content outside class by reading, discussing in their groups, doing homework, and so forth. In class they learn how to apply this content to open-ended design problems. The professor needs to be sure that the printed learning materials are good. If the textbook is not clear by itself, then additional notes or study guides must be developed. Some class time needs to be available for answering questions, reviewing the homework, making class assignments and so forth.

Once the guided design period starts, you, as the professor, are a guide and coach, not a lecturer. The first challenge is to form groups and to get the groups off to a good start. Group assignments are discussed in Sections 7.2 and 9.1.2. During the project the professor and the TA can circulate among the groups. If a group is functioning well, the professor or the TA can just listen and then briefly provide some positive feedback such as, "This is a great discussion. Keep it up." Some groups will need help getting started. Ask the group questions about the project or about group processing. If necessary, appoint a leader and a recorder. The behavior of students is often markedly more focused if they have an assigned role. In order to provide proper feedback to the groups, a professor or a TA should be available for every twenty-five to thirty students.

There are a variety of ways to assign group project grades. One approach is to have the groups assign the grade that they think they have earned. If their grade is higher than what you think they have earned, make them redo their project and their report until they have earned the higher grade (Eck and Wilhelm, 1979). This procedure is most appropriate for long projects.

The results reported for guided design have been impressive (Eck and Wilhelm, 1979; Feldhusen, 1972; Stager and Wales, 1972; Wales and Nardi, 1982). The instructor spends much more time with the students on high-level cognitive tasks. And they show better retention, higher grades both in the guided design course and in follow-up courses, increased confidence, and greater motivation. The classes show more cooperation and better group dynamics than other design classes. Students rate guided design classes higher than they do other design classes. However, it is not uncommon to have one or two poor groups which do not function well. The members of these groups do not fully benefit from the guided design experience.

9.1.6. Design Clinics

Actual practice in engineering is obviously beneficial to students. A design clinic is one way of providing an internship activity. Other approaches to providing industrial experience are industrial cooperative programs and summer internships. The following are advantages of the design clinic approach (Harrisberger, 1986a,b):

- Students have a significant industrial experience.
- Students make contacts with practicing engineers.
- Students become confident and more professional.
- The design clinic can fit into the normal course structure.
- Students need no extra time for graduation.
- The design clinic can be controlled by faculty members.
- The design clinic can be self-supporting.

There are models of the design clinic approach at Harvey Mudd College and at the University of Alabama. Our discussion is based on the University of Alabama model (Harrisberger, 1986a,b).

The design clinic assumes that basic technical knowledge has been covered in other courses. In the clinic students first learn a variety of skills and then apply them in a supervised professional practice working on a real industrial problem. Students take a three-credit skills course in the first semester of their senior year which consists of two seminars and one three-hour lab every week. In the seminars the students have lectures, take diagnostic tests such as the Myers-Briggs Type Indicator, make and critique presentations, listen to panels, and so forth. The content is concerned with the practical aspects of engineering instead of technical content. Thus students learn skills for presentation, listening, writing, record keeping, teamwork, leadership, project planning, creative problem solving, design methodology, retrieving and finding information, persuasion, and assertiveness.

The laboratory portion of the skills course consists of group projects in which students have the opportunity to practice the skills covered in the seminars. The laboratory is also used to introduce them to the solution of open-ended design problems. The projects done in the laboratory consist of an ideation exercise, a management simulation game, an extensive guided design project, and an extensive competitive design study.   Note that the class starts with significant guidance in solving open-ended problems and then reduces the amount of structure.

During the second semester students take an internship course. Groups of three work on company-sponsored problems. The companies are expected to pay all direct costs, which average less than $1000 per project plus a clinic fee of $350 to cover administrative expenses. The design group visits the company for an initial visit to learn about the problem and for a final written and oral presentation. Other visits may be scheduled if needed. All companies are within a four-hour drive of the campus, and student groups are selected so that at least one member has a car. The companies are expected to provide the necessary information and to have an engineer work with the students as needed.

Every group meets with a faculty coach once a week for twenty minutes. This coaching helps keep the students from procrastinating and keeps them focused on solving the problem on time. Each group presents a midterm progress report in the clinic. A dress rehearsal is presented in front of a faculty jury before the final presentation to the company. All faculty in the department are modestly involved by coaching two design teams, which takes about one hour per week. Administrative details of running a clinic are discussed by Harrisberger (1986b).

Although the design clinic idea has not been widely adopted, it does appear to be a cost-effective way of providing an industrial internship for all engineering students. In addition, design clinics do not require that the faculty have extensive industrial design experience to teach design.

9.2. LABORATORY COURSES

More than any other topic in this book, teaching laboratory courses in engineering is specific to the field of engineering and the type of laboratory. Since we must avoid discipline specificity, this section is an abstract discussion of the most concrete part of engineering education—the laboratory.

9.2.1. Purposes of Laboratory Courses

Laboratory courses can have a variety of different purposes, many of which are explored by ASEE (1986), Eastlake (1986), Jumper (1986), Kersten (1989), and Radovich (1983). Since the laboratory and the course structure depend upon the purposes of the laboratory course, these objectives should be decided upon first. No laboratory can be optimal for all purposes. The goals for the course can include:

Experimental skills. Students can learn a variety of skills involved in doing experimental engineering work. These can include certain psychomotor skills, planning an experiment, recording, analyzing and interpreting data, and using modern measuring instruments.

Real world. Students can learn to function in a real-world environment where the theory may or may not work and the equipment occasionally malfunctions. They can learn to distinguish reality from theory. They can also experience working in a climate of uncertainty and can learn the manifold meanings of Murphy's law.

Build objects. Students can actually build and test their designs. A sense of craftsmanship can be gained. They can learn to use working models to solve engineering problems (Hills, 1984). Models are used in many industrial settings but are often ignored in the education of engineers.

Discovery. Students can discover results which can improve theory and reinforce their ability to predict the results of using complex devices.

Equipment. Students can work with modern equipment, which adds a concrete aspect to an otherwise abstract education. While working with equipment, students can also learn about the importance of safety.

Motivation. "The theoretical work was difficult—some of it exceedingly so—but the physical *doing* made it seem worthwhile" (Florman, 1987, p. 8).

Teamwork. Many laboratories are team efforts, and students can learn to function as part of a team. This can include an opportunity to be the team leader.

Networking. In addition to teamwork, students may have to find information from a variety of sources including industrial contacts, professors not connected with the laboratory, technicians, and so forth. This is an appropriate experience before accepting their first industrial position.

Communication. Both written and oral communication skills can be emphasized through preparation, progress, and final reports.

Independent learning. Since all the knowledge needed for laboratory classes will not be at their fingertips, students will have to independently review old material and learn new material. This can help prepare them for the real world where independent learning is important.

We have not tried to be encyclopedic, and there are obviously other purposes for laboratory courses.

9.2.2. Laboratory Structure

The structure of the laboratory should depend upon the major purposes of the course. It can range along a continuum from a totally structured, cookbook-type approach to a partially guided experience to an unstructured class (Alexander et al., 1978). A cookbook approach can be satisfactory if the purpose is to develop psychomotor skills and the ability to use measuring instruments. These purposes have become less important as easy-to-use digital instruments have replaced analog instruments which often required considerable expertise. However, learning to use instruments or tools is still a legitimate purpose for a laboratory course. A cookbook approach may be used when the purpose is to reinforce theory. Unfortunately, this does not tend to be extremely convincing, and a discovery approach is more effective.

In an unstructured laboratory students are given fairly general instructions or goals. For example, the goal may be to design and build a new logic circuit, to survey a new subdivision, or to scale up a chemical process. The students must decide what needs to be done and how best to do it. An unstructured laboratory might ask students to explore a phenomenon such as the effect of pH and temperature on a biochemical reaction. No other directions are given. Unstructured laboratories are certainly appropriate for seniors who are mature enough to handle the uncertainty and who need the experience in planning and decision making before graduation.

Lower-division students may be lost in an unstructured laboratory. A partially guided experience is appropriate. A student is given some guidance in setting up the experiment and told what to do first. For later parts of the experiment much of the detail is left to the student. For example, a student can be told to look at the effect of several temperatures in a given range but not be told how many or which temperatures to use. In addition, the student would not be told what to expect although he or she might be told to predict the behavior.

Laboratory experiments appear to be most effective when the solution is not known ahead of time (Jumper, 1986). Measuring an orifice coefficient when fifty other students have already done so is not the stuff of a marker event. As a professor you need to be creative. Assume, for example, that the method of measuring an orifice coefficient is important in a fluids laboratory. The method will be learned much better if the student is given a noncircular hole as the orifice. Where does one look up the orifice coefficient for ellipses, rectangles, parallelepipeds, and triangles? What about five- or six-pointed stars and quarter moons? By varying the dimensions and the shapes, each student group can do a unique experiment, and the groups will not be able to dry-lab the results. In addition, this sort of "research" can eventually result in a technical note. Being the coauthor of a technical note or presentation (even if it is in a student magazine or at a student convention) will make the laboratory a marker event for the students. If time is available, this type of laboratory experiment can be made even more useful by asking students to predict the behavior of their orifice ahead of time.

Laboratory classes can be structured to reinforce lectures not with cookbook exercises but with the scientific learning cycle (see Section 15.1.) Do the laboratory work before the topic is covered in lecture and have the students explore the phenomenon. Let them discover many of the characteristics of the device. For instance, in the orifice example the students can determine the general form of the equation relating velocity to pressure drop. Then in lecture the theoretical development will be much more believable and would already have been partially verified. The students will be more likely to appreciate the power of theory to include additional terms without needing additional experimentation. The lecture would be the term introduction step in Figure 15-1. For concept application students can use their data to determine the orifice coefficient and solve additional problems.

Design laboratories are often unstructured. Students may be asked to design a large-scale apparatus. The purpose of the laboratory is to determine certain coefficients or efficiencies needed for the design. The students must determine what must be measured and must allocate their time between laboratory experimentation and design calculations. A design laboratory can also be used to design, build, and then test something. Hills (1984) suggests having students design and build simple working models, while Balmer (1988) believes that they should solve real industrial problems and test their solutions in the laboratory. Williams (1991) requires students to design and then build microcomputer boards. Many electrical and mechanical engineering problems can fit into these types of design laboratories.

9.2.3. Nitty-Gritty Details

A number of decisions must be made in any laboratory course. Should the laboratory be part of a lecture course or should it be a stand-alone course? Both arrangements have their advantages. If the purpose of the laboratory is to reinforce the theory and allow students to discover results, then a laboratory attached to a theoretical course makes sense. Scheduling is easier, and the connection between experiments and theory will be more obvious to the students. If the purpose of the laboratory is to synthesize several theory courses and have students design or build something, then a stand-alone course with appropriate prerequisites makes sense. In either case, the laboratory workload should be congruent with the credit granted (Radovich, 1983). If students are supposed to be able to finish laboratory experiments and reports in the laboratory, then it needs to be structured so that at least the better groups can do this.

Should students work individually or in teams? Although there are a number of reasons why teamwork is beneficial to students, the decision may be made on the basis of availability of apparatus. Equipment availability often determines team size, but most schools seem to have settled on two students for bench scale equipment, and three or four students per group for larger equipment. If teams are used, how should they be selected? This question is discussed in detail in Section 9.1.2. It is better to make a rational choice than just to continue what has been done for many years.

Students should be required to plan their experiments in advance. Many laboratory courses require students to pass an oral readiness quiz before they can go into the laboratory. This is a good safety precaution which encourages students to think before experimenting. In a design laboratory with projects lasting four weeks, we found it useful not to allow students to collect any experimental data during the first class. This time was spent in planning.

What types of records should students keep, and how should they report their results? Laboratory notebooks are commonly used in industry to support possible future patent claims. Experience in keeping a neat laboratory notebook which follows industrial practice is appropriate in an engineering laboratory (McCormack et al., 1990). Since communication is often an important goal of the laboratory (and all too often of *only* the laboratory), both oral and written reports are often required. The best feedback for oral reports can be provided by videotaping student presentations and having them watch their tapes (see Section 8.1.3). For written reports the most improvement in writing will occur if students receive prompt feedback and then rewrite the report for a grade. This obviously requires proper scheduling of the laboratory session and diligence on the part of the instructor.

The quality of the equipment in the laboratory is a never-ending problem (ASEE, 1986), and obsolete equipment and poor maintenance are often problems when programs are accredited. We do not see any substitute for modern instrumentation. Components such as resistors and transistors and major pieces of equipment such as nuclear reactors, distillation columns, or jigs do not have to be new, but the analytical instrumentation does. Mechanical balances, for example, are now obsolete and should be retired. If the purpose is discovery, much of the equipment can be simple and homemade. If the purpose is to familiarize the student with industrial equipment, then it is better to use commercial equipment. There is no substitute for a planned and funded maintenance and equipment replacement program. Safety should be a primary concern when equipment is repaired and when new equipment is purchased. Safety needs to be stressed with undergraduates (and with TAs). Stern measures are taken in industry when workers fail to follow safety rules, and stern measures should be taken with students who do not follow safety rules.

Teaching assistants may try to avoid laboratory assignments because they are often more work than the grading of papers in other courses. The department needs to be sure that the workloads for all TA assignments are appropriate and roughly equal. Laboratory TAs usually have significant contact with the students; thus, they should be able to communicate well. TAs often need to be trained, and a convenient time to do this is the week before classes start.

Group grading needs to be carefully considered. It is appropriate in laboratory courses to foster both interdependence and individual responsibility (see Section 7.2.2). Each student's grade should be partly based on the team effort and partly on the individual effort. Groups should be encouraged to make the laboratory a group effort, not merely a leader with two drudges. Professors and TAs should make a regular practice of circulating through the laboratory and observing the groups at work. After a few weeks of casual observation, it is usually clear who the malingerers are. This regular observation and a perusal of laboratory notebooks also help to discourage dry-labbing. Students can also be asked to assign part of the grade to the other students on their team. This procedure can work, but abuses can occur.

9.2.4. Advantages and Disadvantages of Laboratory Courses

From the student's point of view, laboratory work can provide a concrete learning experience where principles can be discovered. The chance to design and possibly build equipment can serve as a marker event in the student's undergraduate career, and friendships developed in laboratory teams may last for years. In addition, a student may get to know his or her laboratory instructors better than any other professors, and the student will rely on the laboratory professor for advice and letters of recommendation.

Of course, everything is not always this ideal, and there can be disadvantages. The laboratory may be an incredible time sink as an overzealous professor tries to have the students learn everything about engineering in one course. The equipment may not work or may be obsolete. Files may be readily available, and drylabbing of cookbook experiments may be rampant. A student's group may malfunction, leaving him or her with all the work and only one-third of the rewards. The professor may be absent, and the TAs may not speak English. Other than tradition, the reason for a laboratory course may be unclear.

The professor, whose task is to make the reality closer to the ideal, can have significant student contact and a chance to make a real difference in students' careers. Design laboratories often require a synthesis of the material from several courses. This helps the professor stay current in areas other than his or her research specialty. Working with real equipment can also help the professor be a better teacher of theoretical concepts.

Grading can be a chore when a number of long reports are turned in. It helps to have someone trained in English available to grade the communication aspects of the reports and to work with students on their communication skills. This reduces the burden on the engineering professors and provides the students with better instruction. Unfortunately, the workload is often heavier in laboratories than in other courses, and less credit may be given for teaching laboratory courses. This unfair workload has been criticized by ASEE (1986).

From the departmental point of view excellent laboratories are a source of pride. If you don't believe this, visit a department with an excellent undergraduate laboratory and note the attitude of the professor who guides you through the laboratory. Excellent laboratories also help produce well-prepared engineering graduates. And excellent laboratories are an advantage at accreditation time. Of course, the department gets what it pays for. Excellent laboratories require money for equipment, maintenance, a technician, and dedicated professors, who will remain dedicated only if suitably rewarded. Departments which use the laboratory as a way to save money when the budget is tight will pay the price of less-than-excellent laboratories fairly quickly.

9.3. CHAPTER COMMENTS

It should be clear that we believe that design and laboratory classes are important. We also believe that there are a variety of nontechnical skills which are critical for the successful practice of engineering. These include communication skills, management skills, and interpersonal skills. More engineers are removed from positions because of a deficiency in these skills than because of a lack of technical ability. Design and laboratory courses provide an opportunity for teaching these skills. Students learn by doing. However, the doing is more effective for learning if it is initially guided and supervised. Thus, we have included teaching procedures which specifically guide the student and provide feedback.

We enjoy teaching laboratory courses. The extra student contact makes up for the burden of grading laboratory reports. In addition, our school has done an adequate job of financing the laboratory and rewarding the participation of professors. Since we enjoy teaching laboratory classes, most students don't mind taking them from us.

9.4. SUMMARY AND OBJECTIVES

After reading this chapter, you should be able to:

• Discuss what design and laboratory work add to the education of engineers. Discuss the problems inherent in teaching design and laboratory courses.
• Develop a plan to incorporate design throughout the undergraduate engineering curriculum.
• Compare and contrast the different ways to teach design. Highlight the advantages and disadvantages of each method.
• Describe how you would select groups for a design project or laboratory experiment. Justify your method.
• Explain the appropriate laboratory structure for students at different levels.

HOMEWORK

1 Determine what roles design and laboratory classes play in the curriculum at your school. Do they meet the spirit of the ABET requirements? If not, what can be done to improve them? Or, why do you think the ABET requirements are irrelevant?
2 Develop a plan to include design throughout the engineering curriculum at your school.
3 Choose one of the methods of teaching design. Outline how to incorporate this method into one of the design courses at your school. Explain how this method would help students achieve the course objectives.
4 Assume one of the design groups in your class is not functioning well. Develop an intervention strategy to help get this group back to healthy functioning.
5 Select appropriate objectives for a laboratory course at your school. Outline a structure to help students meet these objectives.

REFERENCES

ABET, *Criteria for Accrediting Programs in Engineering in the United States*, Accreditation Board for Engineering and Technology, New York, 1989.

Alexander, L. T., Davis, R. H., and Azima, K., "The laboratory," *Guides for Improvement of Instruction in Higher Education, No. 9*, Michigan State University, East Lansing, MI, 1978.

ASEE, "Executive summary of the final report: Quality of Engineering Education Project," *Eng. Educ.,* 16 (Oct. 1986).

Baasel, W., "Goals of an undergraduate plant design course," *Chem. Eng. Educ.,* 26 (Winter 1982).

Bailie, R.C. and Wales, C.E., "Pride: A new approach to experiential learning," *Eng. Educ.,* 398 (Feb. 1975).

Balmer, R.T., "A university-industry senior engineering laboratory," *Eng. Educ.,* 700 (April 1988).

Bishop, E. H., and Huey, C.O., Jr., "The administration of an industry-supported capstone design course," *Proceedings ASEE Annual Conference*, ASEE, Washington, DC, 1661, 1988.

Culver, R. S., Woods, D., and Fitch, P., "Gaining professional expertise through design activities," *Eng. Educ.,* 533 (July/Aug. 1990).

Cundy, V. A., Smith, S., and Yannitell, D. W., "Practical creativity—LSU's senior design experience," *Proceedings ASEE Annual Conference*, ASEE, Washington, DC, 472, 1987.

Dekker, D. L., "Designing is doing," *Proceedings ASEE Annual Conference*, ASEE, Washington, DC, 784, 1989.

Durbin, P. T., "Coursework needs for technological literacy," *Proceedings ASEE/IEEE Frontiers in Education Conference*, IEEE, New York, 174, 1991.

Eastlake, C. N., "Tell me, I'll forget; show me, I'll remember; involve me, I'll understand (The tangible benefit of labs in the undergraduate curriculum)." *Proceedings ASEE Annual Conference*, ASEE, Washington, DC, 420, 1986.

Eck, R. W., and Wilhelm, W. J., "Guided design: An approach to education for the practice of engineering," *Eng. Educ.,* 191 (Nov. 1979).

Emanuel, J. T., and Worthington, K., "Senior design project: Twenty years and still learning," *Proceedings ASEE/IEEE Frontiers in Education Conference*, IEEE, New York, 227, 1987.

Emanuel, J. T., and Worthington, K., "Team-oriented capstone design course management: A new approach to team formulation and evaluation," *Proceedings ASEE/IEEE Frontiers in Education Conference*, IEEE, New York, 229, 1989.

Evans, D. L., and Bowers, D. H., "Conceptual design for engineering freshmen," *Int. J. Appl. Eng. Educ.,* 4, 111 (1988).

Evans, D. L., McNeill, B. W., and Beakley, G. C., "Capstone design for engineering freshmen?" *Proceedings Innovation in Undergraduate Engineering Education Conference,* Engineering Foundation, New York, 45, 1990.

Feldhusen, J. F., "Guided design. An evaluation of the course and course pattern," *Eng. Educ.,* 541 (March 1972).

Florman, S. C., *The Civilized Engineer*, St. Martin's Press, New York, 1987.

Green, D. G., "A curriculum approach to teaching engineering design," *Proceedings ASEE Annual Conference*, ASEE, Washington, DC, 1509, 1991.

Griggs, F. E., Jr., and Turano, V. S., "The Merrimack College capstone design program," *Proceedings ASEE Annual Conference*, ASEE, Washington, DC, 1279, 1990.

Harrisberger, L., "Development of human software for industry," *Proceedings ASEE Annual Conference*, ASEE, Washington, DC, 479, 1986a.

Harrisberger, L., "Engineering clinics and industry: The quintessential partnership," *Proceedings ASEE Annual Conference*, ASEE, Washington, DC, 979, 1986b.

Henderson, J. M., "Design in mechanics courses?" *Proceedings ASEE Annual Conference*, ASEE, Washington, DC, 1146, 1989.

Henderson, J. M., Bellman, L. E., and Furman, B. J., "A case for teaching engineering with cases," *Eng. Educ.,* 288 (Jan. 1983).

Herring, S., *From the Titanic to the Challenger*, Garland, New York, 1989.

Hills, P., "Models help teach undergraduate design," *Eng. Educ.,* 106 (Nov. 1984).

Hudson, W. B., and Hudson, B. S., "Special education and engineering education: An interdisplinary approach to undergraduate training," *Proceedings ASEE/IEEE Frontiers in Education Conference*, IEEE, New York, 53, 1991. (This article has names, addresses, and phone numbers for organizations that provide assistive technology.)

Jansson, D. G., "Creativity in engineering design: The partnership of analysis and synthesis," *Proceedings ASEE Annual Conference*, ASEE, Washington, DC, 838, 1987.

Jones, J. B., "Design at the frontiers of engineering education," *Proceedings ASEE/IEEE Frontiers in Education Conference*, IEEE, New York, 107, 1991.

Jumper, E. J., "Recollections and observations on the value of laboratories in the undergraduate engineering curriculum," *Proceedings ASEE Annual Conference*, ASEE, Washington, DC, 423, 1986.

Juricic, D., and Barr, R. E., "Integration of design into mechanical engineering curriculum," *Proceedings ASEE Annual Conference*, ASEE, Washington, DC, 358, 1991.

Kersten, R. D., "ABET criteria for engineering laboratories," *Proceedings ASEE Annual Conference*, ASEE, Washington, DC, 1043, 1989.

Kidder, T., *The Soul of a New Machine*, Little-Brown, Boston, 1981.

Klein, R. E., "The bicycle project approach: A vehicle to relevancy and motivation," *Proceedings ASEE/IEEE Frontiers in Education Conference*, IEEE, New York, 47, 1991.

McCormack, J., Morrow, R., Bare, H., Burns, R., and Rasmussen, J., "The complementary roles of laboratory notebooks and laboratory reports," *Proceedings ASEE Annual Conference*, ASEE, Washington, DC, 1429, 1990.

Magleby, S. P., Sorensen, C. D., and Todd, R. H., "Integrated product and process design: A capstone course in mechanical and manufacturing engineering," *Proceedings ASEE/IEEE Frontiers in Education Conference*, IEEE, New York, 469, 1991.

Manning, F. S., Wilson, A. J., and Thompson, E. E., "The use of industrial interaction to improve the effectiveness of the senior design experience," *Proceedings ASEE Annual Conference*, ASEE, Washington, DC, 620, 1988.

Middlebrook, R. D., "Low-entropy expressions: The key to design-oriented analysis," *Proceedings ASEE/IEEE Frontiers in Education Conference*, IEEE, New York, 399, 1991.

Miller, L. S., Papadakis, M., and Nagati, M. G., "Design content in traditionally non-design courses," *Proceedings ASEE Annual Conference*, ASEE, Washington, DC, 6, 1989.

Myers, D. D., "Need for case studies: New product development," *Proceedings ASEE Annual Conference*, ASEE, Washington, DC, 89, 1991.

Overholser, K. A., Woltz, C. C., and Godbold, T. M., "Teaching process synthesis—The integration of plant design and senior laboratory," *Chem. Eng. Educ.*, 16 (Winter 1975).

Paris, J. R., "Professional software in process design instruction: From why to how to beyond," *Proceedings ASEE Annual Conference*, ASEE, Washington, DC, 1161, 1991.

Peterson, C. R., "Experience in the integration of design into basic mechanics of solids course at MIT," *Proceedings ASEE Annual Conference*, ASEE, Washington, DC, 360, 1991.

Pierson, E. S., "A team-based senior-design sequence," *Proceedings ASEE/IEEE Frontiers in Education Conference*, IEEE, New York, 221, 1987.

Radovich, J. M., "What is needed for a good laboratory program?" *Eng. Educ.*, 749 (April 1983).

Riffe, W. J., and Henderson, B. P., "A second year mechanical engineering design course," *Proceedings ASEE Annual Conference*, ASEE, Washington, DC, 980, 1990.

Ring, S. L., "Don't overlook the cities for engineering design labs," *Proceedings ASEE/IEEE Frontiers in Education Conference*, IEEE, New York, 272, 1982.

Sloan, E. D., "An experimental design course in groups," *Chem. Eng. Educ.*, 38 (Winter 1982).

Smith, C. O., and Kardos, G., "Need design content for accreditation? Try engineering cases!" *Eng. Educ.*, 228 (Jan. 1987).

Stager, R. A. and Wales, C. E., "Guided design. A new concept in course design and operation," *Eng. Educ.*, 539 (March 1972).

Stern, H., "Team projects can offer incentives," *Proceedings ASEE Annual Conference*, ASEE, Washington, DC, 394, 1989.

Sullivan, W., and Thuesen, J., "Integration of economic principles with design in the engineering science component of the undergraduate curriculum," *Proceedings ASEE Annual Conference*, ASEE, Washington, DC, 525, 1991.

Wales, C. E., and Nardi, A., "Teaching decision-making with guided design," Idea Paper No. 9, Center for Faculty Evaluation and Development, Kansas State University, Manhattan, KS (Nov. 1982).

Wales, C. E., and Stager, R. A., *Educational Systems Design*, Morgantown, WV, 1973. (Published by authors.)

Wales, C. E., and Stager, R. A., *Guided Design*, Part I, Morgantown, WV, 1977. (Published by authors.)

Wales, C. E., Stager, R. A., and Long, T. R., *Guided Engineering Design*, West Publishing Company, St. Paul, MN, 1974a.

Wales, C. E., Stager, R. A., and Long, T. R., *Guided Engineering Design, Project Book*, West Publishing Company, St. Paul, MN, 1974b.

Warfield, J. N., "Design science: Experience in teaching large system design," *Proceedings ASEE Annual Conference*, ASEE, Washington, DC, 39, 1989.

Whittemore, O. J., "Patents: A tool for teaching design," *Eng. Educ.* 229 (Jan. 1981).

Williams, R. D., "A project-oriented class in microcomputer system design," *Proceedings ASEE Annual Conference*, ASEE, Washington, DC, 1514, 1991.

Woods, D. R. (Ed.), "Using troubleshooting problems," *Chem. Eng. Educ.*, 88 (Spring 1980), 130 (Summer 1980).

Woods, D. R., "Workshop in using troubleshooting problems for learning," *ASEE Annual Conference, Session 3516*, June 22, 1983. (This paper is not in the proceedings.)

CHAPTER 10

ONE-TO-ONE TEACHING AND ADVISING

In a perfect world professors would have the time to get to know every one of their students as individuals and would be able to tutor them when they had difficulties. Although this is seldom feasible, professors do have significant one-to-one contact with students. One-to-one contact occurs when a student asks a question and the professor makes eye contact while answering the question. It also occurs when a student asks a question after class or in the hall, and when a student comes to the professor's office to ask questions. Although brief, these encounters have a considerable impact on his or her rapport with students; thus one-to-one contact has a major effect on the professor's effectiveness as a teacher. Advising and counseling usually involve significant one-to-one contact with students. The area where many professors have the most contact with individual students is in serving as research advisers for graduate students.

The one ability which is common to all these examples is skill in listening. Actively listening to people and responding so that they know you understand is a necessary skill for excellent one-to-one teaching and advising. Unfortunately, this ability is often neglected. Listening skills will be discussed first, and then particular one-to-one teaching and advising situations will be considered.

10.1. LISTENING SKILLS

Everyone who writes about listening laments the lack of skill in this important communication area. Professors who take the time to listen to students will benefit, and their students will greatly appreciate the rare chance to be heard. This will increase the professor's effectiveness as a teacher significantly. However, learning to listen can be very difficult for professors since many of them really do like to talk. Listening skills are also critical for effective advising and tutoring. If one of the goals of your department is to improve the communication skills of the engineering graduates, then it may be appropriate to teach students to improve their listening skills. Exercises that do this can easily be incorporated into laboratory and design courses.

Listening is a skill that can be learned, but practice is required. Listening skills are discussed in many counseling books (e.g., Bolton, 1979; Brammer, 1985; Edwards, 1979; Hackney and Nye, 1973), in many books on teaching (e.g., Eble, 1988; Lowman, 1985; McKeachie, 1986), and in articles in the engineering education literature (e.g., Katz, 1986; Miller, 1980; Root and Scott, 1975; Stegman, 1986; Wankat, 1979, 1980). Reading about listening skills can be a first step in improving these skills, but long-term gains require practice.

10.1.1. Setting the Climate

As the professor you must first create a climate so that listening can occur. To become known as someone who listens, you must be available, and the easiest time to be available for the largest number of students is before and after class. Students must come to class anyway, so the barrier to talking to the professor is significantly less than in coming to the professor's office. Make a point to come to class five or ten minutes early. Not only will this give you a chance to make sure that the room is ready for class, but it will send a subtle message that you are interested and looking forward to the class. It also gives students a chance to talk to you. Early in the semester it is useful to walk around the room with your class list, talking to students and learning their names. Later in the semester students will come up to you to talk.

Students often have questions after a class; by staying a few minutes you can further develop a rapport with them. To do this you may have to avoid scheduling a meeting immediately following the class. If the after-class period is too rushed, you might consider finishing class five minutes early. Since you don't want to delay the start of the next class, it helps to be available for short questions in the hall. Office hours are useful for longer discussions and for dealing with private concerns of students (see Section 10.2).

Professors and students are not equal. As the professor you have significantly more knowledge and experience in the subject area. In addition, you have power over the student. These inequalities in power and status inhibit some students (nothing inhibits other students). You can facilitate student interaction by making the environment more equal. Reduce barriers: Step from behind the podium and take a few steps toward the students. Wander in the audience to solicit interactions with students. Be relaxed and nonverbally encourage students to talk. Sitting down on the edge of a table or desk indicates that you are relaxed and have time to talk. Rearrange your office so that the desk is not a barrier between you and the students. (If you are new to academe and feel a bit insecure, you might want to have the desk between you and the students.)

A professor's attitude is important. Those who want to help students telegraph this attitude to them. Generally speaking, people who are classified as feeling types on the Myers-Briggs

Type Indicator (see Chapter 13) will have an easier time conveying to the students the impression that they want to listen. Thinking types need to think about the students' feelings. Perceptives tend to enjoy the uncontrolled give-and-take of discussions with students, while judging types need to schedule this time for students. By knowing yourself, you can adjust to be available and to listen to students. A note of caution: If you don't particularly like students but love the content, don't try to fake being the students' friend. They will see through your facade. Aloof professors who are content experts can be good teachers (see Table 1-1).

To encourage interaction with students, professors need to be nonjudgmental: There are no "dumb" questions. There are questions which show a lack of understanding, and there are questions you don't understand. The purpose of listening is to clarify your understanding of the questions so that you can help students understand the material. It is also helpful to avoid being defensive. This can be difficult when students are angry and are attacking a test, and may be attacking you. Although there are no dumb questions, there are hostile ones. Sometimes acknowledging a student's feelings (see Section 10.1.2) will calm her or him so that he or she can listen to facts. Sometimes humor is useful in deflecting the hostility. If no progress is made, offer to talk to the student privately after class. Discipline problems are discussed in detail in Chapter 12.

Being nonjudgmental does not mean "anything goes" or that there are no standards. Instead, it means that actions and behaviors are evaluated, not the inherent worth of the student. There are times "when a student is rationalizing about the difficulties and needs to be told bluntly to make an attitude adjustment and work harder (or more efficiently)" (Herrick and Giordano, 1991). When being blunt with a student, tell him or her the probable consequence of actions or inactions, but without a character analysis.

10.1.2. Focus

A professor's first focus should be on the student. Make eye contact, move or lean toward him or her, offering nonverbal encouragement. Listen to what the student says completely without trying to formulate your response before he or she is finished. Use your brain's "free time" to ask yourself questions about what the student is trying to say. What is the underlying message that may be hidden in the student's response? (Katz, 1986). A useful technique is to paraphrase the question briefly after the student is finished. This ensures that you have understood it, and in a classroom situation ensures that everyone has heard the question. Repeating the question also gives you a little more time to formulate an answer to the question.

Your focus should be on the student's problem being considered. The best atmosphere for a class is one where the professor is present to help the students master the objectives. Unfortunately, in many classes the professor is the enemy. Anything that you can do through one-to-one contact to help students feel that you are there to help them learn helps improve the atmosphere in the classroom. How much of the student's problem should be solved by the professor and how much by the student is a judgment decision which is discussed in Section 10.2. Another trick for focusing on a speaker's message is to take notes. This may help you listen and pay attention at faculty and committee meetings, seminars, after-dinner speeches, and so forth. It is also appropriate to encourage both undergraduate and graduate students to take notes in meetings with you.

There should also be some focus on emotions. Emotions are always present, and if not dealt with directly may prevent communication and learning. This is particularly appropriate in private, but can also be appropriate in a classroom. In class, it is usually sufficient to acknowledge the emotion and then move to the content of the question. For example, "This appears to be an emotional issue for you. Let's look at it from another angle"; or, "I see that you are upset about the grading of this test. Let me answer the question now and then we can discuss the grading after class." In private, more time can be spent exploring the student's emotions (see Section 10.3).

Although it is appropriate to focus on the student's emotions, it is usually not appropriate to focus on your emotions as the professor. Try to remain rational and nondefensive. This is particularly true in class where an emotional outburst can do significant damage to your standing and credibility. Unwind later by talking to a friend.

10.1.3. Responses

Individuals make nonverbal, minimally verbal, and verbal responses to others. An additional response is silence. All these responses should be congruent. Students receive a confusing mixed message if your words do not agree with the nonverbal signals. This is one reason why most people cannot fake interest or caring for long periods.

Nonverbal messages include facial expressions, eye contact or lack of eye contact, interpersonal distance, hand gestures, and body language (Axtell, 1991; Miller, 1980). In Western cultures direct eye contact with occasional breaking and reforming of the contact is expected. Leaning forward is usually interpreted as a sign of interest, as are nods and encouraging hand gestures. An open stance or sitting position is interpreted as signifying openness, whereas crossing one's arms suggests a closed position. Clenched fists are often interpreted as anger, as are angry facial expressions. These are powerful signals which most individuals raised in a Western culture transmit and receive unconsciously (Axtell, 1991). The signals are often so powerful that words are ignored if they are incongruent with the message.

An individual can change the nonverbal messages he or she is sending. Changing behavior often changes the individual's feelings. For example, if you find that you have your arms tightly crossed and you are resisting listening to a message at a meeting, purposely opening your arms and relaxing will probably result in your being more open to listening. Since changing behavior often changes underlying emotions, it is useful to monitor and change the nonverbal clues you are sending to your students. One problem with nonverbal messages is that they have to be interpreted, and thus they may be misinterpreted. For example, the tightly crossed arms in the previous example may simply mean that the person is cold, while it is interpreted as being closed to an idea. In addition, the nonverbal messages of different societies are different (Axtell, 1991). In India shaking one's head signifies agreement, not disagreement as it does in Western society. The appropriate degree of eye contact and comfortable interpersonal differences are very different in different societies. If you are listening to one of your students and the nonverbal and verbal messages appear to be incongruent, it may be that you are misinterpreting her or his nonverbal messages. And he or she may be misinterpreting your nonverbal messages.

Minimal verbal messages are sounds like "uh" and "uh-huh" and words like "oh," "yeah," and "OK" which do not convey much meaning but encourage the person to keep talking. Minimal verbal messages sent by the listener usually imply that he or she is paying attention and understands the speaker. These messages are often used in private conversation, although they are also appropriate when a student is talking in class. If speakers often act as if they don't know whether you are listening, you may need to increase your use of minimal verbal messages. However, faking minimal verbal messages when you aren't listening will get you into trouble fairly quickly.

Verbal messages are an important part of the active listening process. Probes are questions or directives which ask the speaker to tell more. Probes can be nonspecific, "Elaborate on that" or "Tell me more"; or quite specific, "What would one observe if the weld was bad?" or "You are confused about the application of Kirchhoff's laws in this situation." Probes are often more effective if they are open-ended questions or directives which cannot be answered with a simple yes or no response. If you ask closed-ended questions and get yes or no responses, then change the questions to make them open-ended.

Paraphrasing what the person has said in your own words is a useful method for letting him or her know that you understand. With student questions it is useful to rephrase the question, and with both paraphrasing and rephrasing it is appropriate to ask if your interpretation is correct. Summarizing long statements in both classroom and private discussions is another useful active listening technique: "What I heard you say is . . . ." Again, it is important to check with the speaker that the summary is correct.

Silence is not golden if it does not encourage communication or becomes threatening. Use silence to encourage communication, not to punish students. Professors usually do not pause long enough after asking questions. A period of silence is necessary to allow students time to respond. In class this will be less threatening if you do something useful during the period of silence. For example, ask a question, clean off the blackboard, and then turn back to the students for an answer. Silence, perhaps punctuated with a nonverbal response, is also an appropriate response when a student is clearly processing information and is not ready for more communication.

Silence can also be very useful when students are trying to manipulate the professor. A common ploy is for a student to tell all the reasons why he or she will have trouble handing in an assignment on time or taking a test when it is scheduled, but never make a direct request for a postponement. Since no request has been made and no question has been asked, there is no need to respond. Silence is an effective counterploy since it forces the student to be honest about the request. An alternative response is to use a probe such as "Well, what are you going to do about it?" There is a final use for silence. When a student breaks down and starts crying in your office (yes, this can happen), one appropriate response is to offer a tissue and be silent until he or she has regained control.

TABLE 10-1   COMPARISON OF LISTENING AND NON-LISTENING BEHAVIORS (WANKAT, 1979)
Reprinted with permission from *Chem. Eng.* (Oct. 8, 1979). © 1979, McGraw-Hill.

	Non-listening behavior	Listening behavior
Time limitations	Does not mention time limitations, but shows in obvious ways he/she is busy	Honest about time limitations
Climate	Defensive— 1. Evaluates and judges 2. Tries to control speaker 3. Uses strategy 4. Is neutral and avoids feelings 5. Shows superiority 6. Is certain and dogmatic	Open and Supportive— 1. Non-evaluative and non-judgmental 2. Problem-oriented 3. Is honest and spontaneous 4. Accepts and shows feelings 5. Sets up an equal environment 6. Tentative about conclusions
Focus	1. Internal – self-conscious 2. External – on other work 3. Other – does not watch speaker 4. Interaction – on mechanics of the conversation	1. On speaker 2. On speaker's topic 3. Looks directly at speaker 4. On what is being communicated
Non-verbal behavior	Non-attending – closed posture, expressionless face, faces away from speaker	Attending – open posture, shows expression on face, looks at and leans toward speaker
Dialogue	No dialogue – either silent or monopolizes conversation	Dialogue – reflects and summarizes what speaker has been saying, clarifies unclear points, asks relevant questions
Overall attitude conveyed	Not interested	Interested

10.1.4. Comparison Between Listening and Nonlistening Behavior

Table 10-1 presents a brief comparison of listening and non-listening behavior. This can serve as a useful checklist for monitoring your behavior or for helping students improve their listening skills.

10.2. TUTORING AND HELPING STUDENTS

We are using an inclusive definition of tutoring to include helping students before and after class, during office hours, in special help sessions, in the halls and on the telephone. We rejected the idea of calling this section "Office Hours" since only a fraction of the students in a class come to see a professor during office hours. A majority of students can receive individual attention and at least minimal amounts of tutoring when the professor broadens her or his availability.

10.2.1. Tutoring Locations

Right before and right after class are the most efficient times for tutoring because many students ask questions then but are not tempted to visit with the professor. Coming to class early and staying late also shows accessibility and interest in the students. This technique is one of the few methods which is efficient and effective for both students and professors, and we strongly recommend that you try it. Since there are minimal barriers to the students, the hall can also be an effective place for informal student contacts. Professors who are open, and friendly and know the names of their students are often asked questions in the hall. Many of these questions can be answered immediately. For questions requiring more time or the use of a blackboard, you can make an appointment with the student or invite her or him into your office immediately. Taking a student with you into your office is one way to encourage students who otherwise would never come on their own.

Office hours are useful for the "regulars" who will use them. Unfortunately, many students, particularly introverts, who could benefit from help do not take advantage of a professor's office hours. Encourage the whole class to visit both you and the TA during office hours. (Of course, be sure that both you and the TA keep office hours.) Private notes on returned homework and tests asking students to come in and see you or the TA can also be effective. In lower division courses it may be appropriate to tell struggling students that they must come in. This fits in with our general strategy of being more directive to beginning students. Some professors require all students to stop in early in the semester as a way of getting to know them. This also reduces the barrier to students' coming to see you.

Telephones can also be used for long-distance tutoring. For television courses at remote sites, telephone contact with students is indispensable. Usually, a specified time is set aside when the professor will be available for phone calls about the course. Most students never use this service, but the existence of the service is important psychologically. Similarly, set-aside hours with the TA or the professor available to answer phone calls can be used for on-campus students. This service is particularly valuable for commuters who might find it difficult to come in for scheduled office hours.

Should you give your home telephone number to students and encourage them to call you at home? This is your decision. Some professors do this and some do not. If you do, it is appropriate to set limits on when they can call.

10.2.2. Advantages and Disadvantages of Tutoring

Tutoring and lecturing can fill complementary functions, as shown in Table 10-2, but they also differ in their ability to satisfy some of the basic learning principles listed in Section 1.4.

TABLE 10-2 COMPARISON OF LECTURING AND TUTORING

Item	Lecture	Listening behavior
Purpose	Transmit information	Troubleshooting
Where done	Lecture hall	Anywhere
Focus	Entire class	One student or small group of students
Coverage of material	Broad – Use of material may not be obvious to student	Narrow – immediately useful
Emotions	Not dealt with	Can be dealt with
Personal attention	Very little	Lots
Barriers to student use	None	Hesitant to bother professor Professor may not be receptive
Path	Linear/sequential	Branched/multiple
Information transfer	Mainly one-way	Interactive
Efficiency: Professor	High	Low
Student	Low	High
Professor needs	Basic knowledge Ability to organize material Presentation Skills	Basic knowledge Listening skills Troubleshooting skills
Advantages	Good at transmitting information Efficient use of Prof's. time	Individualizes Can work on problem solving

This comparison is shown in Table 10-3. A complete course package of lectures, tutoring, homework, and tests can satisfy all the learning principles; however, it is a good idea to try to satisfy as many of these as possible without the tutoring component since many students will not come in for tutoring.

10.2.3. Goals of Tutoring

The definition of good tutoring depends upon the goals of tutoring. The professor's goals are often to make a student a better problem solver who can become independent of the professor. Other possible goals include getting to know students better, receiving feedback about what they understand and do not understand, having an opportunity to interact more with them, motivating them to learn the material, stretching and challenging them, and minimizing the time spent tutoring.

TABLE 10-3   SATISFACTION OF LEARNING PRINCIPLES

Learning Principles	Lecture	Tutoring
1. Active learner	Often no	Usually yes
2. Feedback	Usually no	Can include
3. Knows objective	Can transmit in lecture	Could check on – but not usually done
4. Motivate learner	Can happen – often doesn't	Reinforces motivation
5. Individualize	Difficult	Yes
6. Important types of learning:		
Concepts	Yes	Can clarify
Apply principles	Sometimes	Yes
Illustrate problem solving	Can be	Can give practice
7. Problem solving requirements:		
Acquisition of knowledge	Yes	Can clarify
Practice	Usually no	Yes
8. Structured hierarchy material	Yes	No
9. Thought-provoking questions	Can do	Can do
10. Professor enthusiastic	Can show	Can show

Students often have a different perspective. Many want an answer to their current difficulty and are not concerned about overall development as a problem solver. However, there are students who are genuinely concerned about learning the course content and want to become better problem solvers. Some students use tutoring as a short-cut to finding information that they could clearly obtain on their own. Overly dependent students often want to check that they are doing everything correctly. They want reassurance. Students with a high need for affiliation may want to get to know the professor and use office hours for this purpose.

The dilemma for you is to satisfy your goals and at the same time satisfy enough of the student's goals. If none of the student's goals are satisfied, he or she will not return. This minimizes the time you spend tutoring but does not satisfy any learning objectives. Various methods can be used to improve tutoring.

10.2.4. Methods for Improving Tutoring

Tutoring is an art. A tutor must continually make decisions about what will be most helpful for a student at a particular time. Sometimes only a short answer or a pat on the back is needed.

In other cases significantly more time is required. Most of the suggestions in this section apply to these longer contacts.

One main advantage of tutoring is that a tutor can individualize the instruction for a particular student. Since different students want and need different things, vary your approach and responses. Observe and listen closely throughout the semester. When you meet these wants, the student will be happy and motivated. Unfortunately, this may not satisfy the student's needs. For example, someone who has trouble generalizing solution methods wants to see the solution method applied to all possible cases. What he or she needs is to learn to generalize. Good tutoring may consist of showing one additional case to satisfy his or her wants. Then the tutor can show how to generalize from the base case to the new case, and follow this by making the student generalize to another new case.

Students also require different emotional responses. One student may respond to a challenge, while another may require initial hand-holding and encouragement. Some students respond well to the socratic approach; others become flustered and frustrated. What works best also varies from day to day. Right after a hard test is not the time to offer another challenge. By observing and listening to a student, you can get an idea of his or her emotional state. Then respond accordingly.

One important way to improve tutoring is by improving listening skills (Section 10.1). Remember to be nonjudgmental. It doesn't help a student when he or she is yelled at and called stupid. This requires patience. Without jumping to conclusions, try to find what the student's difficulties really are. Ask open-ended questions to help the student find her or his own mistake. Whenever the problem is nontrivial, encourage the student to talk. Some focus on emotions can be helpful. A very short comment such as "I see you're really frustrated" can have a remarkable effect. It frees up the student, helping her or him see you as human and to feel that you understand.

Listening is particularly important when a student has subtle misconceptions. Conceptualize what the student is thinking and compare his or her approach to possible correct approaches. Since words may be confusing, the student should write down equations or draw a figure. This can help you see what he or she is talking about.

Interest in helping students is another important ingredient in excellent tutoring. You have to want to help students, although the reason why you want to help is probably not very important. Interest is important because a student can sense it from many verbal and nonverbal clues. Interest is also important as a motivating force for the professor. Tutoring can be hard, frustrating work. Being interested in helping students provides the patience and energy needed to be an effective tutor.

When a student comes in for tutoring, the passive lecture approach has failed. Make the student do things. The student can explain an approach in detail, write equations on the board, or solve problems on paper. Involve the student in the process instead of giving another minilecture. After you explain something, don't accept the polite but often meaningless phrase, "I understand." For example,

Professor: Do you see it now?
Student: Yeah, I understand.

Professor:	Good, now finish this problem on the board.
Student:	You mean right now?

This forces the student to become actively involved. It also allows you to observe and correct mistakes as they occur. Give minor help, and allow the student to work the problem through to completion. This can provide confidence that he or she can solve the problems.

Another way to make the student be active is with probing questions which eventually lead her or him to the desired solution. Unfortunately, in our experience this does not work with all students. When it does work, it is a fine method for making the student reason his or her way through a problem.

Another method is to have the student explain your answer in her or his own words. Sometimes you'll answer one student's question and then have a second student come in and ask the same question. Ask the first student to explain the answer to the second student. This forces the first student to be active, gives you a chance to check on understanding, and fulfills the learning principle of having students teach.

Students often know concepts but are unable to use them to solve problems. Since problem solving is an art, the one-to-one contact of tutoring can help students improve their techniques. Many students have no idea how they or anyone solves problems. When tutoring, use the problem-solving strategy taught in your course. This may involve only one or two steps of the strategy or may involve going step by step through an entire problem.

For example, when a student is having trouble getting started, going over the define and explore steps is appropriate (See section 5.3).

Professor:	O.K, Let's define the problem.
Student:	Well, it's written down here.
Professor:	Go to the board and draw me a figure.
	Good, now label all knowns.
	O.K., what are you asked to find?
	Good, that defines the problem. What's the next step?
Student:	It's to explore.
Professor:	What does explore mean.
Student:	Look for different possible approaches.
Professor:	Right. Give me five different ways you might be able to solve this problem.
Student:	Five?
Professor:	Uh huh, if you only have one you're not exploring.
Student:	Well I could . . . .
Professor:	Good, now let's go on and work on the most likely approach.
	Which approach is most likely to work?
Student:	I think that the . . . .
Professor:	Fine. Now let's plan how you would do that.

This approach does not produce an experienced problem solver immediately. However, it does guide the student toward improving.

When a student has put in considerable effort but is not converging on the right answer, troubleshooting is called for. Knowledge of typical student mistakes is helpful. For example, beginning students often have trouble with unit conversions or forget to convert units. A brief study of tests and homework will show you what these typical errors are in your subject.

If the difficulty is not a typical mistake, then more subtle errors must be searched for. This is where expert knowledge of the area becomes important. An expert can evaluate different approaches and find subtle errors. Excellent tutors must be subject matter experts, but not all subject matter experts are good tutors. How many times have you heard some variant of "He really knows the material, but he can't get it down to our level." The other ways to improve tutoring appear to be more important than becoming a subject matter expert. This is why students often make good tutors.

Several little tricks can help in tutoring. The first of these is humor. Professors with a good sense of humor are well liked even if they are hard taskmasters. Humor can aid the interaction by defusing anxiety and making learning fun. Groups of students can often be tutored at the same time. Since they often have similar difficulties, students can learn from others' questions. The exception to this is the student who is totally lost and needs individual attention. A final suggestion is to talk to your colleagues. Find out what works for them with particular students.

10.2.5. Tutoring Problems

Time is the number-one problem. How do you find enough time to prepare lectures, tutor, do research, attend committee meetings, and do everything else which needs to be done? Tutoring does require time, and so it helps to be efficient (see Chapter 2). One method which can help you control your time is to set specific office hours for tutoring. Unfortunately, this solution can generate a new problem. What do you do about students who come in at times other than your office hours? One possibility is to respond, "I'll help you this time, but in the future please come during office hours." If a particular student keeps ignoring this request, he or she may be overly dependent.

The overly dependent student is another problem. To become an independent problem solver the student needs to be weaned from the professors. The student is probably getting what he or she wants, but not what he or she needs. One approach is to discuss dependence with the student. This needs to be done in a nonjudgmental fashion. For example, tell the student that employers expect their engineers to be relatively independent, and that you are worried that he or she is not learning this skill. If the dependent behavior persists, you may have to limit the student to a specified number of minutes per week. Fortunately, this extreme behavior is rare.

The opposite problem is the extremely shy student who may not ask for help even when he or she would benefit from it. Encourage the student. If this student does come in, force yourself to listen even if you are very busy. One interpersonal mistake may drive a shy or sensitive student away. For a shy student to go to a professor's office is an act of courage. By being too busy or locking your door you may destroy this courage.

10.3. ADVISING AND COUNSELING

Probably the most neglected area in engineering education is advising, and certainly this is the area where students show the least satisfaction (Anonymous, 1986; Wankat, 1986). Inadequate advising is a commonly cited deficiency during ABET accreditation visits. Eble (1988) calls advising in college "a mess." Hewitt and Seymour (1992) found that inadequate advising was a frequent complaint of students who left enginering. Although there has been an upsurge in interest in it, advising is still tremendously neglected as compared to teaching or research. In this section we will first discuss academic advising and then personal counseling. Professional counselors often draw a distinction between advising and counseling, but we will use these two terms interchangeably.

10.3.1. Academic Advising

Advisors need to be trained to do academic advising (Eble, 1988; Vines, 1986; Wankat, 1980, 1986). This is true for professors, peer counselors, administrative assistants, and professional counselors. Training is required for the specific information needed, in listening skills, and in specific counseling skills.

Recent studies (e.g., Light, 1990) have found that there are significant gender differences in what students want from their academic adviser. These differences are shown in Table 10-4. To some extent they may mirror differences in the percent of men and women who are feeling or thinking types according to the Myers-Briggs Type Indicator (see Chapter 13). The implication for advising is that procedures which have worked for men may not satisfy women. Tannen (1990) reports on the fascinating research on how men and women tend to differ in their conversations. In general, men communicate facts and try to maintain independence, whereas women search for rapport and connections. Within this general framework the pattern of answers in Table 10-4 makes sense.

TABLE 10-4   WHAT STUDENTS WANT FROM THEIR ACADEMIC ADVISOR.
Percent saying "very important" (Light, 1990).

	Men	Women
1. Will take the time to know me personally	30	72
2. Shares my interests so we have something in common	31	58
3. Knows where to send me to get information	48	51
4. Knows the facts about courses	64	43
5. Makes concrete and directive suggestions	66	23

First, an adviser's information must be accurate and up-to-date as to the university's requirements for registration, prerequisites, dropping and adding courses, probation, transferring to a different school within the university, grade appeals, and so forth. The adviser needs to be able to tell a student the consequences of doing or not doing something. If the adviser does not know an answer, he or she needs to know whom to ask to find out. At large universities keeping up-to-date is a nontrivial task since courses, graduation requirements, and rules are continually changing. Advisers probably need to attend a meeting every semester to refresh their memories and to learn about changes.

Each school has a slightly different philosophy about advising. We believe that the responsibility for obtaining an education is the student's. The adviser is there to help, provide information, and explore alternatives with him or her, but the student must retain the ultimate responsibility. Thus, part of the function of an adviser is to help students gradually become more responsible in such things as checking for errors on a schedule, selecting a major and selecting electives, getting to know a few professors well, and eventually deciding upon graduate school or an industrial position.

Since we believe that the ultimate responsibility lies with the student, we recommend a relatively nondirective approach. Listen to the student and let her or him do most of the talking, empathize, probe, tell the student the probable consequences of a particular action, and then let the student make the decision. Of course, there are certain actions such as taking a course without the prerequisites which are not allowed, but if the action is allowed, the student should make the final decision. The adviser's expertise is important in explaining the consequences of a particular action. For example, many students are unaware of the consequences of being dropped from the university. Or, a student may not realize that dropping a particular required course will delay graduation by one year.

Many lower-division students are not ready for the responsibility of conducting their own affairs. A fairly proactive stance is appropriate for lower division students when there are signs of problems, such as excessive absences, D and F grades, probation, failure to register on time, and so forth. Phone calls, letters, and personal visits can help prevent a student from getting into serious trouble. Students often just let things go because they do not understand the probable consequences of their actions. A formal written contract with the student may be appropriate (e.g., to get a tutor). Since such students are often irresponsible, it is important for the adviser to keep notes on what has happened and what agreements have been made.

With freshmen and sophomores it is appropriate to discuss academic skills if students are having problems. Engineering professors often assume that students know how to study and understand the tricks of taking tests. Many students have not learned these skills. Study methods, problem solving, test taking tricks, relaxation methods, and methods for budgeting time can all be useful to such students. These academic survival skills can be covered informally in small groups, and students can be encouraged to form study teams. Many universities have these types of programs available through a counseling office or the psychology department.

Upper-division students should be more mature, and with them a more laissez-faire approach is appropriate. There are skills which these students probably have not learned which will be useful. In particular, being interviewed and decision making are of considerable interest to seniors. These can be taught in seminars or small groups. Individual attention with a professor who is familiar with a given industry or with several graduate schools can be extremely helpful to students. Although students want to hear the professor's opinions, it is still important to listen to them and let them make the decision.

Advisers often have to communicate with students. At large universities it is surprisingly difficult to be sure that all or even most students understand what they are supposed to do for registration, dropping classes, company interviews, and other official tasks. Try a variety of different modes of communication. Seminars, announcements in class, bulletin board announcements, letters, catalogs, advertisements in the student newspaper, and individual discussions are different modes each of which will reach a few students that the others won't. The failure of a student to know the rules is not an excuse, but try to reach as many students as possible. Since this problem reoccurs every year, you must patiently keep trying to communicate.

Advisers can spend a great deal of time on routine matters such as registration. There are ways to make this and other routine matters more efficient. Records should be computerized to remove this burden from the advisers. Registration can be handled efficiently by the combination of a group seminar to provide information and individual sessions with the counselor for final course selection. Individual sessions are important to avoid losing students and to give the adviser an opportunity to question the student's course selections (Grites, 1980). Some students sign up for completely inappropriate courses for a variety of reasons. The adviser can catch this by asking questions.

Hiring a professional counselor, empathic administrative assistant, or peer adviser (an upper-division student) can be a cost-effective way of reducing the burden on engineering professors while simultaneously increasing the effectiveness of advising. The routine bookkeeping, processing of forms, and enforcing of discipline do not have to be done by professors. They can better spend their time advising students on professional decisions involving the choice of electives, graduate school, or industrial jobs.

At large universities many students seldom have the opportunity to speak individually with a professor—with the lone exception of engineering professors who serve as advisers. It is important for advisers to be open and take advantage of this opportunity. Use of listening skills and empathy may lead the conversation from a rather mundane presenting problem to more serious concerns. The term "presenting problem" refers to a problem which is raised by the student but which may conceal another problem—the one needing to be addressed. Oftentimes an adviser is the only official person at the university who has really taken the time to listen to the student. This illustration of caring can make a major difference in the student's career, and many times is what keeps a student in school. Doing this requires interest, time, and counseling skills which are the subject of the next section.

10.3.2. Counseling Skills for Personal Problems

Advising students on personal problems is not a major role of engineering professors, but it is sometimes required. The procedures used for dealing with personal problems are often useful for academic and career counseling. A simple crisis intervention model useful for short-term interventions will be presented (Edwards, 1979; Wankat, 1980). Professors should always aim for short-term intervention with students with one or at most two sessions. If the professor has a student in class, then great care should be taken in counseling him or her on personal problems. The roles of teacher and personal counselor are in many ways incompatible (Lowman, 1985). All professors should be aware of the resources on campus so that they can refer students to additional help.

An ABCF model can be used to help students deal with a crisis (Edwards, 1979; Wankat, 1980). The four steps in this crisis intervention model are:

A. Acquire information and rapport. Students who come in to talk to an adviser very often have a presenting problem which is the first thing they talk about. The presenting problem is real and often is the only problem. However, there may be other problems hiding behind the problem which the student would like to talk about if given the opportunity. Advisers who are open and use listening skills give students the opportunity to discuss these deeper problems. In addition, there are often other signs of serious problems which can sometimes be noticed by casual observation: excessive absences, sudden plummeting of the quality and quantity of work, the smell of alcohol on the person, slurred speech, and so forth (Civiello, 1989).

Once it is clear that there is probably a problem, the adviser uses active listening skills to acquire information and establish rapport with the student. Empathy, which may be crucial for gaining rapport, involves knowing what it is like to walk in that person's shoes. It can be obtained by focusing on feelings since the feelings of sadness, anger, fear, happiness, and so forth are universal. The adviser may not have been in a situation similar to the student's, but he or she will have felt similar feelings. With this focus on feelings some students may become quite emotional and start crying. Signify that this is OK, give the student a tissue, and let him or her cry. The short book by Mayeroff (1971) is useful for insights into empathy and caring.

Individuals often skirt subjects that are taboo, but they probably want to talk about these issues. It is permissible to bring up issues such as death, AIDS, suicide, poverty, and broken relationships. (Note that this can be difficult for the teacher because he or she usually has to evaluate performance in class.) If you bring up a topic which is not the problem, the student will correct you. Probe, but encourage the student to do 90 percent of the talking.

Often counseling should stop at the acquiring step. People, particularly women, may want a confirmation of feelings, not problem solving (Tannen, 1990). This behavior is more likely to occur with peers who are of equal status than with students who are in a subordinate position.

B. Boil the problem down. Sometimes the student knows what the problem is and sometimes he or she has not accepted what the problem is. If it seems clear to you that the student knows what the problem is, it is OK to ask, "What's going on here?" Follow this with silence to give the student time to collect her or his thoughts. When the problem is unclear, an important function of a counselor is to help the student clearly state the problem. While in the acquiring stage you can hypothesize about the problem. Then explore if this is a possible problem by probing with open-ended questions. When there is sufficient evidence, you can formulate a problem statement and check it with the student. Do this in a tentative fashion. "It seems that the underlying problem is . . . ." If the student agrees or modifies the statement slightly, then you are ready to move on to step C.   If not, return to step A.

C. Coping—help the student cope. If we think of working with students on personal problems as problem solving, then steps A and B are the define stage. Step C consists of the explore and plan stages. The adviser helps the student devise a plan to cope with the problem. Since it is the student's problem, it is not helpful to give advice. Explore alternatives and try to get the student to set up an action plan. This may be difficult.

There are problems in which the situation cannot be changed but must be accepted. For example, a death in the family or a parent who is dying of cancer are not problems which are amenable to action. However, there are actions which may help the student, such as joining a support group, seeing a professional counselor, or taking incompletes in a few classes. These possible actions can be all be explored.

As a counselor you can serve as a resource person during the coping step. Students often do not know what resources are available on campus or in the community. They can be referred to the university counseling center, the office of the dean of students, the student hospital, the financial aid office, the local crisis center, or whatever else is appropriate. It may be helpful to call the referral office and make an appointment for the student with her or his permission.

F. Follow-up. The student needs to go out and actually carry out the action plan. It is sometimes appropriate to schedule one follow-up session to check on his or her progress and offer encouragement. This follow-up can be suggested informally ("Stop in and see me when you've gotten this resolved") or formally ("Can you come in and see me at this time in two weeks?") Whether or not a follow-up is appropriate depends on your judgment.

One paradox of helping is that the more severe the crisis, the less training the adviser needs. Natural caring and empathy are often sufficient for acute problems. Students with long-term chronic problems and dysfunctional students require trained professional counselors, social workers, or psychologists. Refer these students to the appropriate professionals.

10.4. RESEARCH ADVISERS

An extremely important type of one-to-one teaching involves being a research adviser. This role is closer to the tutoring role than that of an academic adviser or personal counselor, although it has elements of all these roles. Research advisers have two major objectives: to help students develop and become competent researchers, and to do research and publish the results. Although usually complementary, there can be conflicts between these two objectives. The resolution of these conflicts depends upon the relative importance each adviser places on the two objectives.

10.4.1. Undergraduate Research

A research experience can be extremely valuable for undergraduates, and strong arguments can be made that a senior thesis should be required (Prud'homme, 1981); or if classes are large and resources limited, undergraduate research should be an encouraged option (Fricke, 1981). Research can give students the opportunity to explore a particular topic in greater depth than would be possible in class. The student can receive individual attention from one professor and get to know this professor, "try out" research to see if graduate school should be considered, or improve his or her efficiency, time management skills, and ability to schedule complex projects. If teams are used, the student will learn how to function in a team.

Most undergraduate engineering students in the United States have little experience in working independently on a large project. Thus, initial supervision needs to be fairly structured. An undergraduate may need to be taught laboratory skills which professors normally assume graduate students know. Ideally, the professor will have time to teach the undergraduates the appropriate skills. If the professor is busy, an advanced graduate student can be assigned the duty of helping them get started. Making a graduate student the supervisor of an undergraduate provides for much of the day-by-day assistance to the undergraduate. In addition, the experience at supervision is useful for the graduate student. With this arrangement professors can meet with the team once a week and review progress and provide ideas.

The selection of projects for undergraduate research is critical. Real problems which push the knowledge level of the student are appropriate. However, the project must be doable in a finite amount of time. Team projects can be significantly more complex than individual projects (Masih, 1989). Reporting of results in both written and oral form should be part of the project requirements. If possible, it is very motivating to list the undergraduate as a coauthor of any papers resulting from the research.

10.4.2. Graduate Student Research

In most schools the major goal of research programs, particularly the Ph.D. program, is a research project and the thesis which results. Then the role of the research adviser becomes critical. This role really has no equivalent in undergraduate education, and in many ways it is essentially unchanged from that of medieval tutors. At some schools the research adviser essentially controls when the graduate student graduates, how long the graduate student is funded, what the project is, whether or not the student goes to conferences and so forth. Once the student signs on for a given project, he or she may lose many rights. Since there will always be a few professors who will abuse such a powerful relationship, the department needs to consider formal institutional controls. Since disgruntled students always know how other students are being treated, it is helpful to have uniform policies which research advisers must follow. New professors need mentoring in becoming research advisers since they have probably had only their own adviser as a role model.

In engineering most graduate students doing research receive support. Policies on how long the student will be supported to the M.S. and to the Ph.D. are useful to ensure uniformity. Such policies also help prevent advisers from keeping students too long and prevent students from abusing the system because they enjoy graduate school. Some leniency in the cutoff date is useful for exceptional cases, but the extra period of support should be limited.

The selection of a research adviser is probably the most important decision an individual makes as a graduate student. Unfortunately, the process is often random and the student does not ask the right questions (Amundson, 1987). A better approach would be to train students in selection processes before they listen to faculty presentations and interview faculty members. Such training could include a discussion of differences in personality type (see the description of the Myers-Briggs Type Indicator in Chapter 13). In addition, students could be shown various decision methods and perhaps a list of generic questions considered important by the faculty. Since mistakes are made in the selection of a research adviser, a uniform policy on allowing students to switch advisors should be followed. Some schools require all students to interview other potential advisers at some specified point such as on completion of the M.S. Students who are satisfied with their current advisers attend these interviews in a cursory fashion. This policy protects the students and removes any stigma from switching advisers.

Palmer (1983) lists the philosphical needs of students as *openness, boundaries,* and an *air of hospitality*. Openness is a sense that there are few barriers to learning. Admit students to the community of scholars and expect them to learn. Firm boundaries help create an open space where students can choose interesting, meaningful research, but not so open that they run wild. Research and learning can be very painful processes since they do not represent linear paths. Therefore, the learning space must be hospitable. Both new ideas and failures must be treated gently so that the student has permission to keep trying.

10.4.3. Masters' Research

Masters students following a thesis option do research projects which in some ways are longer versions of undergraduate research. The project must be doable in a finite amount of time, which means that the professor probably needs to define most of the problem for the student. Most masters students have just made the transition from being undergraduates. Graduation does not suddenly make them mature, self-starting individuals. Many of these students need the same initial structure as undergraduates; however, graduate students should be made more independent fairly quickly. Thus, it is appropriate to define the problem for the student but require her or him to control the execution of the project. Asking a senior graduate student to train new graduate students in laboratory skills and safety practices is still a good idea. But it is probably not a good idea to use an advanced graduate student as a research supervisor. New graduate students deserve the attention of the professor.

New graduate students may not be very efficient, may not know how to do a scientific literature survey, and probably do not know how to schedule a long project. All these things need to be taught. Thus, there is a strong tutoring aspect to starting new graduate students. At the beginning of the graduate student's tenure as a student, it is useful to have regularly scheduled meetings. Even though there may be no research as yet, there are many other things to talk about. Shortly after being assigned an adviser, the student should be asked to develop a project plan for when he or she wants to graduate. This makes the student think about what must be done to graduate. The plan can be revised on a regular basis as the student learns more about the time demands of the project. The student also needs to learn to balance the work on courses where there are immediate demands and research where there may be no immediate demands. Research also requires accepting delayed gratification.

Some discussion with the student about background and goals is appropriate. This will help develop rapport with the student. It is also useful to know where the student has come from (figuratively speaking) and where he or she is going to help design methods to motivate the student. Use active listening skills to get the student to talk about her or himself.

Regular meetings of the graduate and undergraduate students in a research group can help foster a sense of belonging. A senior graduate student can be assigned to organize the meetings. The presentation of research results to the group is a useful practice on a regular basis and makes students more polished when they make formal presentations. New students can be asked to make presentations on papers from the literature. Discussion of the presentations should be critical yet friendly. The professor can ensure that this happens. New ideas should be greeted with the PMI or another positive approach (see Section 5.6.3).

10.4.4. Ph.D. Research

When they graduate, Ph.D.'s in engineering are expected to be independent researchers. This means that they should be able to analyze a situation, define the problem, outline a plan of attack, and then conduct the research. Many companies assume that a new Ph.D. can supervise the work of technicians. In academia the new Ph.D. will be expected to generate new research ideas, supervise the research of graduate students, write proposals, review papers, teach, serve on committees, and publish. If the Ph.D. is to be able to do all these at graduation, then he or she needs guided practice in doing these or similar things while a graduate student. This need for practice guides our advising of Ph.D. students.

Since the professor usually has research money for a specific area, the general area for the Ph.D. student's research is usually set. However, within this broad area the student needs to define the problem he or she will work on. This is a nontrivial task. The easy part involves doing a literature search to see what others have done. The professor can help the student by steering her or him to the appropriate tools in the library including computer searches. During this period, attendance at a professional meeting can help the student see what the current hot areas are.

The hard part for most students is the intellectual development often required to be in charge of the research. Determining what is important and what research should be done requires that he or she be relatively mature (see Chapter 14). Most students have always been told what to do, and they find the freedom of Ph.D. research frustrating. The period of frustration can easily last a year while they work to determine "What they want." Duda (1984) notes that many students have misconceptions about graduate research, not realizing that research is a problem-solving method and that a straight, linear path seldom works. The backtracking, deadends, and lack of obvious progress can be frustrating.

Amundson (1987) suggests meeting with students on a regular basis until it becomes necessary to turn them loose. Then they need to be told to work on their own until they come up with something that is new or surprising. It must be made clear to the student that the value of a coefficient is not a surprise even if it is new, especially at the Ph.D. level. Most students respond amazingly well to this ploy and develop independent ideas. A few start their graduate programs already independent and appreciate and use the freedom to develop research on their own. A few are unable to cope and probably should be encouraged to look outside the Ph.D. program for career opportunities.

When a student does present a new idea, it needs to be accepted but with the challenge that it be further developed. Ask the student to develop the idea and then prove it theoretically or experimentally. Suggest that he or she develop twenty or fifty alternatives to the idea. Refuse to judge the idea since the student needs to learn how to do this.

While conducting research, a student needs to learn to do a host of other tasks. These can be paced over a long period so that he or she is not overwhelmed at any one time. And the student needs plenty of freedom to learn to think; otherwise, he or she may escape from thinking into mindless work on other tasks. Foremost among these tasks is learning to communicate. Encourage the student to review the literature and then write a review article with you. This is efficient since it accomplishes the necessary review process, helps the student learn how to write, and earns the student a publication early in her or his graduate student career. The review article can be completed before the student is challenged to become independent. Communication includes oral presentations. Students can be required to make oral presentations in group meetings and later at professional society meetings. These presentations need to be evaluated to help the student improve. Videotaping some of the group presentations is a good way to encourage growth. Students should be given the opportunity to review papers and proposals. This should first be a supervised activity where the professor also reviews the paper or proposal and the reviews are compared and contrasted. Then, the student can review papers and proposals independently.

After a student has learned to conduct research independently, he or she can help to supervise other students. This supervision itself should initially be supervised. That is, have meetings with the Ph.D. student, both individually and with the undergraduate student, to discuss the progress of the undergraduate student's research. Your supervision can then be slowly reduced.

After a graduate student has completed some research, he or she will want to write a research paper. Ideally, the first paper should be written after the student's M.S. thesis. Take an active role in outlining this paper and in the necessary revisions to the paper. When the reviews arrive, show them to the student. The reviews often provide for reality testing. The student should then help with revisions of the paper on the basis of the reviewer's comments. Once the student has started doing independent research, he or she needs to become independent in writing papers. After experiencing the first flush of success in publishing a paper, the student may suddenly write several papers. If these are not well done, return them with a note indicating that they are not up to professional standards, but do not provide detailed comments. Rewriting is a necessary part of writing, and the student needs to master this art.

Although it may not be appropriate to place a student completely in charge of writing proposals, he or she can certainly help prepare them. Let the student share the joys or frustrations of submitting research proposals. Discuss with her or him the strategy you use for obtaining research support.

Students who intend to follow an academic career need additional teaching experience beyond being a TA. A course or self-study on teaching methods is useful. The chance to do supervised teaching of a class or seminar is also an excellent experience which far too few students receive.

10.5. CHAPTER COMMENTS

The ways that a professor interacts with students in one-to-one situations obviously depend heavily on the professor's personality. Some suggestions have been provided which have been found to be useful.

The section on being a research adviser is likely to be controversial. Many professors use very different procedures as research advisers. Obviously, there is no one best way to advise research students. We have clearly stated our value judgments and then suggested an advising procedure which follows these value judgments. If readers do the same, then Section 10.4. will have achieved its purpose.

10.6. SUMMARY AND OBJECTIVES

After reading this chapter, you should be able to:

- Explain and use methods to improve listening.
- Improve tutoring of students and become more effective in helping them learn.
- Improve both academic and personal advising skills.
- Outline your personal value structure for advising research students, and then develop procedures to improve your advising of these individuals.

HOMEWORK

1 Set up a role play to practice listening skills. This requires that you have a partner to take the role of a student, and a facilitator to observe the interactions and record both their positive and negative aspects. The observer needs to watch the climate, the focus, and the responses (nonverbal, minimal verbal, and verbal). Several role plays should be done. They can include the following situations.
   a A student needing help after class.
   b A student needing help during office hours.

c A student needing academic advising.

d A student with a personal problem which is causing academic difficulty.

e A Ph.D. student who is having difficulty getting started on research.

2 List the rules and regulations for undergraduate students at your university as far as registration for classes is concerned.

3 What is the purpose of Ph.D. education in your engineering field? Based on this purpose discuss what the ideal thesis adviser would do. Then develop a program to make your own advising more closely approach the techniques of your ideal adviser.

REFERENCES

Amundson, N.R., "American university graduate work," *Chem. Eng. Educ.*, 21, 160 (Fall 1987).

Anonymous, "Engineering utilization study findings on engineering education," *Eng. Educ. News,* (Jan. 1986).

Axtell, R. E., *Gestures: The Do's and Taboos of Body Language around the World*, Wiley, New York, 1991.

Bolton, R., *People Skills*, Prentice-Hall, Englewood Cliffs, NJ, 1979.

Brammer, L.M., *The Helping Relationship. Process and Skills*, 3rd ed., Prentice-Hall, Englewood Cliffs, NJ, 1985.

Civiello, C.L., "Identifying and assisting students with serious problems," *Proceedings ASEE Annual Conference*, ASEE, Washington, DC, 915, 1989.

Duda, J.L., "Common misconceptions concerning graduate school," *Chem.Eng.Educ.*, 18, 156 (Fall 1984).

Duda, J.L., "Graduate Studies. The middle way," *Chem. Eng. Educ.*, 20, 164 (Fall 1986).

Eble, K.E., *The Craft of Teaching*, 2nd ed., Jossey-Bass, San Francisco, 1988.

Edwards, R.V., *Crisis Intervention and How It Works*, Charles C. Thomas, Springfield, IL, 1979.

Fricke, A.L., "Undergraduate research: A necessary education option and its costs and benefits," *Chem. Eng. Educ.*, 15, 122 (Summer 1981).

Grites, T.J., "Improving Academic Advising," Idea Paper No 3, Center for Faculty Evaluation and Development, Kansas State University, Manhattan, KS, 1980.

Hackney, H. and Nye, S., *Counseling Strategies and Objectives,* Prentice-Hall, Englewood Cliffs, NJ, 1973.

Herrick, R. J. and Giordano, P., "EET counselor takes on student role," *Proceedings ASEE/IEEE Frontiers in Education Conference*, IEEE, New York, 434, 1991.

Hewitt, N. M. and Seymour, E., "A long discouraging climb," *ASEE Prism,* 1(6), 24 (Feb. 1992).

Katz, P. S., "Listening: The orphan of communication," *Proceedings ASEE Annual Conference*, ASEE, Washington, DC, 955, 1986.

Light, R. J., *The Harvard Assessment Seminars,* Harvard University Press, Cambridge, MA, 1990.

Lowman, J., *Mastering the Techniques of Teaching*, Jossey-Bass, San Francisco, 1985.

Masih, R., "The importance of research projects for undergraduate students," *Proceedings ASEE Annual Conference*, ASEE, Washington, DC, 1165, 1989.

Mayeroff, M., *On Caring*, Harper and Row, New York, 1971.

McKeachie, W. J., *Teaching Tips*, 8th ed., D.C. Heath, Lexington, MA, 1986.

Miller, P. W., "Nonverbal communication: How to say what you mean and know what they're saying," *Eng. Educ.*, 71, 159 (Nov. 1980).

Palmer, P. J., *To Know as We Are Known: A Spirituality of Education*, Harper-Collins, San Francisco, 1983.

Prud'homme, R.K., "Senior thesis research at Princeton," *Chem. Eng. Educ.,* 15 , 130 (Summer 1981).

Root, G. and Scott, D., "The interpersonal dimensions of teaching," *Eng. Educ.,* 184 (Nov. 1975).

Stegman, L., "Listening pays dividends: Improve student learning through listening techniques," *Proceedings ASEE Annual Conference*, ASEE, Washington, DC, 1019, 1986.

Tannen, D., "*You Just Don't Understand,*" Ballatine Books, New York, 1990.

Vines, D. L., "Mentors," *Proceedings ASEE/IEEE Frontiers in Education Conference*, IEEE, New York, 326,1986.

Wankat, P. C., "Are you listening?," *Chem. Eng.*, 115 (Oct. 8, 1979).

Wankat, P. C., "The professor as counselor," *Eng. Educ.*, 153 (Nov. 1980).

Wankat, P. C., "Current advising practices and how to improve them," *Eng. Educ.*, 213 (Jan. 1986).

CHAPTER 11

TESTING, HOMEWORK, AND GRADING

For many students, grades constitute the number-one academic priority. Tests, or any other means professors use to determine grades, are the number-two priority. Because of this concern about grades, tests and scoring of tests generate a great deal of anxiety which can translate into anxiety for the professor. It is easy to deplore students' excessive focus on grades; however, this excessive focus is at least in part the fault of the professor. In addition, a student's focus on grades and tests can be used to help the student learn the material.

Testing and homework can help the professor design a course which satisfies the learning principles discussed in Section 1.4. Homework and exams force the student to practice the material actively and provide an opportunity for the professor to give feedback. With graduated difficulty of problems, the professor can arrange the tests so that everyone has a good chance to be successful at least initially. This helps the professor approach the course with a positive attitude toward all the students, which in turn helps them succeed. The desire to achieve good grades can help motivate students to learn the material, particularly if it is clear that the tests follow the course objectives. Anxiety and excessive competition can be reduced by using cooperative study groups. Thought-provoking questions can be used both in homework and in exams to use the students' natural curiosity as a motivator. Students can be given some choice in what they do in course projects.

Although testing and homework can help the professor satisfy many learning principles, they also can serve as a barrier between students and professors which inhibits learning. It is difficult for students to truly use the professor as an ally to learn if they know he or she is evaluating and grading them (Elbow, 1986). Perhaps the ideal situation would be to completely separate the teaching and evaluation functions. One professor would teach, coach, and tutor students so that they learn as much as possible. Then a second professor would test and grade them anonymously. An alternate method with which to approach this ideal can be obtained with mastery tests and contract grading (see Section 7.4). If these alternatives are not possible, there will always be tension between learning on the one hand and testing and grading on the other. In the remainder of this chapter we will assume that you have resolved to live with this tension.

Why does one test and how often does one test? What material should be included on the test? What types of tests can be used? How does one administer a test, particularly in large classes? These are the questions we'll consider in this chapter. Then our focus will shift to scoring tests and statistical manipulation of test scores. Homework and projects will be explored. How much weight should be placed on homework? How does the professor limit procrastination on projects? Finally, the professor's least favorite activity, grading, will be considered from several angles.

11.1. TESTING

Testing requires careful thought. Fair tests which cover the material can increase student motivation and satisfaction with a course. As long as a test is fair and is perceived as being fairly graded, rapport with students will not be damaged even if the test is difficult. Unfair and poorly graded exams cause student resentment, increase the likelihood of cheating, decrease student motivation, and encourage aggressive student behavior.

11.1.1. Reasons for and Frequency of Testing

There are many educational reasons for having students take tests. Tests motivate many students to study harder. They also aid learning since they require students to be active, provide practice in solving problems, and offer feedback. Tests also provide feedback for the professor on how well students are learning various parts of the course.

Tests are stressful since they are so closely associated with grades. Stress and pressure are part of engineering. Mild stress can actually increase student learning and performance on tests, but excessive stress is detrimental to both learning and performance for students and practicing engineers. In addition, exams can be stressful for the professor because they are so tightly coupled with grades. What can be done to harvest the benefits of tests while simultaneously reducing the stress they induce?

Give more tests!

Giving more tests reduces the stress of each one since each exam is less important in deciding the student's final grade. Courses with only a final or a comprehensive exam make the test enormously important and thus very stressful. If there are four tests during the semester, each one is significantly less important. If there are fifteen quizzes throughout the semester, then each quiz has a modest amount of stress associated with it. Having frequent tests or quizzes also allows professors to ignore an absence or discard the lowest quiz grade.

Frequent testing spreads student work throughout the semester, which increases the total amount of student effort and improves the retention of material. The more-frequent feedback to the students and to the professor is beneficial. Both the students and the professor know much earlier if the material is not being understood. The increased forced practice, repetition, and reinforcement of material aids student learning. Because stress is reduced, frequent testing serves as a better motivator for students. The net result is improved student performance (Johnson, 1988). One of the advantages of PSI and mastery courses is that they require frequent testing (see Chapter 7). Frequent exams also provide a more valid basis for a grade since one bad day has much less of an effect.

Frequent tests do have negatives. The considerable amount of class time required may reduce the amount of content that can be covered; however, the content that is covered will probably be learned better. A considerable amount of time may also be required to prepare and grade the frequent examinations. At least some of this time is available since less homework needs to be assigned when there are frequent exams. Perhaps the most important drawback of frequent tests in upper-division courses is that they do not encourage students to become independent, internally motivated learners.

We have adopted the following compromise solution to the question of how frequently to test. In graduate-level courses we give infrequent tests (two or three a semester) but usually have a course project which represents a sizable portion of the grade. In senior courses we use slightly more tests (three or four). In junior courses, despite the great deal of material to be covered, we increase the number to six or seven during the semester. In sophomore courses where there is often little new material to learn but students need to become expert at applying it, we have gone as high as two quizzes per week (and no homework). For these courses one quiz per week seems to work well. This frequency may also be appropriate for computer programming courses. Frequent quizzes ensure that students are practicing the material and are receiving frequent feedback.

What about finals? There are very mixed emotions about finals (for example, see Eble, 1988; Lowman, 1985; McKeachie, 1986). Finals do require students to review the entire semester and to integrate all the material. They can also be useful for slow learners and for those who initially have an inadequate background since they allow these students to show that they have learned the material. Finals are also useful for assigning the course grade. Unfortunately, they are very stressful for students and are almost universally disliked. In addition, feedback to the professor is too late to do any good in the current semester. To the students it is almost nonexistent. Many students look only at the final grade and do not study their mistakes on the test.

A professor choosing to give a final has several interesting options which can reduce the stress. If other tests have been reasonably frequent during the semester, students can be told that the final can only increase but not decrease their grade. When this is done, it may make sense to tell students their current earned grade and then make the final exam optional. In PSI and mastery courses an optional final can be used as one way to improve students' final grades with no risk. Another option is to give a required final but tell students that their grades will automatically be the higher of their composite grade for the entire course or their grade on the final. The reasoning behind this strategy is that it makes sense to give high grades to students who prove at the end of the semester that they have mastered the material, but having only a final is too stressful. In this way you are also rewarding them for what they know at the end of the term instead of penalizing them for deficiencies they may have had at the start of the semester. Feedback can be made more meaningful by going over the final in a follow-up course the next semester.

Many universities have a scheduled finals period. If the professor decides not to have a final, this time may be used for other purposes. In a course with projects, the final examination period is an excellent time for student oral reports on projects. This period can also be used for a last hour examination which is not a final. One advantage of using the finals period for an hour examination is that more time is usually allotted for the final, and students taking an hour examination during this period have sufficient time to finish even if they work slowly.

One additional type of quiz is the unannounced, surprise, or "pop" quiz. Some professors like to give several of these during the semester. After answering questions the professor announces there will be a pop quiz. Once the students' groans subside, a short quiz is administered. The advantages of pop quizzes are that they help keep students current and they reward attendance. The major disadvantage is that they increase stress. This increase in stress can be controlled by:

1 Noting in the syllabus that there will be unannounced quizzes.
2 Making the quizzes a small fraction (2 to 3 percent) of the course grade.
3 Giving some points for the student's name (i.e., rewarding attendance).
4 Throwing out the lowest quiz grade. This helps students who miss a class which happens to have an unannounced quiz.
5 Making the quizzes short (five to ten minutes).

11.1.2. Coverage on Tests

How does a professor decide what to put on a test? If objectives have been developed for the course, the decision is relatively simple. The important objectives are tested. At what level in Bloom's taxonomy (see Chapter 4) should the test be? If at the higher levels, then the test questions need to be evaluated for appropriateness.

An effective method for ensuring that the test covers the objectives appropriately is to develop a grid (Svinicki, 1976) as illustrated in Figure 11-1. For each objective or topic, think of a question or problem which allows you to test at appropriate levels of Bloom's taxonomy. It may not be necessary to have any problems which are solely at the knowledge or comprehension levels since these levels are usually included in higher-level problems.

Once the preliminary grid has been developed, you can check it to see if the proposed test satisfies your goals for a particular section of the course. Since not all objectives or topics can be included at all levels of the taxonomy in a single test, you need to make some compromises. Is the coverage of topics on the test a fair representation of the coverage during lectures and of the homework? If not, the exam probably is not a fair test of the course objectives, and students are likely to think it is unfair. Although not all topics can be covered, one should try

FIGURE 11-1   EXAMPLE GRID FOR TEST PREPARATION

Objectives or Topics	Level					
	Knowledge	Comprehension	Application	Analysis	Synthesis	Evaluation
1	X		X			
2		X		X		
3	X	X			X	
4		No problem for this objective				

to have reasonably wide coverage. If a topic is discussed in two separate parts of the course, it might be reasonable to include it in one test and not the other. The levels of the questions also need to be considered. If higher-level activities are important, they need to be included in homework and in tests. Without a conscious effort, it is highly likely that only the three lowest levels will be used since questions at these levels are the easiest to write (Stice, 1976). For the grid shown in Figure 11-1, the instructor has decided not to test for objective 4 or to include any questions at the evaluation level on the test.

Should the test be open book or closed book? The argument in favor of open book tests is that practicing engineers can use any book they want to solve a problem. Open book tests also reduce stress. One argument against them is that too many students use the book as a crutch and try to find the answer in the book instead of by thinking. Another opposing argument involves logic. The practicing engineer argument relies on a false analogy because the purpose of the open book is different: Unlike students, these engineers are not being tested on their knowledge. One problem with closed book tests is that students may be forced to memorize equations which they would always look up in practice. Closed book tests may encourage memorization of all content and not just the equations.

Some compromise arrangements are between the extremes of open book and closed book tests. The instructor can prepare a sheet of important equations for students to use during the exam and hand this sheet out to them before the test so that they know what will be available for the test. When the exam is administered, each student receives a clean set of equations. The advantage of this compromise is that the professor has control over the information each student has available during the test. Another compromise is to allow each student to bring a key relations chart (see Section 15.1) on one piece of paper or an index card. The advantage of this procedure is that students benefit from preparing the chart and often do not glance at it during the test.

11.1.3. Writing Test Problems and Questions

How does one write the problems or questions for tests? What style of questions is appropriate? This section discusses some general rules for writing exams and then explores specific formats for questions.

In writing examination questions, avoid trivial questions even when testing at the knowledge level. Avoid trick questions also since they do not test for the student's understanding and ability in the course. Problems should be as unambiguous as possible unless you are explicitly testing for the ability to do the *define* step of problem solving. To test for clarity have another professor or your TA read the test and outline the solutions. The time required for the exam can be estimated by taking the time you require to solve the problems and multiplying by a factor of about 4. The number of points awarded for each problem should be clearly shown on the test so that students can decide which problem to work on if time is short.

Solve the problems before handing out the test. This aids in grading and helps to prevent the disaster which will occur if an unsolvable problem is on the exam. (If you want the students to perform a degree of freedom analysis to determine if the problem is solvable, then it is reasonable to have an unsolvable problem on the test. However, warn them ahead of time that this may happen; otherwise, they will assume all problems are solvable.)

If tests are returned to students (which is a useful feedback mechanism), then you should assume that files exist on campus for all old exams. Even if you require students to return tests after they have seen their grades, you should assume that at least rudimentary files exist. Since the purpose of a test is to determine how much a student has learned and not who has the best files, you should write new tests. If exams are given frequently, this is a considerable amount of work. Once a large number of questions, particularly of the multiple-choice variety, have accumulated, you can recycle a few questions on each test. Old test questions do make good homework problems, and students appreciate the opportunity to practice on real test problems. Since some students have files, many professors provide files of old tests so that everyone has equal access to information. Most university libraries place test files on reserve. Another more drastic solution to the file problem is to periodically revise the curriculum and reorganize all the courses.

Although it may sound contrary to the previous advice, we suggest that every once in a while a homework problem should be put on a test. This rewards students who have diligently solved problems on their own and is a clear signal to students that they should work on the homework.

How does the professor generate interesting problems which test for the objectives at the correct level but are not clones of textbook or homework problems? One way is to take an existing problem and do permutations of which variables are dependent and which are independent. Changing the independent variable often changes the solution method remarkably. Brainstorm possible novel problems. Use problems from other textbooks (but if this is done consistently, some students will catch on). Set up an informal network with friends at other universities to share test problems and solutions. As part of their homework assignments have students write test problems. The occasional use of one of these will reward the student who made it up. (In our class on teaching methods the second test is based entirely on student-generated questions.) Don't wait until the last minute to start generating problems. It is often productive to generate ideas throughout the semester. Then, the details of the problem and the solution can be worked out when the exam is made up.

Test problems usually fit into one of the following categories: short-answer, long-answer, multiple-choice, true-false, and matching. Since true-false and matching have scant use in engineering, they will not be considered here but are discussed elsewhere (Canelos and Catchen, 1987; Eble, 1988; Lowman, 1985; McKeachie, 1986).

Short-answer. Short-answer problems include problems requiring identification of a principle, a brief essay, and short problems. In engineering, short problems are the most common. As long as complete long problems are also employed, short problems are an excellent way to determine if students have mastered certain principles. These problems are set up so that three to five lines of calculation give the desired answer. The problem is tightly defined so that the student is tested for application to a single principle.

Short-answer problems can also be used to develop students' skills as problem solvers. The problem focuses on one or two stages in the problem-solving strategy. For example, students can be asked to define the problem clearly but not solve it. Or, they can be given a "solution" to the problem and asked either to check the solution or to generalize it. Students need instruction in doing this type of short answer problem since they always want to calculate.

Long-answer. Long-answer problems include essay and complete long problems. In engineering, complete problems are probably the most common type of test problem. They are necessary to determine if students can find a complete solution. Unfortunately, an exam consisting entirely of a few long problems cannot test for all the objectives covered in the course. Thus, a mix of both long- and short-answer problems is often appropriate. Long-answer problems can also be difficult to score for partial credit (see Section 11.2.1).

Multiple-choice. With the regrettable but probably inevitable increase in class size at many engineering schools, multiple-choice examinations will become increasingly popular. They are easy to grade and, if properly constructed, can be as valid as short-answer questions (Kessler, 1988). Unfortunately, proper construction of the classical type of multiple-choice question is more time-consuming than constructing a short-answer question. Thus, the professor transfers some of her or his time from grading to test construction. This trade makes sense only with large classes.

General rules for constructing classical-style multiple-choice questions are given by Eble (1988), Lowman (1985), and McKeachie (1986), while examples for particular engineering courses are presented by Canelos and Catchen (1987) and Leuba (1986a,b). The stem, which is the question itself without the choices, should be complete, unambiguous, and understandable without reading the choices. The correct answer and the incorrect answers (the distractors) should be written as parallel as possible. Thus, all possible answers should be grammatically correct and about the same length. There should be no "cues" which allow a good test taker who is unfamiliar with the material to discard any of the distractors or to pick the right answer. Most authors suggest a total of four choices, all of which should appear reasonable. The instruction should ask the student to pick the "best" choice so that arguments with students can be minimized.

In writing a multiple-choice question, the professor usually starts with a short-answer problem. The correct answer is then obvious. Indicate that the answer is a number within a given percentage (say, 1 percent). The challenge lies in choosing distractors. If a similar short-answer question has been used in the past, look at the students' solutions to find common errors. Then construct the distractors so that the numerical answer follows from these common student mistakes. Most authors suggest that "none of the above" is an improper distractor or answer. Once the distractors have been written, randomly assign the answer and the distractors as a, b, c, and d.

When questions have numerical answers, there is a clever alternate type of multiple-choice question (Johnson, 1991). For each question, list ten numbers in numerically increasing order. Tell the students to select the choice nearest to their calculated answer. If the calculated answer is the average of two adjacent choices, tell them to select the higher choice. The effort in writing distractors is thereby reduced. Now all you have to do is to pick choices over a feasible range at reasonably narrow intervals. This procedure also reduces the probability of a guess being correct. With the usual type of multiple-choice question the student who doesn't get one of the listed answers knows that he or she has made a mistake, but this procedure does not provide this clue. In addition, if you initially make a mistake solving the problem or there is a typographical error in the problem statement, all is not lost. As long as the problem is solvable, one of the choices is correct.

One of the advantages or disadvantages of multiple-choice questions (depending upon your viewpoint) is that there is no partial credit. Students who know how to do the problem but who make an algebraic or numerical error will receive the same credit as students who have no idea how to do the problem. Since numerical and algebraic errors cause loss of all credit, we suggest that multiple-choice questions be used only to replace short-answer questions and not long problems. Both multiple-choice and one long-answer problem can be included on a test. This will significantly reduce the grading in a large class without significantly decreasing the validity of the test.

Tests are stressful for students. This stress can be reduced by providing space on the examination for student comments. Tell the students the purpose of this space and explain that the comments will not affect their grades. Then, when you read a comment which says "This problem stinks," you will realize that the student is just letting off pressure.

11.1.4. ADMINISTERING THE TEST

The first part of administering a test occurs the class period before it is given. Discuss the exam with the students. Clearly state the content coverage by telling them which book chapters and which lecture periods will be covered. Explain the type of test and show a few old problems as examples. Discuss the ground rules, such as staggered seating, closed book or open book, time requirements, and so forth. Particularly for lower-division students, it is helpful to give a few hints on studying and test taking.

Many instructors find optional help sessions useful. If you plan to have an optional help session, set the rules for the session first. We hold help sessions in which students must ask questions. When the student questions stop, the help session is over. If a student asks a question which is very similar to a test problem, the best idea is to answer the question in exactly the same manner as you answer other student questions.

McKeachie (1986) suggests making up about 10 percent extra exams. It is easy for the secretary to miscount or to collate a few exams with blank pages. The extra copies allow you to rectify these problems quickly. Take reasonable precautions to safeguard the test copies, such as locking them up in a briefcase or desk in a locked office.

To students the exam is one of the most important parts of the class, so plan on being there if it is at all possible. As the professor, only you can answer student questions properly and help students understand what they are supposed to do. In addition, if a student finds a typographical error, only you can make last-minute changes to correct the problem. Professors usually have better control of the class than do TAs.

Come early and have the TAs come early. This gives you time to check the lighting, straighten up the chairs, and start to arrange the students in alternate seats. Plan to pass out the tests as quickly as possible to give everyone equal time. In very large classes put a cover sheet on the exams and tell the students not to open them until given the signal to start. Have them put their names on the test immediately. Then have them count the questions to be sure they have a complete test.

If your school does not have an honor code, it is traditional to proctor the examination. It is also helpful to have someone present to answer student questions. A circulating proctor can do wonders in reducing the desire students might have to cheat. A TA standing discretely in the back of the room can also be a major deterrent. It is much better to prevent cheating than to deal with it after it has occurred (see Chapter 12).

Periodically write on the board the time remaining. Then state, "You have two minutes, please finish your papers." When the time is up, stop the class firmly and collect the papers. It is best to give tests where there is effectively no time limit, but this is often difficult to schedule.

As soon as the examination is over, count the tests. Then check them in against the student roster. It is best to know immediately if a student has not handed in a test or was not present. Students have been known to occasionally complain that their test was lost.

11.2. SCORING

We will draw a distinction between scoring tests, which has a feedback function essential for the student's learning, and grading, which is a communication at the end of the semester of how well the student has done in the course. Grading will be discussed in Section 11.5. Unfortunately, both of these activities are often called grading.

11.2.1. Scoring Tests

Extra effort taken while preparing an examination is recovered when the tests are scored. Multiple-choice tests can be machine-scored or with a homemade stencil. In fact, the attractiveness of multiple-choice tests for large classes lies in the ease of scoring.

For other tests an answer sheet and a detailed scoring sheet should be prepared by you as the professor. Evaluation is difficult, and a professor can do a better job than a TA in preparing both the answer sheet and deciding the breakdown of points. The scoring sheet should be developed for the "standard solution." The TA should be instructed to show you unique solution paths. Occasionally, a student develops a creative solution path but makes a numerical error and gets the wrong answer. To avoid dampening creativity, it is important that you carefully consider these alternate solutions.

Whoever scores the test should do so without looking at the name. Students should receive the score that they earn, not the score that the grader thinks they should earn. Extremely important tests such as qualifying examinations should probably use a code letter for every student instead of a name. It is best to grade every test for one problem before grading a second problem. This procedure helps to ensure that grading is uniform. For a series of short-answer questions it might be feasible and faster to grade the entire sequence on each test paper before proceeding to the next. After one problem has been graded on all tests, review the scoring, particularly of the first few tests that were graded. Be sure that the scoring is uniform.

For long problems it is often useful to look at a few sample tests before grading everything or before giving the tests to the grader. The sample tests may show a common mistake that will require adjustment of the grading scheme, or they may indicate a second correct solution path. If a grader is available, sit down with him or her for a few minutes and go over both the solution and the scoring sheet. Indicate the type of feedback you want put on the tests. Give the TA or the grader a reasonable deadline for return of the exams as well as some hints on how to grade that type of test. Tell the TA to bring in any nonstandard solutions so that you can check them over.

We believe in awarding partial credit for long problems. Crittenden (1984) presents the opposite viewpoint that partial credit should either be given sparingly or not at all. Our reason in favor is that students can often demonstrate understanding of how to solve a problem and not have the correct solution because of a relatively small error in technique, an algebraic error, or a numerical error. On the other hand, students also need to realize that engineers must be accurate. Problems without partial credit can be given as short-answer or multiple-choice questions.

If partial credit is to awarded, develop the scoring sheet for the standard solution. Do this in advance and then adjust it after looking at a few tests. You can determine partial credit by awarding points for parts of the solution that are correct or by subtracting points for parts that are wrong or missing. In long problems these two approaches often result in different scores, and if a scoring sheet is not used will certainly result in different scores. For the highest reliability use a scoring sheet and calculate a score by adding positive items and subtracting negative ones. Discrepancies in the results obtained are a signal that the scoring needs to be reconsidered.

In addition to scoring the exam, provide written feedback and marks on the test or instruct the TA to do so. Correct parts of the test can be indicated quickly with check marks, while incorrect parts can be crossed out. Be sure that there is some mark on each page, including empty pages, so that the student will be sure that every page has been seen. Both positive and negative comments should be written on the test. Comments which explicitly correct the student's work are much more useful than writing "wrong" or "incorrect" without explaining why. Positive comments such as "good" or "clever derivation" serve as motivators.

To be effective, feedback must be prompt. Ideally, feedback would be given immediately after the student has finished the test. This procedure is used in some PSI classes (see Section 7.4). In large classes it takes longer to grade tests, but there is no excuse for taking a month or longer to return tests. If possible, hand them back the next class period. If that is not possible, be sure to return them within one week. Tell the TAs in advance which weeks there will be tests so that they can arrange to have sufficient time to grade the exams quickly. Ericksen (1984, p. 119) believes "this business of immediate feedback is overdone." He suggests taking more time to do detailed critiques and evaluations.

If it is to be useful, students must pay attention to the feedback. There are several methods that can be used to ensure that this happens.

1 Hand back the test and discuss it in class. A variant of this is to have small groups discuss the exam. This procedure is useful since it can reduce student aggression.

2 Before discussing the solution, assign one of the test problems as a homework.

3 Give one or more of the problems on a second test.

4 Ask students who obviously do not understand the material to see you privately. Student scores on exams are private, privileged information. Write the score on the inside or fold the test over when returning the test papers. If grades are posted, use student numbers or code letters instead of names.

11.2.2. Data Manipulation and Critiquing the Test

After the test fix up any problems which are not quite perfect for later use as homework or in that book you will write someday. Correct any typographical errors on all copies of the test you keep and in your computer files. If some students misinterpreted the problem, reword it so that this will be less likely to occur in the future. Perhaps one of the misinterpretations will give you an idea for an alternate test problem which can be used next year. Write the idea down and put it into your test file for future use.

It is easy to determine if an exam problem discriminates between students who do well on the test and those who do poorly. Johnson (1988) suggests a simple procedure for doing this. Separate out the tests of the ten (or fifteen in large classes) students with the highest scores on the test and of the ten (or fifteen) students with the lowest scores. For problems where no partial credit is given, let H = number of top ten students who got the problem correct, and L = number of bottom ten students who got the problem correct.

The test has positive discrimination if $H - L > 0$, and negative discrimination if $H - L < 0$. If a problem has negative discrimination, the better students are having more difficulty. These problems need to be rewritten. The sum of correct scores, $H + L$, can also be looked at. Johnson (1988) suggests that this figure should be between 7 and 17 (except in mastery courses where 20 may be reasonable). If partial credit is given, the discrimination of each item can be determined by looking at the sum of scores for the ten best and for the ten worst students.

In large classes (more than twenty students), standard scores can be useful for comparing student scores on different tests and for deciding final grades (Cheshier, 1975). Calculate the mean test score $\overline{x}$ for each student (N = number of students, x_i = test score),

$$\overline{x} = \frac{\sum x_i}{N} \tag{1}$$

and the standard deviation s,

$$s = \frac{1}{N} \sqrt{N \Sigma \left(x_i^2\right) - \left(\Sigma x_i\right)^2} \tag{2}$$

Then the z_i score is

$$z_i = \frac{x_i - \overline{x}}{s} \tag{3}$$

The z_i score is a normalized score for each student which has a mean of zero and a standard deviation of 1. The z scores can be converted to T scores where the T score has a mean of 50 and a standard deviation of 10.

$$T_i = 10 \, z_i + 50 \tag{4}$$

The standardized scores are easily calculated with a calculator or computer. If the class follows a normal distribution, which does not always happen, then the z and T scores are shown in Figure 11-2. The z or T scores for each student can be averaged and then compared to other students' scores. Doing this for raw scores is not statistically valid since both the means and the standard deviations vary from test to test. A very simple example may help to clarify the use of standard scores. Consider Debbie who has the following scores on three tests: 60, 40, 80. Her corresponding z scores are 0, +1, and —1, while the T scores are 50, 60, and 40. Compared to the class, her lowest grade is the last one which looks highest on the basis of raw scores.

There can be problems with the use of standard scores. First, in small classes they are not statistically valid and should not be used. Second, scores of 100 or 0 do not remain 100 or 0 when translated to T scores. Extreme scores can become negative or greater than 100. Thus, T scores can be misleading for these extreme scores. Third, the usual interpretation of the meaning of one standard deviation is valid only for normal distributions. T and z scores can still be used but must be interpreted with care. Cheshier (1975) highly recommends the use of standard scores, but McKeachie (1986) does not think they are worth the effort. You get to choose. If you do use standard scores, it is important to spend a few minutes explaining them to the class. Of course, in a class which uses statistics or discusses error analysis, the use of standard scores can be a useful part of the course objectives.

11.2.3. REGRADES

Allow regrades! If handled properly, regrades make the professor seem fair, reduce student aggression, force some students to reexamine the test problems, and do not take much time.

In small classes regrades can be handled informally by discussions between the students and the professor. In large classes a more formal procedure is necessary (Wankat, 1983). Regardless of the method used, the regrade procedure should be discussed with the class when the first test is returned. Students are ready to listen at that time.

If the scoring error the student wishes to correct is the incorrect addition of points, then we encourage the student to see the professor immediately following the class. In large classes

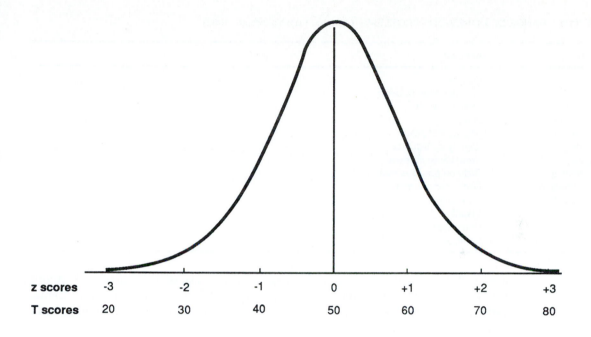

| z scores | -3 | -2 | -1 | 0 | +1 | +2 | +3 |
| T scores | 20 | 30 | 40 | 50 | 60 | 70 | 80 |

FIGURE 11-2   DISTRIBUTION OF T AND Z SCORES FOR NORMAL DISTRIBUTION OF SCORES

there will be several students clustered around the professor at this time. Thus, it is a good idea to collect the tests to allow time to check the addition.

The second type of scoring error is a mistake in the scoring where the student believes he or she deserves more points. In large classes we require a written regrade request. Students are told to make no additional marks on their tests. On a separate sheet of paper the student is asked to logically explain why he or she deserves more points. The emphasis here is on "logical," not the plea "I deserve more points." For example, a student who uses a different solution path than the standard solution may claim that his or her path was correct but that the answer was incorrect because of an algebraic or numerical error. The student can then rework the problem by using his or her path and show that the correct solution is obtained. Based on this type of argument, we have occasionally given a student a large increase in a test score. Quite often while trying this procedure the student finds that the path really does not work, and no regrade is requested.

Students are told that there may be an increase, no change, or a decrease in their test score. We ask for the entire test back but seldom regrade the entire exam. The advantage of getting the entire test back is that the professor can tell if extra pages have been inserted since the original pages will have additional staple holes in them. Some professors regrade the entire test (Evett, 1980), but this policy seems designed to prevent students from asking for regrades instead of being for the educational benefit of the student.

Give students a deadline (one week is sufficient) for regrade requests. This prevents last minute "grade grubbing" by students. Once the regrade requests have all been collected, sit

TABLE 11-1   RANGE OF HOMEWORK PROBLEMS (Adapted from Yokomoto, 1988)

Concrete	Abstract
Simple	Complex
Linear solution	Simultaneous solution
Linear solution	Trial-and-error
Short	Long
Answer given	Answer not given
Very clearly defined	Slightly ambiguous
Data given	Need literature data
Self Contained	Build on previous material
Forward solution	Backward solution
Hand calculation	Computer
Written	Visual
Logical	Brainstorm
Numerical	Symbolic

down with the TA and discuss them. The purpose of this is to ensure that grading is uniform. It is poor policy to give complainers higher scores just because they complain. Chronic complainers can be controlled if the professor carefully checks the TAs scoring before returning the tests.

11.3. HOMEWORK

What is the purpose of assigning homework? To keep students off the street and out of trouble? Or to help them learn the material? While doing homework the students are active and have a chance to practice the skills being taught in the course. A modest amount of drill can be useful since students learn how to perform certain operations quickly and accurately. Of course, the value of this practice depends on the timeliness of the feedback about the homework. To be effective, this feedback should consist of both positive and negative comments. Homework problems also provide students with a fair chance for success, yet some should also be challenging since both success and curiosity are motivating. The use of study groups should be encouraged since these groups are beneficial for extroverts and field-sensitive students. Homework is beneficial since there is a strong correlation between effort on the homework and test scores (Yokomoto and Ware, 1991).

Homework problems should cover all levels of Bloom's taxonomy and all levels of the problem-solving taxonomy (see Chapters 4 and 5). A gradation of problems, from easy to difficult, to cover many different aspects should be used. At least some of the homework problems should be at the same level of difficulty as the tests. Homework problems can focus on various aspects of the problem-solving strategy such as defining the problem, brainstorming possible solutions, and checking with an independent solution method. Other dimensions which need to be considered are discussed by Yokomoto (1988) and are shown in Table 11-1. Computer problems should emphasize the use of software tools.

How often should homework be assigned and how many problems should be given? Students need an activity for every week whether it is a test, homework, or a project. These activities should be due on different days. By working around your test schedule, you can determine when homework needs to be done. With a large number of tests or quizzes students do less homework. The number of problems obviously depends upon length. Following the need for a range of problems as shown in Table 11-1, the professor can make some assignments consisting of five small problems and other assignments consisting of a single long problem.

However, assigning a significant amount of homework involves scoring the homework and providing adequate feedback. This is particularly significant if the professor does not have a TA. One solution is to score only selected problems. Tell the students ahead of time that not all problems will be scored. If the problems to be scored are randomly selected, the final homework grade will be proportional to the total amount of homework the student does during the semester. Students need to have solutions available for problems which are not scored. An alternative is to score the homework in class by having students score someone else's homework (Mafi, 1989). With a large number of quizzes it is not necessary to have students hand in homework. They will soon come to believe the professor when he or she tells them that students who do the homework do better on the quizzes. This realization comes quickly if a homework problem is occasionally used on a quiz.

What percentage of the course grade should be based on the homework? If the percentage is low, students will tend to ignore the homework unless a special effort is made to illustrate the correlation between homework effort and test results. If the percentage is high, many students will be encouraged to copy others' work or to cheat in other ways. A reasonable compromise seems to be 10 to 15 percent. This is low enough that you can encourage students to work in groups, but require each to hand in an individual homework paper.

Late submissions can be a difficulty. In industry, late work is accepted grudgingly and does not earn promotions or handsome raises. We suggest telling students this and then following industrial practice. Accept late work grudgingly and take off some percentage based on how late it is. We accept no homework after the solution has been posted.

Reading assignments pose somewhat different problems. Students will read the assignments if they see that reading leads to success in homework and tests. The professor's task is to ensure that the reading contributes directly to the student's success. And if the reading does not help the student achieve success, why is it assigned? Be sure that a good textbook or other readings have been selected. Skip certain material in lecture but make it clear to the students that it is material they are expected to learn from the readings. Refer to this material in the lecture, but do not cover it. Ask questions in class based on the readings. Reiterate to the class that it is important material. Then assign homework based on the material and include test problems based on it. It doesn't hurt if one of the homework problems is a clone of an example in the textbook.

11.4. PROJECTS

Design and laboratory reports can be considered special types of project reports; however, since they were discussed in Chapter 9, they will not be considered further here. Projects are most common in smaller classes such as senior electives and graduate courses. Long projects are not appropriate for first-year students since they are not ready to pick topics of special interest and need the discipline of more frequent assignments (Erickson and Strommer, 1991).

Projects can fulfill some educational objectives which are diffiuclt to fulfill with lectures and tests. A project can allow a student to explore in depth a topic of her or his choice. This choice of project gives the student some control over her or his education which is often missing from other courses. Consequently, students sometimes become very strongly motivated and continue the project long after the class is over. Projects also provide an opportunity for students to work on communication skills through written and oral progress and final reports. In the ideal case, through the project the professor empowers the student to work on an area in which he or she is intensely interested, and the professor encourages the student to develop a meaningful project. Our experience in requiring graduate students to do projects is that approximately 20 percent comes close to this ideal.

Projects must have a deliverable. In a library research project the deliverable is a paper. In engineering it is often preferable to require the student to produce "something" such as a computer program, an integrated circuit, a teaching module, a laboratory experiment, a novel solution to a mathematical problem, and so forth. Then the deliverables are a short report describing the something in addition to that something. The more choices the student has on what to deliver, the more likely he or she is to become excited about the project.

Students naturally want to know what the professor wants. Explaining what is desired (i.e., the objectives for the project) can be surprisingly difficult. However, at some point the professor does decide what he or she wants and grades accordingly. Waiting until the projects are graded to decide what one wants leaves the professor open to student complaints: "Oh, so he does know what he wants; he just won't say until it's too late" (Starling, 1987). As a professor you need to explain what it is that you really want. If you can't, then analyze the projects from the last time you taught the course to determine what you wanted when you graded them.

How big should a project be? If it is only 5 percent of the grade, students will treat it as a homework project. If it is 50 percent of the grade, students will feel very anxious about it. Projects that are about 25 percent of the grade have worked well for us.

Procrastination is the biggest problem involved in student projects. To limit but not eliminate procrastination, set up a series of deadlines. First, introduce the project in lecture relatively early in the semester. Describe the project and the evaluation procedure as clearly as possible. List the dates of all intermediate and final deadlines. In a small class, both written and oral progress reports are useful since they allow for early feedback and make students do at least some work throughout the semester. Individual meetings with each student can also help prevent procrastination.

Evaluation of projects is time-consuming, which is one reason they are most commonly used in small classes. If communication skills are included in course objectives, then projects should be evaluated on organization and writing ability. Professors who are very serious about this can correct about a third of the report and have the student correct the entire report before it is evaluated for a grade. The written report should also be graded for content, including correctness, depth, and creativity. Although projects are usually turned in late in the semester, many students seriously study the feedback because they have become involved with the material. The best feedback method for oral reports is a videotape of the student presentations. Since the project represents a sizable portion of the course grade, instructors should use their discretion in accepting late reports.

11.5. GRADING

The best advice we can give before a professor decides on the final course grades is to go into an empty room and repeat out loud, "I can't satisfy everyone. I can't satisfy everyone. I can't satisfy everyone." This will help create the proper frame of mind for awarding final course grades.

11.5.1. Purpose of Grades

There are very diverse views in the literature about the purposes and suitability of the typical grading methods used in colleges. These range from grades being indispensable to worthy of being abolished. The writers who defend grades, at least moderately, include McKeachie (1976, 1986), Johnson (1988), and Lowman (1985). The purposes they note for grades include:

1 Reward or penalty for student accomplishment.
2 Communication to others about what the student has accomplished.
3 Predictor of future performance.

Grades certainly do serve as rewards or penalties, but for feedback purposes they come much too late in the semester to provide any motivation. As rewards, grades are often used to determine who will receive honors, scholarships, and so forth. As penalties, they are used to place students on probation and to drop students.

Using grades as a communication tool is often confusing since there is no generally agreed-upon definition of what a grade means. Professors who arbitrarily change the meaning of grades are not communicating well because those who see the grade interpret it differently. However, some communication exists since there is general agreement that an A or a B means the student has learned more than a student who receives a D or an F.

Grades are also used as predictors. Students use grades as a predictors of how well they will do in the rest of their college careers, and in this sense good grades may motivate a student to continue. Professors also use grades to predict who will do well in later courses and who might do well in graduate school. Since grades are reasonably good at predicting grades in future courses (Stice, 1979), this use of grades is somewhat reasonable. Even in this case grades can be misused since they do not predict who will be good at research, which is a major part of many graduate school programs.

The most controversial use of grades is as predictors of success in life after school. Many employers use grades as part of their selection procedures. Unfortunately, many studies agree that there does not seem to be any correlation between grades and success after school (Eble, 1988; Stice, 1979). This is true regardless of how one defines success. What this means is that engineers who graduate are good enough in whatever it is that grades are measuring to be a success, and that other variables become important. These other variables can include drive, motivation, inherited wealth, common sense, communication skills, interpersonal skills, who the person knows, and luck. The supporters of grades note that since grades are part of the selection criteria, one cannot expect them to be a major predictor of success since the sample is already fairly homogeneous (McKeachie, 1986).

11.5.2. Grading Methods

Some intelligent and thoughtful people state that one should select a grading method that subverts the grading system (Eble, 1988; Elbow, 1986; Smith, 1986). The Keller plan for PSI is an example (see Chapter 7). We will assume that, being new to teaching, you are not ready or willing to subvert the system (yet). Thus, the remainder of this chapter will be on grading hints and methods.

Regardless of the grading method used, the more scores you have, the easier it is to give grades. When there are many scores, and the students know what these scores indicate, there are fewer conflicts with students when grades are awarded. If a student does complain about a grade, it is appropriate to listen to what he or she has to say. But unless a grading error has been made, it is unwise to change grades. Once a grade is changed, the word gets around and many students want their grades changed.

TABLE 11-2   TYPICAL GRADING SCALES FOR DIFFERENT CLASS COMPOSITION IN TERMS OF T SCORES (© 1975, American Society for Engineering Education)

Letter Grade	Class Made Up Primarily of			
	Poor Students	Average Students	Exceptional Students	Graduate Students
A	69 and above	63 and above	57 and above	55 and above
B	59 - 68	53 - 62	47 - 56	45 - 54
C	49 - 58	43 - 52	37 - 46	35 - 44
D	39 - 48	33 - 42	27 - 36	25 - 34
F	38 and below	32 and below	26 and below	24 and below

The most appealing grading method uses an absolute standard. Although appealing, this is often difficult. Contract or mastery grading is another way to use a standard, but this method has its detractors since the grade no longer means what most people think it means (communication), and the grade is no longer a good predictor of grades in a "standard" class. However, Eble (1988), who is not fond of grading, states that the predictive capabilities of grades should not be taken too seriously.

Travers (1950) originally suggested a standard grading scheme which has been echoed by McKeachie (1986) and Johnson (1988). The major and minor objectives of the course are clearly defined. Then the grade is a communication to the student and to others of what fraction of the course objectives has been achieved. The meaning of the grades can be defined in a manner similar to the following:

A: achieved all major and minor objectives
A⁻: achieved all major and most minor objectives
B⁺: achieved most major and most minor objectives
B: achieved most major objectives and many minor objectives
B⁻: achieved most major objectives and some minor objectives
C: acceptable performance
D: student is not prepared for advanced work requiring this material
F: failed

Even with a system like this the professor needs to decide if the student meets an objective, and subjective decisions will have to be made. Normative grading, commonly known as grading on a curve, is often used because the professor does not have to develop and correctly test for absolute standards. Instead, students are compared to each other and the grade curve is broken up into A, B, C, D, and F. This has the unhealthy effect of increasing competition. In addition, a student performance which earns an A in one class may earn a C in another class merely because student competition is better. Because of this effect of class quality, a professor should never force grades into predetermined percentages.

Many professors grade on a curve but slant the curve to take into account student quality. Cheshier (1975) suggests the grading scales shown in Table 11-2, where the grades listed are the average T score for each student. The professor needs to decide what the average quality of the class is and then use these ranges as a guide. Since a T score of 50 is the average, Cheshier is suggesting that the average student in a poor class receive a low C while the average student in a good class or a graduate-level class receive a B.

An alternate procedure is to list all the students' total scores or average scores for the semester. Then decide first where the average grade in the course should be. Many professors believe that the average grade in upper-division courses should be higher than in freshmen courses since the poorest students have dropped or transferred out of engineering. Thus the average student might receive a B or a B⁻. Then look at the distribution and decide upon cutoff points. One method for assigning the cut point for F's is to see what the score of a good but not exceptionally brilliant student is (usually the second or third best student in the class). Then any student who receives less than half this number of points fails the course. If no one is that low, then everyone passes. With the grade of the average student chosen and the F's chosen, the other grades can be selected. It is convenient to look for natural gaps in the grade distribution and put the A-B and B-C, boundaries there. When this has been done, look at the grades that are on the edges. Should they be moved up or down? Many professors prefer to give students the benefit of a doubt if their scores have been increasing throughout the semester. This can also be accomplished by giving greater weight to later tests. Try to apply wisdom at the boundaries. We have seldom been wrong when we have found reasons to be generous.

11.6. CHAPTER COMMENTS

Much more could be said about testing. Tests can be analyzed for validity and reliability by statistical methods. Multiple-choice tests in particular are easy to analyze. We do not think that this type of analysis is particularly valuable in engineering education and doubt that many engineering educators would use these procedures. In any case, most large universities have a testing service which machine-scores multiple-choice tests and calculates the appropriate statistics.

Students usually consider exams to be the most important part of a course since they are the main determiner of their grades. Professors can lament the students' values, but these values are difficult to change. Complaints about tests can be decreased by making them as fair as possible and by having enough tests so that one test will not completely determine a student's grade. Because they consider exams to be so crucial, some students will be tempted to cheat, and cheating is another fact of life which must be faced (see Chapter 12).

11.7. SUMMARY AND OBJECTIVES

After reading this chapter, you should be able to:

• Discuss the advantages and disadvantages of different types of test questions. Write test questions using each of the major test question styles.
• Develop a grid to determine the course material to be covered on a test.
• Explain to a TA how to score a test fairly. Score a test fairly.
• Determine the discrimination of test questions and calculate z and T scores for students.
• Develop a scheme for using homework and projects as part of a course which satisfies learning principles.
• Develop a personal value system for giving course grades.

HOMEWORK

1 Assume you will write a test for this chapter.

a Develop a test grid to decide on the coverage of the test.

b Write at least two long-answer questions, two short-answer questions, and two multiple-choice questions for the test.

c Select the questions for the test so that it can be given in a regular fifty-minute class period.

d Write the solutions and the grading scheme for this test.

2 Do a project on teaching engineering material. The project must involve both content (engineering subject matter) and teaching method. You must have a deliverable such as a videotape, CAI, a self-paced module, a laboratory experiment, a National Science Foundation proposal for curriculum development, a student handbook for commercial software, or course demonstrations.

REFERENCES

Canelos, J., and Catchen, G. L., "Test preparation and engineering content," *Proceedings ASEE Annual Conference,* ASEE, Washington, DC, 1624, 1987.

Cheshier, S. R., "Assigning grades more fairly," *Eng. Educ.*, 343 (Jan. 1975).

Crittenden, J. B., "Partial credit: Not a God-given right," *Eng. Educ.*, 288 (Feb. 1984).

Eble, K. E., *The Craft of Teaching*, 2nd ed., Jossey-Bass, San Francisco, 1988.

Elbow, P., *Embracing Contraries: Explorations in Learning and Teaching*, Oxford University Press, New York, 1986.

Ericksen, S. C., *The Essence of Good Teaching,* Jossey-Bass, San Francisco, 1984.

Erickson, B. L. and Strommer, D. W., *Teaching College Freshmen,* Jossey-Bass, San Francisco, 1991.

Evett, J. B., "Cozenage: A challenge to engineering instruction," *Eng. Educ.*, 434 (Feb. 1980).

Johnson, B. R., "A new scheme for multiple-choice tests in lower-division mathematics," *Amer. Math. Mon., 427 (May 1991).*

Johnson, G. R., *Taking Teaching Seriously: A Faculty Handbook*, Texas A&M University, Center for Teaching Excellence, College Station, TX, 1988.

Kessler, D. P., "Machine-scored versus grader-scored quizzes—An experiment," *Eng. Educ.*, 705 (April 1988)

Leuba, R. J., "Machine scored testing, Part I: Purposes, principles and practices," *Eng. Educ.*, 89 (Nov. 1986a).

Leuba, R. J., "Machine scored testing, Part II: Creativity and item analysis," *Eng. Educ.*, 181 (Dec. 1986b).

Lowman, J., *Mastering the Techniques of Teaching*, Jossey-Bass, San Francisco, 1985.

Mafi, M., "Involving students in a time-saving solution to the homework problem," *Eng. Educ.*, 444 (April 1989).

Masih, R. Y., "Perfecting the engineering education and its exams," *Proceedings ASEE Annual Conference*, ASEE, Washington, DC, 200, 1987.

McKeachie, W. J., "College grades: A rationale and mild defense," *AAUP Bull.*, 320 (Autumn 1976).

McKeachie, W. J., *Teaching Tips*, 8th ed., D.C., Heath, Lexington, MA, 1986.

Smith, K. A., "Grading and distributive justice," *Proceedings ASEE/IEEE Frontiers in Education Conference*, IEEE, New York, 421, 1986.

Starling, R., "Professor as student: The view from the other side," *Coll. Teach.,* 35 (1), 3 (1987).

Stice, J. E., "A first step toward improved teaching," *Eng. Educ.*, 394 (Feb. 1976).

Stice, J. E., "Grades and test scores: Do they predict adult achievement?" *Eng. Educ.*, 390 (Feb. 1979).

Svinicki, M. D., "The test: Uses, construction and evaluation," *Eng. Educ.*, 408 (Feb 1976).

Travers, R. M. W., "Appraisal of the teaching of the college faculty," *J. Higher Educ.,* 21, 41 (1950).

Wankat, P. C., "Regarding tests: A chance for students to learn," *Eng. Educ.*, 746 (April 1983).

Yokomoto, C. F., "Writing homework assignments to evoke intellectual processes," *Proceedings ASEE Annual Conference*, ASEE, Washington, DC, 579, 1988.

Yokomoto, C. F. and Ware, R., "The seven practices—Persuading students to do their homework," *Proceedings ASEE Annual Conference*, ASEE, Washington, DC, 1767, 1991.

CHAPTER 12

STUDENT CHEATING, DISCIPLINE, AND ETHICS

In universities, as in society in general, cheating is a fact of life, but as a new instructor you can drastically reduce the incidence of cheating in your classes by taking simple precautions. The aim of this chapter is to help you prevent incidents of cheating. Related issues about ethics and student discipline are also considered.

12.1. CHEATING

Most engineering professors agree with Gregg (1989) that cheating "must not be tolerated in any form." Despite this, many graduates admit that they have cheated sometime during their college career. In one study 56 percent of a graduating class of engineering students admitted to having cheated (Todd-Mancillas and Sisson, 1986). It is much better to prevent cheating than to have to deal with it after the fact.

12.1.1. Prevention of Cheating

The best method for reducing large scale cheating is to create an atmosphere which is not conducive to cheating (Eble, 1988; Kibler et al., 1988). When good rapport exists between students and professor and among students themselves, cheating is drastically reduced. It is much easier to cheat when a professor is cold and aloof. A student who feels like a number and knows that the professor does not know her or his name finds it easier to cheat than a student who is known to the professor by name. Students cheat significantly less in a class with shared objectives, and where there is an obvious excitement in learning. Any class where students feel that the professor is a partner in learning will have a low incidence of cheating.

Professors who develop a reputation of writing fair tests and of grading fairly will have less cheating on their tests than professors with a reputation of writing unfair tests or of being "superhard" graders. Students must be challenged, not overwhelmed.

It is important to discuss the rules of cheating and plagiarism with students (Evett, 1980; Walworth, 1989). Just as different cultures define the act of sharing answers with a friend in different ways, professors need to share their definitions and rationale with the class. Many students simply do not know the rules about plagiarism, so this discussion is particularly important. To make the discussion positive, present it in a positive form by including it in a larger discussion on the importance of engineering ethics (see Section 12.3). The appropriate time for this discussion is immediately before the first test or the first assignment which is to be done independently. A little bit of humor can help get the point across and make the discussion less threatening for the students.

Reducing anxiety on tests also decreases cheating (Kibler et al., 1988). The pressure of any one test can be reduced by giving numerous quizzes or tests. Equal access to test files reduces the urge to cheat on the part of those without access. Access to the professor or to TAs for help and a help session immediately before the test make the course seem fairer and help reduce pressure. Open book exams or tests with equation handouts or key relations charts help to reduce pressure and eliminate the use of illegal cheat sheets.

Before a test is administered, some commonsense security measures can reduce the temptation to cheat. You may want to adopt some of the following suggestions to help prevent both casual cheating and the deliberate stealing of tests. Make up a new test shortly before the test date. Have a secretary, not a work-study student, do any typing and make the copies. Any waste copies should not be thrown away (students have been known to search waste baskets) but should be kept with the other tests and discarded after the test. Any computer files which are not secured by a password should be cleared or the disk should be locked up. Test copies should be locked up, taken home, or craftily hidden. Even a normally honest student will be sorely tempted if a copy of the test is sitting out in plain view on a desk. Make up extra copies of the test, but number the copies so that you will know exactly how many have been distributed.

The most common forms of cheating occur during the administration of a test. Long-answer tests with many calculations are the most difficult to cheat on. Multiple-choice problems are the easiest to cheat on, and if other precautions are not taken may invite cheating. Short-answer problems are intermediate in cheatability.

In large classes both the professor and the TAs should proctor the test. Proctoring has a major deterrent effect on cheating (Kibler et al., 1988). The proctoring can be done in such a way that it is clear that the proctors are alert so that they can help students with questions. Stationing a TA at the back of the room is an effective deterrent in large lecture halls since students cannot easily keep track of the proctor's location.

If at all possible, have the students sit in alternate seats since this drastically reduces cheating of the wandering eyes variety (Evett, 1980). If a large enough room is not available, consider using two rooms simultaneously. Assign students to each room in advance and have each room proctored. Another possibility is to use alternative test forms which have either the questions or the answers in different order, or to use different values in calculations.

Before the exam starts have students place books underneath their desks unless the test is open book. Since many calculators can now store significant amounts of alphanumeric data, the calculator has become a possible cheat sheet. The temptation to use it in this way can be reduced if you stroll about the room while looking for students who need help. Students should not be allowed to share calculators unless a TA clears the calculator first. The significance of cheat sheets is eliminated if every student is allowed to bring one in or if the test is open book.

Since the purpose is prevention not proof of cheating, take action as soon as something suspicious happens. Standing near the student (while waiting to answer the questions of other students) may be a sufficient deterrent. Asking the student if he or she has a question is a subtle way of letting the student know that you are watching. If the situation persists, ask the student to move to a less crowded spot. If the student prefers to stay put, suggest that you prefer that he or she move. Some professors announce to the class that students should not look around the room. This can be effective, particularly if the professor looks at the suspicious student, but it is a bit distracting for the class.

In very large classes where the professor does not know each student by name and may not even recognize some faces, it is fairly common to use picture IDs to prevent "ringers" from taking the examination for someone else. The IDs can be placed on the corner of the desk or they can be shown to the TA as the student turns in the test.

The chaos of test turn-in time also invites cheating (Felder, 1985). Students see someone else's solution or the professor's solution and then quickly change their answers. This can be prevented by making everyone stop writing at a particular time. The professor's solution can be guarded until after all tests have been turned in. The professor can have the TAs collect the test while watching for suspicious activity.

As soon as the tests have been collected, log them in. In this way you know immediately who did not take the test. Tests need to be kept secure after the exam is over. Students have been known to steal a large number of tests so that the exam cannot be graded. Both the professor and the TAs need to be careful not to lose any tests. Losing a student's test creates major difficulties.

Be sure that the graders make a mark on every page of the test so that students cannot claim that the grader did not see a page. It is best to use bound examination books so that pages cannot be inserted after the test has been returned. If a stapled test is used, carefully staple all the pages together when the test is returned. It is extremely difficult to add a page without making an extra staple hole in the other pages.

Any students who are suspected of being dishonest should receive extra care in grading. You can spend a little extra time going through their tests in detail to be sure that the grading is correct. Make a copy of the suspect's test before it is returned. Who is a suspect? Any student who has been caught cheating previously, who has been suspected of cheating, or who has received significant points by having a previous test regraded. It is best to handle these cases quietly without the help of the TAs since the student may well be innocent.

Procedures for homework, projects, and take-home tests are similar, but it may be harder to catch cheaters. Making any of these assignments a large part of the grade, particularly in a class where there is no rapport between the professor and the students, is asking for trouble. If a take-home assignment is to be done independently, this needs to be discussed in class and it needs to be stated in writing on the top of the assignment. Despite this, collaboration on take-home assignments is very high (Todd-Mancillas and Sisson, 1986). Even graduate students collaborate on take-home tests. The rules for plagiarism of papers need to be clearly spelled out. A student who believes that he or she will receive more credit by properly citing sources is less likely to plagiarize. The easiest way to decrease cheating on take-home assignments is to make them a small percentage of the course grade and then encourage students to collaborate (Felder, 1985).

12.1.2. The Cure for Cheating

Once cheating has been detected, resolving the situation can be very painful and time consuming. Cheating must be fully documented. If possible, have someone witness your proof. If there is reasonable doubt that cheating has occurred, the best course is to put the student on your suspicious list and be more vigilant the next time.

If the proof of cheating is clear, then obtain a copy of your university's regulations and read them very carefully. Courts have upheld the principle that some form of due process must be followed in academic discipline cases [see Kibler et al. (1988) for citation of the court cases]. Follow your university's regulations. Most universities have developed regulations that provide students with appropriate due process. If you make allegations of cheating in good faith and follow your university's regulations, then you will be well protected from personal liability even if the student is found not guilty (Kibler et al., 1988). However, you will be liable if the student is found not guilty and you impose penalties anyway.

Some universities allow the professor to discuss the case with the student, and *if* the student confesses, the professor can decide the penalty. This can range from a zero on the test to a lower grade in the course. Kibler et al. (1988) suggest that this informal procedure without reporting the case is somewhat dangerous. If the student later recants and claims that he or she was coerced into confessing, then the professor may be liable even though the student signed a confession. It is safer to go through the formal university channels. The university committee also has access to records which may show that the student is a chronic cheater, which will result in a more severe penalty. Some professors lower the student's grade without discussing the allegations with the student. This is unwise since due process has clearly been denied the student and the professor may be liable. We repeat, it is much better to prevent cheating than try to deal with it once it has occurred.

12.2. OTHER DISCIPLINE PROBLEMS

Although cheating is the most prevalent discipline problem, there are other discipline problems which the professor must learn to deal with. Once again, prevention is the best policy.

Some professors have fewer problems with students than do others. These are the professors who develop rapport with students, are fair and accessible, are excited about the material they are teaching, and try to function as an ally to the student in learning the material. In addition, these professors know where to draw the line. Students are similar to children in that they test the professor. Just as a parent must know when to stop this testing, the professor must be able to tell students that their request is unreasonable. This can be achieved if the professor is friendly but keeps a certain professional distance (see Chapter 17).

Discipline problems can occur in class. They include talking, reading newspapers, and wearing headphones. State both the rules and the reasons for the rules during the first class period. Talking and newspaper reading disturb other students, and a student wearing headphones cannot, even by accident, pick up anything of value from the lecture. Offenders can be asked politely to stop. If the activity continues, the offender can be called on to contribute to the class. If the offense still continues, the student can be called in for a private discussion.

Late arrivals are mildly disruptive. And a latecomer who then asks questions about material which has already been covered can be quite disruptive. Some professors lock the door when the bell sounds. This approach seems extreme. Talk to a student who is chronically late but do so in a nonthreatening manner (see Chapter 10). Perhaps there is a good reason for the tardiness, and some sort of special arrangement may be appropriate, such as transferring the student to another section. At the least you can request that the latecomer save questions about material which has already been covered until after class.

Hostile students are another problem. Hostility is most prevalent following a test, but some students start the semester hostile. Hostility following a test can usually be deflected by having a fair regrading procedure and by asking the student to talk to you after class. Since this type of hostility usually decays rapidly, a good strategy is to give the student time to cool off and then listen to her or him. A chronically hostile student is a different matter. Look at the student's file and talk to professors who have had him or her previously. This information may give a hint of how to proceed. If there is no hint, you can call the student in for a chat. Try to be nondefensive and listen to her or him. Don't expect miracles but see if an accommodation can be worked out for the semester.

Students with excessive absences cause problems since they skew the curve downward on tests, are usually late with or don't turn in homework, and often complain about the course. In some courses such as laboratories and seminars, it is appropriate to require attendance and reduce the student's grade accordingly for absences. Many students, and some professors, think that attendance in lecture courses should be optional and what the student learns should determine his or her grade. Keep track of attendance and point out the excellent correlation between attendance and learning. Refusing to grade on a curve prevents excessively absent students from skewing the grading. If you want to grade on a curve, one solution is to plot the scores of students who have attended at least some minimum number of classes and use this curve to set the course grades. Then on the basis of this curve, students with excessive absences receive whatever grade they have earned. If you do this, be sure to explain the grading procedure clearly in advance.

Students do procrastinate and assignments are often turned in late. Accepting late assignments at full credit does not seem fair to students who have done the work on time, and it rewards students turning assignments in late for bad behavior. On the other hand, following a policy of never accepting late assignments seems overly rigid. We accept overdue assignments but penalize students a given percentage for each day the assignment is late.

Students sometimes miss tests, with excuses ranging from oversleeping (probably true) to being sick (maybe true) to the death of a grandmother (doubtful). The easiest policy for dealing with absences is to automatically discard the lowest test score (preferably based on the T score) that the student receives during the semester. A test missed for any reason becomes a zero and is discarded. A student who protests can be offered an opportunity to take the test for practice. A second policy that some professors use is to allow makeups only for illness with a signed form from a medical doctor. A third possible policy is to write a makeup test. The procedure to use is up to you.

Students argue about grades after each test and at the end of the semester. A formal regrade policy (see Section 11.2.3) is useful. For arguments at the end of the semester, students should be shown the courtesy of being listened to. At some universities this is also the first step in a formal grade appeal procedure. Changing student grades unless a mistake has been made in recording the grade or in adding points will drastically increase the number of complaints the professor gets in the future.

On rare occasions one hears stories of students trying to buy grades with money, gifts, or sexual favors: "Professor, I'd do anything for a B in this class." Since the offers are usually not explicit, the best response is to act as if nothing unethical was intended: "Here's a study schedule with ten hours a week on this course plus an hour a week of tutoring with the TA. If you follow this you will be sure to improve your current grade."

A different type of student discipline problem involves the graduate student who does not appear to be working or performing any research. Unfortunately, it can be difficult to tell if the student is working or performing. Help the student set realistic goals. Sometimes graduate students stop working because they do not have a job lined up after graduation. If this is the case, you may be able to help with a postdoctoral position. If the student is supported by research funds, no work—no pay is a realistic policy. Often the student can continue to be paid if he or she continues working on research and just delays turning in the thesis. In extreme cases stopping payment is appropriate.

12.3. TEACHING ETHICS

Teaching students to become ethical engineers is important. This subject is particularly appropriate for this chapter because part of becoming an ethical engineer consists of behaving ethically as a student. It is improbable that students who cheat their way through school will suddenly become ethical engineers upon graduation. A general discussion on ethics is a good way to start a discussion on cheating since the students are less likely to feel accused.

Professors cannot assume that students are automatically ethical. Ethics must be instilled in students (Walworth, 1989)—most effectively by including the subject throughout the curriculum instead of adding an ethics course or lecture at the very end of the student's career.

TABLE 12-1   CODE OF ETHICS

Fundamental Principles

Engineers shall uphold and advance the integrity, honor, and dignity of the engineering profession by:

1. using their knowledge and skill for the enhancement of human welfare;
2. being honest and impartial and serving with fidelity the public, their employers, and clients; and
3. striving to increase the competence and prestige of the engineering profession.
4. supporting the professional and technical societies of their disciplines.

Fundamental Canons

1. Engineers shall hold paramount the safety, health, and welfare of the public in the performance of their professional duties.
2. Engineers shall perform services only in areas of their competence.
3. Engineers shall issue public statements only in an objective and truthful manner.
4. Engineers shall act in professional matters for each employer or client as faithful agents or trustees, and shall avoid conflicts of interest.
5. Engineers shall build their professional reputations on the merits of their services.
6. Engineers shall act in such a manner as to uphold and enhance the honor, integrity, and dignity of the engineering profession.
7. Engineers shall continue their professional development throughout their careers, and shall provide opportunities for the professional development of those engineers under their supervision.

Reprinted with permission of Accreditation Board for Engineering and Technology, Inc.

Ethics can be introduced in a "just-in-time" format in every engineering class. When the ethical issue comes up, discuss the ethics involved.

A variety of methods can be used to instill ethical behavior. A few of these methods are:

1 Model ethical behavior at all times. The ethics of being an engineering professor are discussed in Chapter 17.

2 Before the first test in every course discuss the need for ethical behavior in engineers. Note that you expect students to start practicing that ethical behavior right away. Then discuss the rules for honesty in taking a test.

3 Seminar classes on professionalism should have at least one session on ethics. The appropriate code of ethics should be distributed to all students and then discussed. A simple code is the one shown in Table 12-1 which is the ABET code of ethics and part of the codes of many engineering societies. The new IEEE code is discussed by Singleton (1991). Unfortunately, it is easy for students to read any code without thinking about its ramifications.

4 In discussions of ethics the professor can usefully play the role of devil's advocate. Florman (1987) presents an interesting hypothesis that laws have taken the place of self-policed ethical codes and thus most of the codes are obsolete. He recommends a commonsense approach to ethics. His first postulate is:

Don't break the law.

Since most engineering failures are due to human error or sloppiness, Florman's second postulate is:

Be conscientious, that is, careful, hardworking, dedicated, and innovative.

Florman notes that whistle blowing is usually unnecessary. He suggests:

Try to influence events without becoming excessively disruptive. Work within the system.

5 Case studies with class discussion or a class debate are excellent for involving students in a discussion of ethics, either as part of a regular class or as part of seminar classes. For example, in a senior design class the ethics of environmental problems can lead to a lively hour of discussion. Or the students can discuss or debate whether an engineer working on offensive weapons in the defense industry is satisfying fundamental principle 1 of the code of ethics. Students in a senior seminar can become very involved in a discussion of the ethics of interview trip expenses and accepting a job. The explosion of the space shuttle Challenger serves as an interesting case study for any engineer (Florman, 1987; Singleton, 1991). Cooley et al. (1991) present a case study which includes the results of a trial in which they had the students role play different roles.

Ethics is a dry subject only if the professor makes it a dry subject. A little creativity can make the ethics portion of a class lively and interesting.

12.4. CHAPTER COMMENTS

In a class on teaching, the material in this chapter can be fun to teach. Every student and professor knows stories about the zany things students have done to cheat or to escape doing work. A little humor can make this interesting and counter the seriousness of the topic. As for cheating, we cannot emphasize too much that prevention is better than a cure. One additional method, which is outside the scope of this chapter, is to develop a student-run honor code. Schools with well-functioning honor codes have a significantly lower incidence of cheating

12.5. SUMMARY AND OBJECTIVES

After reading this chapter, you should be able to:

• Define methods to prevent cheating.
• Discuss the appropriate methods to handle cheating at your university.
• Develop methods to handle other disciplinary issues.
• Introduce ethics into your engineering course in short segments at the appropriate time.

HOMEWORK

1 Obtain a copy of the regulations for handling cheating at your university. Compare the policies of your university to the more general discussion in this chapter.

2 From newspapers or professional publications find a current news item which involves ethical issues. Develop a five-minute presentation to include this issue in an engineering class. Develop a plan for student discussion based on Table 12-1.

3 List additional cheating methods and how to handle them.

REFERENCES

Cooely, W. L., Klinkhackorn, P., McConnell, R. L., and Middleton, N. T., "Developing professionalism in the electrical engineering classroom," *IEEE Trans. Educ.*, 34, 149 (May 1991).

Eble, K. E., *The Craft of Teaching,* 2nd ed., Jossey-Bass, San Francisco, 1988.

Evett, J. B., "Cozenage: A challenge to engineering instruction," *Eng. Educ.,* 434 (Feb. 1980).

Felder, R. M., "Cheating—An ounce of prevention . . . or the tragic tale of the dying grandmother," *Chem. Eng. Educ.*, 12, (Winter 1985).

Florman, S. C., *The Civilized Engineer*, St. Martin's Press, New York, 1987.

Gregg, N. D., "Ethics in the teaching profession," *Proceedings ASEE Annual Conference*, ASEE, Washington, DC, 341, 1989.

Kibler, W. L., Nuss, E. M., Paterson, B. G., and Pavela, G., *Academic Integrity and Student Development: Legal Issues. Policy Perspectives,* College Administration Publications, Asheville, NC, 1988.

Singleton, M., "The need for engineering ethics education," *Proceedings ASEE/IEEE Frontiers in Education Conference*, IEEE, New York, 145, 1991.

Todd-Mancillas, W. R. and Sisson, E., "Cheating among engineering students: Some suggested solutions," *Eng. Educ.*, 757 (May 1986).

Walworth, M. E., "Professionalism and cheating in the classroom," *Proceedings ASEE Annual Conference*, ASEE, Washington, DC, 316, 1989.

CHAPTER 13

PSYCHOLOGICAL TYPE AND LEARNING

How do individuals learn? What can teachers do to aid learning? Why do different teaching methods have different effects on individuals (or why doesn't everyone learn the same way I do?). Complete answers to these and related questions are not known despite years of intensive research; however, what is known can be helpful to professors in understanding the learning and teaching processes. Chapters 13 through 15 explore these questions and suggest ways of incorporating them in engineering education. Chapter 14 examines two theories of cognitive development, Piaget's and Perry's, and offers implications for engineering education. Piaget's theory leads to the learning theory known as constructivism, which is dealt with in Chapter 15, along with Kolb's learning cycle and, finally, motivation.

This chapter focuses on the natural differences among students that need to be considered in the planning of instruction and for handling interpersonal relationships. By accounting for such differences we can not only retain the students we have, but also attract the nontraditional groups that are underrepresented in engineering. As Sections 13.2 through 13.4 show, the personality instrument discussed in this chapter, the Myers-Briggs Type Indicator (MBTI), has proven to be a successful tool in engineering education for recognizing and accommodating these differences.

The essence of the theory behind the MBTI is that "much seemingly random variation in behavior is actually quite orderly and consistent, being due to basic differences in the way individuals prefer to use their perception and judgment" (Myers and McCaulley, 1985, p. 1). "Perception" refers to the ways that we process information or become aware of the world around us. "Judgment" has to do with the ways we make decisions on the basis of what has been perceived. These ideas are based on the theories of the Swiss psychologist Carl Jung (1971) and their application and extension by Katherine Briggs and Isabel Briggs Myers. The MBTI has been used in education and industry as well as career and marriage counseling to help identify personality types in order to improve communication and open the possibilities for learning. Such knowledge is very important for professors and beneficial for students.

An indicator, not a test, the MBTI is a self-reporting instrument which offers a forced-choice format between equally valuable alternatives. In answering the questions or responding to certain word-pairs, we can discover our preferred way of dealing with, and living in, the world. Intrinsic preferences, though possibly inborn, aren't always available to our conscious minds. The way we are raised and the situations we confront may force us to react in ways opposed to our inherent preferences. The MBTI allows us to arrive at a "reported" type and then examine the conclusion in light of our experiences, beliefs and feelings, all the time being free to accept or reject the result. Over the years, the MBTI has become the most widely used personality measure for non-psychiatric populations (Myers and Myers, 1980). What the MBTI does not do, however, is merely classify people. Type does not refer to something that is fixed, permanent. Each type has its own ways of reacting to situations, but no one is true to type all the time. As McCaulley et al. (1983) point out, good type development often involves responding in ways that one does not spontaneously prefer. "The word *type* as used here refers to a dynamic system with interacting parts and forces. The characteristics and attitudes that result from the interactions of these forces do differ, but the basic components are the same in every human being" (p. 397).

13.1. FROM JUNG TO THE MBTI

The seminal work on type theory was done by Carl Gustav Jung, the Swiss psychologist and contemporary of Sigmund Freud. His study *Psychological Types* was published in 1921 after almost twenty years' work in treating individuals and discussing problems and solutions with colleagues, as well as "from a critique of [his] own psychological peculiarity" (Jung, 1971, p. xi). In his difficult yet eminently readable book, Jung looks at the problem of type in the history of classical and medieval thought as well as in biography, poetry, philosophy, and psychopathology.

An excellent biography of Katherine Cook Briggs and her daughter, Isabel Briggs Myers, is Frances Saunders' book (1991); and *Gifts Differing* (Myers and Myers, 1980) gives an excellent discussion of their theory. Katherine Cook Briggs was a lifelong student of the differences among individuals and how they relate to the way in which one functions in the world. In part her interest in personal differences grew out of her desire to be a writer and create fictional characters, and for this reason she was particularly interested in Jung's treatment of biography in his book.

Discovering Jung's work in the English translation in 1923, Briggs is alleged to have said, "This is it" (Saunders, 1991, p. 59). Unfortunately, she was so impressed with Jung's work and terminology that she burned her own notes and adopted the latter's terminology. She shared this interest in personality with her daughter Isabel (later Isabel Briggs Myers) who continued studying type. With the onset of World War II, Myers desired "to do something that might help people understand each other and avoid destructive conflicts" (Lawrence, 1980). She decided to find a way to put the theory to practical use and from this came the idea for a "type indicator." Her first task was to develop an item pool that would reflect the feelings and attitudes of the differing personality types as she and her mother had come to understand them. The first period of development involved item validation with friends and family and then collecting data on 5000 high school students and 5000 medical students. The second period began in 1956 when the Educational Testing Service (ETS) became the publisher. After the 1962 publication of the MBTI manual and form F, the popularity and use of the MBTI began to grow slowly. In 1975, Consulting Psychologists Press took over publication, and since then the use of the instrument has expanded greatly. It has now been translated into Japanese and Spanish, and is also being used in England and Australia. According to McCaulley (1976), the fact "that similar career choices by the same types occur in disparate cultures suggests that Jung's theory taps some fundamentally important human functions that cut across cultural teaching."

13.2. PSYCHOLOGICAL TYPE: ATTITUDES AND FUNCTIONS

In *Psychological Types,* Jung postulated that everyone has a basic orientation to the world which indicates the directions in which energies or interests flow: to the outer world of people and events (extroversion, E) or to the inner world of ideas (introversion, I). He referred to this as an attitude toward the world. Either type, in the conscious aspects of life, processes information either through the senses (S) or by intuition (N) and makes decisions on the basis of this information either by logical, impersonal analysis (thinking, T), or on the basis of personal, subjective values (feeling, F). Jung regarded both thinking and feeling to be rational processes and so the term "feeling" here does not carry the common connotations associated with emotions. As to why there are four functions (S, N, T, F), not more or less, Jung (1971, pp. 540–41) says he arrived at that number on purely empirical grounds. Through sensation we establish what is present, with its meaning determined through thinking. Feeling tells us its value, with possibilities delineated by intuition. The jungian dichotomies are as follows:

Direction of energy/interest:	E or I
Perceiving functions:	S or N
Judging (decision-making) functions:	T or F

To these three Jungian pairs, Katherine Briggs added a fourth: a judging (J) or perceptive (P) orientation to the world. In her research she discovered that individuals tend to function primarily in either the perceiving or the judging mode. That is, some people (P) like to gather more and more information and adapt to situations as they arise; others (J) prefer to lead a more structured, ordered existence, making lists, and trying to control events.

A person's preferences are indicated by these dichotomies, but each person is free to use sensing or intuition, and, similarly, thinking or feeling. As with handwriting, anyone can write a signature using either the left or the right hand; however, most have a preference for one over the other and tend to develop the skill in one more than in the other. The four dichotomies (EI, SN, TF, JP) can be arranged in a four by four table or matrix, giving sixteen personality types from interactions. Table 13-1 summarizes the attitudes and functions, and Figure 13-1 gives an example of the matrix, with breakdowns of percentages for engineering disciplines. Then

TABLE 13-1   THE FOUR MBTI PREFERENCES

Direction of energy or interests		
E	(Outer world of people)	Extroversion
I	(Inner world of ideas and actions)	Introversion
Preference of perception		
S	(Immediate and practical experience)	Sensing
N	(Possibilities and meanings of aspects of experiences)	Intuition
Preferences for decision making		
T	(Logical, objective)	Thinking
F	(Subjective, personal, value-based)	Feeling
Orientation to outside world		
J	(Ordered, planned)	Judgment
P	(Spontaneous, adaptive)	Perception

brief general characteristics of each type will be given (Myers and McCaulley, 1985; see also Singer, 1972). McCaulley, Macdaid, and Kainz (1985) discuss ways of estimating type frequencies among the general population.

Orientation to Life: Extroversion (E) and Introversion (I).   The first pair, extroversion and introversion, focuses on how one approaches the world. Ever since Jung first posited these descriptors of behavior, the terms have become part of the language and as used here they carry the standard psychological connotations.   The outer-directed extrovert enjoys social contact and depends on interaction with others for personal satisfaction. The inward-looking introvert, on the other hand, tends to withdraw from such interactions, preferring quiet for concentration rather than the quick action of the extrovert.   The easy communication of the extrovert is a problem for the introvert, who, preferring ideas, may have trouble communicating.   In problem-solving, the extrovert tends to place greater weight on the situation and other people's views, whereas the introvert tends to focus more on the conceptual framework of the problem (McCaulley, 1987).   No one is purely extroverted or introverted, though some individuals clearly may represent extremes of each type. Instead, the terms refer to preferred orientations. Anyone can exhibit both introverted and extroverted behavior.   For example, an introverted teacher may approach a class with some fretfulness, mustering up all of his or her energy to begin the class, but once settled into the course, may feel comfortable and act the complete extrovert—within the confines of the class. Yet, the preferred orientation is that of a cautious introvert. Each person carries the capability of developing both orientations but by preference tends to develop one of them. The same is true of the other three function pairs. This is an important point to remember while reading about the MBTI. Myers and McCaulley (1985, p. 5) caution that the Indicator is no substitute for good judgment and that the proper way to use it is as a stimulus to the user's insight.

Extrovert (E). (roughly 70 percent of the general population; about 33 percent of the engineering student population)

- Likes people.
- Likes action.
- Acts quickly.
- Communicates easily.
- Is applications-oriented.
- Feels energized by interaction with others.

Introvert (I). (30 percent of the general population; about 67 percent of the engineering student population)

- Prefers quiet for concentration.
- Likes ideas and concepts.
- Has trouble communicating.
- Relies on inner illumination.
- Prefers to work alone and is energized by doing so.

Perception or Becoming Aware: Sensing (S) and Intuition (N).   The second pair, sensing (S) and intuition (N), characterizes the perceptive function, or how one becomes aware of, or perceives, the world. The sensing person leans toward working with known facts rather than looking for possibilities and relationships as the intuitive person often prefers to do. He or she also tends toward step-by-step analysis and prefers to work by established methods. Intuitives favor inspiration and may work in bursts, quickly jumping to conclusions or solutions. Unlike sensing individuals, they are impatient with routine and may appear to be more imprecise. Using their imaginations, they see possibilities, whereas sensing individuals use their senses and work through the powers of observation. To a sensing type, soundness, common sense, and accuracy characterize real intelligence, which for an intuitive is shown by flashes of imagination and insight in grasping complexities. Attitudes characteristically developed from the preference for intuition include a reliance on sudden insight, an interest in the new, and a preference for learning through an intuitive grasp of meanings (McCaulley, 1978). A synopsis of the two types shows the following.

Sensing (S). (70 percent of the general population; 53 percent of engineering student population)

- Uses senses and powers of observation.
- Works through step-by-step analysis.
- Likes precision.
- Prefers established methods.
- Is patient with routine.
- Works steadily.

Intuition (N). (30 percent of the general population; 47 percent of the engineering student population)

- Is imaginative, sees possibilities.
- Relies on inspiration.
- May be imprecise.
- Jumps to solutions (is quick).
- Works in bursts.
- Dislikes routine.

Decision Making: Thinking (T) and Feeling (F). Once all the data are in, whether by sensing or by intuition, one must then decide how to process the information and come to a decision. A person who prefers to be logical and analytical, weighing facts impersonally and objectively, shows a preference for thinking (T) as the mode of decision making; someone who bases decisions on subjective, personal values and standards uses feeling (F). Both poles are accessible to everyone, and often most individuals move freely between them; however, each person has a preferred mode.

Thinking (T). (60 percent male/40 female in general population; 74 percent of engineering students: 77 percent male/61 percent female)

- Is objectively analytical.
- Works through cause and effect.
- Tends to be logical.
- Tends to be tough-minded.
- Tends to be impartial.

Feeling (F). (40 percent male/60 percent female in general population; 26 percent of engineering students: 23 percent male and 39 percent female)

- Understands people.
- Desires harmony.
- Stresses interpersonal skills.

Living in the World: Judgment (J) and Perception (P).   The fourth preference pair is used to identify the way an individual functions in the world. The previous sections considered the attitudes (E and I) and the functions (S, N, T, and F) which Jung used to categorize conscious mental processes. In an elaboration of Jung's ideas, Briggs and Myers added a further dimension: the attitude a person takes toward the world. This attitude is based on the person's relationship to or preference for the functions of perceiving and judging. An individual who prefers to use a perceiving function (S or N) to run his or her life tends toward being open to new perceptions, adapting to situations, and in general taking in information. This flexibility often leads to minimal planning and organization. For someone who uses a judging function (T or F) to conduct his or her outer life, the impetus is toward planning, organization, and closure. Thus, the JP preference indicates how an individual prefers to live in the outer world. If you are curious as to which of these applies to you, just think about the way you plan a vacation. Are you content to fly somewhere and then to take it from there, making plans as you go (P)? Or are you appalled by the thought of such a trip, preferring to schedule hotels, routes, stopovers, and so forth, well beforehand (J)? Do you find yourself taking in more and more information before finally writing that report—often at the eleventh hour (P)? Or do you plan it and work on it section by section, day by day (J)? As with all the pairs, both ways of living in the world are of course accessible to the individual. And even someone given to doing jobs at the last minute may find him- or herself having to be very much the schedule maker and planner in structuring family plans. So both choices are available. The dynamic interplay of all of the preferences (EI, SN, TF, JP) leads to sixteen combinations or types (see Figure 13-1).

Judging (J). (50 percent of the general population; 61 percent of engineering students)

- Prefers to live in a planned, orderly way.
- Likes to regulate and control events.

Perceptive (P). (50 percent of the general population; 39 percentof engineering students)

- Prefers to be flexible, spontaneous.
- Likes to understand and adapt to events.

Dominant and Auxiliary Processes .   According to type theory, children are born with a predisposed preference for some functions over others (Myers and McCaulley, 1985). Lynch (1987) maintains that the dominant function is usually reflected by kindergarten age. In engineering terms, they are hardwired for a given type. This preference leads to fuller development of the preferred function and greater competence in it. A preference for sensing, for example, leads to the development of characteristics commonly seen in a practical-minded sensing individual. At the same time, the opposite pole of the preference tends to be ignored; in the above example a sensing child gives less priority to intuition and thus develops along quite different lines from another child who prefers intuition. It is apparent then that environment ("software programming") plays a key role in one's development, either reinforcing or demotivating development along certain lines. This "falsification" of type can lead one to develop a less preferred function but overall still not feel in control or confident in his or her abilities. In good type development each person uses all four processes, but one process becomes the leading or dominant.

In the literature about type, the roles of the dominant and the auxiliary are often compared to those of a general and an aide. In an extrovert, the general (dominant function) is at the forefront making decisions and taking the lead, for all the world to see. As a result, we say that for an extrovert, "What you see is what you get." For an introvert, however, the aide (auxiliary function) stands as an intermediary with the outside world while the general makes plans inside a tent. The introvert, who focuses on the inner world, is difficult to know until one gets close enough to the individual. The dominant function remains hidden, which is why introverts are often misunderstood.

To see how the dominant and auxiliary functions are determined, consider an INFP and an ENFP. For the ENFP, the fourth pair (that is, the choice between judgment and perception which indicates how the person lives in the world), here the P, indicates that this person prefers to conduct his or her outer life in the perceptive mode. So we only have to look back to the perceiving slot (the second letter, here N, intuition) to find the function used by this person in the outer world. If asked to characterize this individual's type, another person would see the intuitive aspects. Now, by definition, extroverts show the world their strongest function; therefore, the N in this case is the dominant function. For the ENFP the dominant is extroverted intuition; the auxiliary is a balancing introverted feeling (F) (introverted because the auxiliary always balances the dominant, which here is extroverted), with thinking as the third and sensing as the fourth or least developed. So for an ENFP:

Dominant:	N
Auxiliary:	F
Third:	T
Fourth:	S

For an INFP the P indicates the person extroverts his or her perceptive function. Thus a judging function is dominant since an introvert's strength is within. The other perceptive function is third, and the fourth, or least developed, function, is judging (T). Thus,

Dominant:	F	(introverted feeling)
Auxiliary:	N	(extroverted intuition—what world sees)
Third:	S	(sensing)
Fourth:	T	(thinking—least developed function)

These individuals trust introverted feeling the most and use it the most in directing their lives, with intuition in support of the thinking. To the world, they appear intuitive. Like most introverts they are easily misunderstood because their strength is inside, not as open to the world as the strength of an extrovert.

Good type development.   Type development is seen as a lifelong process of increasing mastery or command over the functions of perception and judgment that one prefers, and corresponding but lesser development of the less interesting but essential processes. Myers and McCaulley summarize the process (1985):

• Development of excellence in the favorite, dominant process.

• Adequate but not equal development of the auxiliary for balance.

• Eventual admission of the least developed processes to conscious, purposeful use in the service of the dominant process, even though this use may require the dominant and auxiliary to temporarily relinquish control in consciousness so that the third or fourth function can become more conscious.

• Use of each of the functions for the tasks for which they are best fitted.

13.3. APPLICATIONS OF THE MBTI IN ENGINEERING EDUCATION

The differences described by type theory are familiar parts of everyday life, and so the theory can be used for a wide range of applications: education, counseling, career guidance, situations involving teamwork issues, and communication. Any university counseling or psychological center can provide the necessary testing services, or individuals can be certified through the training sessions such as those offered by the Association for Psychological Type (APT), the Center for Applications of Psychological Type (CAPT), or the Consulting Psychologists Press (see Section 13.6 for addresses). Thomas (1989) offers some preliminary results on "rapid MBTI self-classification."

Jensen and DiTiberio (1989) extensively examine its relevance in the teaching of writing. Provost and Anchors (1987) discuss the uses of the MBTI in higher education. McCaulley *et al.* (1983) consider the results of the ASEE-MBTI Engineering Consortium of eight universities (see Figure 13-2). Their results are summarized later in this section. In the MBTI manual Myers and McCaulley (1985) give numerous rankings of students and colleges by means of various preferences. Schurr, Ruble, and Henriksen (1989) look at the effects of different admissions practices on the MBTI and gender types. Several authors discuss the MBTI and problem solving, with McCaulley (1987) offering a jungian model. Yokomoto et al. (1987) discuss improvement of problem-solving performance and also consider student attitudes toward ethical dilemmas (1987). Three ethical dilemmas were presented to students, who were required to make a decision on what further action, if any, might be taken to resolve them. Analysis of the results showed several biases arising from personality differences, with feeling types recommending action more strongly than thinking types in one situation. Campbell and Kain (1990) investigated whether some types prefer certain forms of information presentation in problem solving. They found that the most time-efficient types (N and J) were also the least accurate, similarly for NT's and NF's. S, P, SF, and ST types tended to be more accurate but took longer to achieve their accuracy. Campbell and Kain conclude that type plays a small role in a person's preference for presentation form, but a larger role in the accuracy and time efficiency of problem solving.

Teaching Methods. Lawrence (1984) synthesizes learning style research involving the MBTI. The MBTI can be used to develop teaching methods to meet the needs of different types, especially on the sensing-intuition dichotomy. As McCaulley (1987) points out, S and N types approach problems from opposite directions: S moves from the specific to the general; N from the "grand design to the details." She then makes a telling point: "In fields with relatively equal numbers of S and N students, such as engineering, the faculty have more of a challenge maintaining student interest than in fields, such as counseling, where students and faculty are more similar" (p. 47). Smith, Irey, and McCaulley (1973) found that personality traits influence student attitude and performance in self-paced instruction. They further note that a major weakness in college teaching appears to arise from a teacher's and student's lack of recognition of each other's differences, which gives rise to the need for different learning activities. Self-paced instruction, according to the authors, can be made more effective if instructional modules or packages are designed which fit different styles of student perception and judgment. Provost, Carson, and Beidler (1987) studied a sample of professor of the year finalists to see how outstanding teachers use their type preferences. This limited study doesn't conclude that most outstanding teachers will have a certain preference or be a certain type; however, it does show that type affects teaching style, assumptions one might make about teaching, and attitudes about what aspects of teaching are seen as rewarding. These teachers have been able to relate to other types and to appreciate the inherent diversity. From the students' standpoint, Rodman et al. (1985, 1986) looked at the self-perception of engineering students' preferred learning style and related it to the MBTI. Among other conclusions, their work shows that in engineering education major differences among types occur in the sensing and intuition classifications.

The sensing-intuition (SN) dichotomy is perhaps the most important one for an engineering educator, both from the standpoint of the instructor and from that of the students (especially as sensing relates to mastering a body of knowledge and the corresponding skills central to a field of practice). Intuition has to do with the ability to think complexly and contextually. The percentages of type in the general and university populations alone tell a significant tale. Sensing types predominate in the general population; intuitives, in a university environment. More college professors are intuitive types than sensing types, and they tend to write exams that more frequently fit their own type (Lynch, 1987). If memorization and recall are important, sensing and judging types will perform better; if hypothesizing and essay tests are required, intuitive students will have an advantage. Aptitude tests are also designed to measure knowledge in the domain of introverted intuitives (IN). The data show that introverts consistently score higher than extroverts on the SAT-Verbal. Intuitives also consistently score higher than sensing types. The sensing-intuition differences, according to Myers and McCaulley (1985), are greater than the extroversion-introversion differences.

In the classroom, the thinking-feeling (TF) preference appears to have less importance than the others, but it can be argued that a predominance of thinking types in a class could "freeze out" the few feeling types. Is it possible that feeling types self-select out of engineering because of the more impersonal emphasis of the predominant thinking types in engineering? One colleague has suggested that it might be easier to teach ethics if students were more interested in human motivations (feeling types), rather than being concerned with building the best device (ST) or developing the most elegant theory (NT). Finally, it is important to remember that type theory does not make judgments on intelligence: All types can succeed in any area, and all types are represented in every area. What is important is that every type can learn to survive in the academic world. Paying attention to type differences and taking them into account in teaching goes a long way toward promoting such success. The fact that certain types predominate in certain careers says more about a type's attraction to the field than whether he or she will succeed in it. Once an individual has gotten past the educational barriers to a given field, being different from the prevailing type can be an advantage since he or she will see things that others miss.

Motivation. The MBTI can also be used to help students if the instructor understands the ways that different types are motivated. An instructor can help students gain control over their own learning and thereby reach more students. Even something as simple as a phrasing can be important. For example, feeling types respond better to a question that is phrased "How do you feel about . . . ?" whereas the thinking type prefers "What do you think . . . ?" Also, the quickness of the N types may discourage an S type, and in a classroom the quicker student is often more praised and honored; the "slower" student quickly forms an impression that he or she is lacking what the "best" students have. We use quotation marks to indicate that intelligence is not the consideration here. In the long run, the "slower" but more thorough and accurate S may be more correct and/or successful. And if not demotivated by the instructor, such a student may be a valuable addition to the class.

Curriculum and Materials. The MBTI can be used to analyze curricula, methods, media, and materials in light of the needs of different types. This should be done in conjunction with the other aspects of learning theories, such as that of Kolb (see Chapter 15). And it can be used to provide extracurricular activities that will meet the needs of all types

Interpersonal Relationships. The MBTI can also be used to help teachers and administrators work together more constructively. Type data from one sample show that administrators tend to be heavily J types [86 percent in Lawrence's (1984) sample]. As in other areas, such as personnel cases in industry, awareness of type differences can lead to a more harmonious working environment. On a personal level, knowledge of type can be very helpful in counseling (Provost and Anchors, 1987). Carey, Hamilton, and Shanklin (1985) use type theory to look at the relationship between communication style and roommate satisfaction. The differences between judging types and perceptive types can often lead to conflicts. What a perceptive sees as a strength in the desire to have complete information or knowledge before proceeding, a judging person often sees as procrastination. And what a judging type sees as decisive action, a perceptive may see as close-minded and precipative behavior. Differences on the extroversion-introversion and thinking-feeling dichotomies can also lead to problems. From type theory, interpersonal competence is related to extroversion and feeling (Myers and McCaulley, 1985). The focus of extroverts is on people and the external world; that of feeling types is on the effects of their actions and decisions on themselves and others.

In engineering, a great deal of work is done in teams. Clearly, it's important that the members work together harmoniously. A good preparation for this takes place in undergraduate laboratories. Accounting for type differences and making students aware of each other's different strengths can go a long way toward easing the tension that arises when, say, a perceptive can't put an end to a literature search which his or her judging partner needs for the next day's oral report. Giving and receiving criticism in these situations can also depend on the individual's preferred way of functioning.

Student Retention. Retention and attrition are complex issues which every college or university must face. Godleski (1987) considers use of the MBTI to increase retention of underachieving college students. His preliminary results showed that there was no difference in extroversion or introversion, but a significantly larger number of sensing over intuitive types and perceptive over judging types who were in academic difficulty. Provost (1991) found type patterns among freshmen experiencing first-year difficulties, with analyses showing overrepresentation of TP combinations. Schurr and Ruble (1988) found that high school performance and the judging preference (J) were the best predictors of college performance. This report was a follow-up to their 1986 study which followed an entire entering college class. McCaulley (1976) describes a study at the Fenn College of Engineering at Cleveland State University comparing freshman who wanted to become engineers and seniors who successfully completed the program. The types in the four corners (the TJ or logical, decisive types) of Figure 13-1 increased their percentage from 45 percent as freshmen to 55 percent as seniors. The group showing the greatest loss, from 17.3 percent as freshmen to 9.2 percent as seniors, was the types sharing intuition and feeling (more frequently found in the behavioral sciences and communication). McCaulley offers some reasons for this pattern (p. 397):

1 People learn in different ways. If the faculty teaches one way, they will favor some types over others.

2 Faculty members serve as role models for students, but there appear to be no data to indicate that engineering faculty are appropriate models for engineers in industry. Students may not realize this.

3 Choice of textbooks (and programmed learning courses) can favor the learning pattern of some types and cause difficulties for others.

Staiger (1989) also uses type to identify subsets of electrical engineering students who need special attention from the point of view of retention and maintains that curriculum redesign should include teaching styles that will accommodate diverse learning styles, including guided design, cooperative learning, and developmental instruction. Kalsbeek (1987) offers a conceptual model for understanding student attrition. His comment offers an appropriate close to this section: By relating type data to student attrition, educators can consider how different types of students interact with types of academic environments and thereby respond appropriately to the challenges posed.

Distributions of Types in Engineering   As Figure 13-1 indicates, all sixteen types are represented in all areas of engineering; however, even the quickest glance reveals that certain types self-select into and are retained very markedly in engineering. For example, the corners of the table are strongly over-represented—what has come to be called the "tough-minded"

FIGURE 13-1   DISTRIBUTION OF THE 16 MBTI TYPES AMONG ASEE-MBTI ENGINEERING STUDENTS
(McCaulley, 1990) (© 1990, American Society for Engineering Education)

	ISTJ	ISFJ	INFJ	INTJ
Male	17.39	4.19	2.58	9.95
Female	12.22	6.48	3.53	7.07
Aerospace	20.18	4.39	3.51	20.18
Chemical	17.53	4.04	3.37	9.89
Civil	22.87	5.04	1.55	3.88
Computer	11.82	4.93	1.97	7.88
Electrical	19.10	3.64	2.33	12.54
Geological	10.78	2.94	4.90	11.27
Mechanical	16.80	4.44	1.93	10.62
Petroleum	16.30	3.26	0.54	7.07

	ISTP	ISFP	INFP	INTP
Male	6.83	2.45	3.90	9.24
Female	3.24	3.39	4.12	4.86
Aerospace	2.63	0.88	1.75	7.89
Chemical	5.17	1.35	4.49	7.64
Civil	5.81	4.65	2.33	4.65
Computer	3.94	1.48	6.40	6.90
Electrical	5.54	1.60	3.79	10.20
Geological	7.84	2.45	9.31	10.78
Mechanical	9.46	2.32	2.12	6.76
Petroleum	4.89	2.17	5.43	9.24

	ESTP	ESFP	ENFP	ENTP
Male	4.57	2.16	3.16	7.47
Female	2.21	3.24	6.48	7.22
Aerospace	0.88	1.75	3.51	6.14
Chemical	2.25	2.70	4.04	8.54
Civil	4.26	1.94	5.43	2.71
Computer	3.45	2.46	3.45	8.37
Electrical	3.79	1.75	2.04	5.83
Geological	2.45	0.98	2.94	11.76
Mechanical	5.02	2.32	1.54	6.37
Petroleum	6.52	2.17	6.52	7.61

	ESTJ	ESFJ	ENFJ	ENTJ
Male	12.56	2.80	1.71	9.05
Female	13.55	7.22	3.98	11.19
Aerospace	8.77	0.88	1.75	14.91
Chemical	11.01	3.37	1.57	13.03
Civil	20.93	3.88	2.71	7.36
Computer	16.26	7.88	1.97	10.84
Electrical	15.01	2.33	2.19	8.31
Geological	7.35	3.43	3.92	6.86
Mechanical	15.25	3.67	1.35	10.04
Petroleum	10.87	4.89	3.26	9.24

Note: Numbers preceding bar graphs represent the percent of the sample falling in that type. In bar graphs one inch represents 20% of sample. If percentage exceeds 20%, a + follows the bar.

TJ types; what is missing, relatively, is participation by the feeling types (the two inner columns in the type table). McCaulley (1990) raises the question "Are engineering schools preparing their students adequately for the 'people complexities' of the profession?" That feeling types are in such a minority may indicate that the answer is no. It is these groups which tend to drop or transfer out of engineering. One speculation worth exploring is whether underrepresented groups in engineering, such as minorities and women, tend to fall into type categories that are only slightly present in Figure 13-1. If it is true, as the data to date indicate, that women classify more as being F than T, it could be that they view the heavy T orientation of engineering as cold and unfriendly. If the ranks of engineering are to be filled in the future, and clearly the standard pool of potential engineering candidates of the past is dwindling, it is these groups that educators will have to look to and encourage.

The report of McCaulley *et al.* (1983) with the ASEE-MBTI Engineering Consortium provides data showing the breakdown of engineering students by type preference. Among their results were the following.

1 Engineering students markedly prefer thinking (74 percent) judgment and judging (61 percent) judgment, with the stereotypical engineer falling into the TJ group. In the consortium data, this group accounted for almost half of the sample (males, 49 percent; females, 44 percent; with the males more often introverted, 56 percent). One would expect all majors to have the same proportion of each type (with 25 percent in each group if the distribution was equal); the fact that the opposite is true gives evidence to the usefulness of the theory.

2 Engineering students differ from other college students. Compared with a sample group of college freshmen, engineering students are more often introvert, thinking, and judging types. Only on the SN scale were they similar to their peers.

3 Male engineering students differ in type from female engineering students. In engineering, 77 percent of the males were T, whereas 61 percent of the females were T. The proportions of S and N were about the same.

4 Engineering disciplines attract different types of students. Figure 13-1 bears this out as well. The fields with the highest proportion of extroverts were industrial (56 percent), computer (55 percent), petroleum (51 percent) and mineral (51 percent). Introverts were more frequent in aerospace (61 percent), geological (60 percent) and electrical (59 percent) engineering. The fields with the highest proportion of the practical sensing types were civil (69 percent), industrial (61 percent), mechanical (61 percent), and mining (60 percent). Intuitives were frequent in geological (62 percent), aerospace (60 percent) and metallurgical (54 percent). As noted above, all fields had a majority of T types, with the highest proportions in aerospace (82 percent), electrical (80 percent), mechanical (80 percent) and physics (76 percent). The fields with the lowest proportion of T types were undecided students (68 percent), geological (69 percent), computer (69 percent) and general (70 percent) engineering.

5 All types survived to year two, but atypical types had lower retention rates. Judging types were slightly, but significantly, more likely to be retained (entering students were 63 percent J, retained were 65 percent J, p < 0.01). The practical SJ types were 34 percent of entering students but were 40 percent of those remaining. Note that there were no differences in retention between male and female, and feeling types were as likely to persist as their more analytical T counterparts.

Implications of consortium study   The implications drawn from the consortium study merit serious consideration (McCaulley *et al.*, 1983).

1 Clearly, and one might say expectedly, many logical, analytical, and decisive types of students are drawn to engineering; however, overemphasis on these characteristics tends to result in an underemphasis on skills related to listening, understanding, and getting things done through people. Since a great part of engineering work depends on communication and teamwork, it is important that faculty stress the importance of these skills and even teach them specifically (which of course will be appreciated by the extroverted, feeling, and intuitive types).

2 Less typical engineering students, extroverts and feeling types, learn better if given frequent feedback and appreciation; unfortunately, the types which are attracted to the field are the least likely to give such feedback. So it is up to the faculty to teach and model such behavior, which in turn will encourage students to do the same in their own work. Type knowledge can also help in identifying behavioral patterns and needs that may be beneficial in advising students (see also Lynch, 1987).

3 Since the numbers of sensing (S) and intuitive (N) types are roughly equal, the implication, though debatable, is that half of the students learn best deductively, and about half, inductively. The sensing types benefit from clear instructions, starting with their practical experiences, and with new material presented with a step-by-step approach. Intuitives, on the other hand, prefer theoretical principles first, followed by mastery of details through problem solving. In order to reach the greatest number of students effectively, instructors must keep these differences in mind. Tests and other measures of evaluation should also be varied so that different types are given a fair chance.

13.4. DIFFICULTIES WITH PSYCHOLOGICAL TESTING

The MBTI is prone to the same kinds of problems that plague any psychological test:

1 A student may not understand a question because of phrasing or vocabulary. Though not likely applicable for an engineering student population, the MBTI requires at least eighth grade language skills (for children, the Murphy-Meisgeier Type Indicator is used for grades two through eight).

2 The wrong box may accidentally be marked.

3 Students may mark what they feel they "ought" to think or may try to "psych-out" the tester. Unconscious biases may also affect the results (Hammer, 1985).

4 Current environmental stress may change one's answers temporarily.

5 Results may be misinterpreted. With the MBTI a little learning can be a dangerous thing, for it's easy to turn the occasion into a parlor game and make it little more than a horoscope reading. Accurate interpretation is assured if a qualified tester is present such as a psychologist, a counselor, or someone certified to administer the MBTI.

6 Reliability. The MBTI is reliable, but people can change. Times of stress may lead to differing results, and over a period of years growth may be reflected in a change of type. However, such changes are expected and predicted within type theory. For example, as one enters middle age, it's common for compensatory development to occur in the less preferred functions (Myers and McCaulley, 1985). Although one's type doesn't change, the way it is experienced and reported may change and give different MBTI results. Seventy-five percent of people who have retaken the MBTI after one to six years have not changed or have done so in only one category. More information can be found in the reliability studies reported in Chapter 10 of the manual by Myers and McCaulley (1985). Hammer and Yeakley (1987) conducted a study to investigate the relationship between "true" type and reported type.

7 Validity. The MBTI has good face validity. The results seem true to the test taker. Does the MBTI measure what it is trying to measure? This is a problem with all psychological tests: What they try to measure is usually based on a psychological theory. Thus, is the underlying theory valid? If it is, does the test accurately measure this? Chapter 11 in the manual discusses more than 100 studies relating to validity. There is also high face validity when one person types another person whom they know well (Carlson, 1989).

13.5. CONCLUSIONS

The Myers-Briggs Type Indicator offers engineering educators a workable instrument with which to meet the changing needs of engineering education. Measuring preferences as indicated by the students themselves, it is not meant to measure the strength of a trait, as other psychological instruments do. Consequently, it is fairly simple to implement and interpret without requiring a staff psychologist within an engineering department. Attention to differences also makes tremendous common sense as the diverse needs of a new population of students must be met before they can succeed in engineering. We can increase participation in the field as well as increase productivity. Quite possibly, as McCaulley (1990) points out, use of the indicator may help move students toward greater maturity of cognitive development in Perry's model (see Chapter 14). Finally, to stress that engineering educators must acknowledge that students learn differently, Staiger (1989) concludes: "It would help to have the phrase 'equal opportunity for learning' included in all university admission statements as a constant reminder" (p. 143).

13.6. CHAPTER COMMENTS

There is much to consider in the areas of student types, development, and learning theory. And all interact, thereby further complicating an already difficult equation. In such a complex area, each theory looks only at a small part. The Myers-Briggs Type Indicator offers an excellent starting point for looking at the differences between and among students. More information on the MBTI can be obtained through the organizations primarily involved in its development and dissemination.

> Association for Psychological Type
> 9140 Ward Parkway
> Kansas City, MO 64114
> Center for Applications of Psychological Type
> 2720 N.W. 6th St.
> Gainesville, FL 32609
> Consulting Psychologists Press
> 3803 East Bayshore Road
> P.O. Box 10096
> Palo Alto, CA 94303

13.7. SUMMARY AND OBJECTIVES

After reading this chapter you should be able meet the following objectives.

1 Describe briefly the background of the Myers-Briggs Type Indicator and its development from the ideas of Carl Jung.

2 Discuss the attitudes and functions of conscious thought and how the four preferences interrelate to form the matrix of sixteen personality types.

3 Consider your own preferences in terms of the descriptions for each attitude and function, arriving at a rough estimate of your own type.

4 Explain the importance of considering type differences among engineering students in the development of instructional materials, tests, and evaluations.

HOMEWORK

1 Consider the process of selecting an adviser for graduate work.
  a How close a match between student and adviser is necessary?
  b Which are more important: EI, SN, TF, JP?
  c How can you figure out the professor's type?
2 Think back to your undergraduate courses. Recall a particular teacher who wasn't fully effective because of a preference for one type over another in his or her approach to teaching. What could this person have done to improve his or her teaching?
3 Determine the dominant, auxiliary, third, and fourth functions for the following types:
  a ESFJ
  b ISFJ

REFERENCES

Campbell, D. E. and Kain, J. M., "Personality type and mode of information presentation: Preference, accuracy and efficiency in problem solving," *J. Psychol. Type,* 20, 47 (1990).

Carey, J. C., Hamilton, D. L., and Shanklin, G., "Psychological type and interpersonal compatibility: Evidence for a relationship between communication style preference and relationship satisfaction in college roommmates," *J. Psychol. Type,* 10, 36 (1985).

Carlson, J. G., "Affirmative: In support of researching the Myers-Briggs Type Indicator," *J. Couns. Develop.,* 67, 484 (April 1989).

von Franz, M.-L. and J. Hillman, *Jung's Typology,* Spring Publications, Dallas, TX, 1971.

Godleski, E. S., "Learning style compatibility of engineering students and faculty," *Proceedings ASEE/ IEEE Frontiers in Education Conference,* IEEE, Philadelphia, PA, 1984.

Godleski, E. S., "Using the MBTI to increase retention of engineering students," *Proceedings ASEE Annual Conference,* ASEE, Washington, DC, 333,1987.

Hammer, A.L., "Typing or stereotyping? Unconscious bias in applications of psychological type theory," *J. Psychol. Type,* 10, 14 (1985).

Hammer, A.L., and Yeakley, F. R., Jr., "The relationship between 'true type' and reported type," *J. Psychol. Type,* 13, 52 (1987).

Harrisberger, L., "Capstoning with the MBTI," *Proceedings ASEE Annual Conference,* ASEE, Washington, DC, 330, 1987.

Jensen, G. H., and DiTiberio, J. K., *Personality and the Teaching of Composition,* Ablex Publishing, Norwood, NJ, 1989.

Jung, C. G., *Memories, Dreams, Reflections,* rev. ed. (recorded and edited by Aniela Jaffe, translated from the German by Richard and Clara Winston), Pantheon Books, New York, 1961.

Jung, C. G., *Psychological Types,* Princeton University Press, Princeton, NJ, 1971 (originally published in 1921).

Kalsbeek, D., "Campus retention: The MBTI in institutional self-studies," in Provost, J., and S. Anchors (Eds.), *Applications of the Myers-Briggs Type Indicator in Higher Education,* Consulting Psychologists Press, Palo Alto, CA, 1987, pp. 31-63.

Lawrence, G., *People Types and Tiger Stripes: A Practical Guide to Learning Styles,* 2nd ed., Center for Applications of Psychological Type, Gainesville, FL, 1982.

Lawrence, G., "A synthesis of learning style research involving MBTI," *J. Psychol. Type,* 8, 2 (1984).

Lynch, A. Q., "Type development and student development," in Provost, J., and S. Anchors (Eds.), *Applications of the Myers-Briggs Type Indicator in Higher Education,* Consulting Psychologists Press, Palo Alto, CA, 1987, pp. 5–29.

McCaulley, M., "Psychological types in engineering: Implications for teaching," *Eng. Educ.,* 66, 729 (April 1976).

McCaulley, M., *Applications of the Myers-Briggs Type Indicator to Medicine and Other Health Professions,* Center for Applications of Psychological Type, Gainesville, FL, 1978.

McCaulley, M. H., Godleski, E. S., Yokomoto, C. F., Harrisberger, L., and Sloan, E. D., "Applications of psychological type in engineering education," *Eng. Educ.,* 394 (Feb. 1983).

McCaulley, M. H., Macdaid, G. P., and Kainz, R. I., "Estimated frequencies of the MBTI types," *J. Psychol. Type,* 9, 3 (1985).

McCaulley, M., "The Myers-Briggs Type Indicator: A jungian model for problem solving," in Stice, J. (Ed.), *Developing Critical Thinking and Problem-Solving Abilities,* Jossey-Bass, San Francisco, CA, 1987.

McCaulley, M., "The MBTI and individual pathways in engineering design," *Eng. Educ.,* 537 (July/ Aug. 1990).

Myers, I. B. and P. B. Myers, *Gifts Differing*, Consulting Psychologists Press, Palo Alto, CA, 1980.

Myers, I. B. and McCaulley, M. H., *Manual: A guide to the development and use of the Myers-Briggs Type Indicator*, Consulting Psychologists Press, Palo Alto, CA, 1985.

Myers, I. B., *Introduction to Type*, Consulting Psychologists Press, Palo Alto, CA, 1991.

Provost, J., and S. Anchors (Eds.), *Applications of the Myers-Briggs Type Indicator in Higher Education*, Consulting Psychologists Press, Palo Alto, CA, 1987.

Provost, J. A., Carson, B. H., and Beidler, P. G., "Teaching excellence and type," *J. Psychol. Type*, 13, 23 (1987).

Provost, J., "Tracking freshman difficulties in the class of 1993," *J. Psychol. Type*, 21, 35 (1991).

Rodman, S. and Dean, R. K., "Personality type as it relates to problem-solving skills," *Proceedings ASEE/IEEE Frontiers in Education Conference*, IEEE, NY, 36, 1983.

Rodman, S. and Dean, R. K. and Rosati, P. A., "Learning styles among engineering students: Self report vs. classification by MBTI," *Proceedings ASEE/IEEE Frontiers in Education Conference*, IEEE, New York, 48, 1985.

Rodman, S. and Dean, R. K. and Rosati, P. A., "Self-perception of engineering students' preferred learning style related to MBTI type," *Proceedings ASEE/IEEE Frontiers in Education Conference*, IEEE, New York, 1303, 1986.

Saunders, F. W., *Katherine and Isabel: Mother's Light, Daughter's Journey*, Consulting Psychologist's Press, Palo Alto, CA, 1991.

Schmeck, R. and Lockhart, D., "Introverts and extraverts require different learning environments," *Educ. Leadership*, 54–55 (Feb. 1983).

Schurr, K. T. and Ruble, V., The Myers-Briggs Type Indicator and first year college achievement: A look beyond aptitude test results, *J. Psychol. Type*, 12, 25 (1986).

Schurr, K. T. and Ruble, V., "Psychological type and the second year of college achievement: Survival and the gravitation toward appropriate and manageable major fields," *J. Psychol. Type*, 14, 57 (1988).

Schurr, K. T., Ruble, V., and Henriksen, L. W., "Effects of different university admission practices on the MBTI and gender composition of a student body, graduation rate, and enrollment in different departments," *J. Psychol. Type*, 18, 24 (1989).

Singer, J., *Boundaries of the Soul: The Practice of Jung's Psychology*, Doubleday, New York, 1972.

Smith, A. B., Irey, R. K., and McCaulley, M. H., "Self-paced instruction and college student personalities," *Eng. Educ.*, 436 (March 1973).

Staiger, E. H., "Curriculum redesign: Psychological factors worth considering," *Proceedings ASEE/IEEE Frontiers in Education Conference*, IEEE, NY, 140, 1989.

Thomas, C. R., "Rapid MBTI self-classification," *Proceedings ASEE Anual Conference Conference*, ASEE, Washington, DC, 1074, 1989.

Yokomoto, C., "How the MBTI has influenced how I teach," *Proceedings ASEE Annual Conference*, ASEE, Washington, DC, 324, 1987.

Yokomoto, C. F., Buchanan, W., and Ware, R., "A preliminary study of student attitudes toward ethical dilemmas," *Proceedings ASEE/IEEE Frontiers in Education Conference*, IEEE, New York, 154, 1987.

Yokomoto, C., "Personality models/styles," *Proceedings ASEE Annual Conference*, ASEE, Washington, DC, 986, 1989.

APPENDIX 13A. MBTI MODEL FOR PROBLEM SOLVING

The goals in using the MBTI model of problem solving are to improve the problem solving skills of students and to help them gain respect for others whose minds work differently from their own. The following is a brief and simplified overview of Myers' problem-solving model (Myers, 1991; McCaulley, 1987).

The strategy is to use one process at a time and to use it in its own area. Don't, for example, use sensing for seeking new possibilities or feeling to analyze an equipment problem.

1 Use sensing (**S**) to face the facts, to be realistic, to find what the situation is, to see your actions, and to see other people's actions. Do not let wishful thinking or sentiment blind you to the realities.

2 Use intuition (**N**) to discover all the possibilities, to see how you might change the situation, to see how you might handle the situation differently, and to see how other people's attitudes might change. Try not to assume that you have been doing the only obviously right thing.

3 Use thinking (**T**) to make an impersonal analysis of the problem; to look at causes and their effects; to look at all the consequences, both pleasant and unpleasant; to count the full costs of possible solutions; and to examine misgivings you may have been suppressing because of your loyalties to others or because you don't like to admit you may have been wrong.

4 Use feeling (**F**) to weigh how deeply you care about what your choice will gain or lose; to put more weight on permanent than on temporary effects, even if the temporary effects are more attractive right now; to consider how other people will feel, even if you think they are unreasonable; and to weigh other people's feelings and your own feelings in deciding which solution will work.

It is likely, and natural, that the individual will choose a solution that appeals to his or her favorite process, but such a solution will be more effective or successful if the facts, possibilities, consequences, and human values are considered. What can go wrong if any of these are ignored? Intuitives may base a decision on some possibility without discovering facts which may preclude the conclusion. Sensing types may settle for a faulty solution because they assume none better is possible. Thinking types may ignore human values. Feeling types may ignore consequences. Thus, what can make the process difficult is that the problem solver is asked to use strengths opposite to his or her own.

Using the attitudes:

1 Use extroversion (**E**) to see events in the environment that may influence the problem, to seek people who may have information about the problem, and to talk out loud about the problem as a way of clarifying the ideas.

2 Use introversion (**I**) to consider ideas that may have a bearing on the problem, to look for deeper truths that may be obscured by current fads, to take time to think alone deeply about the problem.

3 Use judgment (**J**) to stay on track and not be diverted, to plan ahead, and to push yourself and others toward a solution.

4 Use perception (**P**) to ensure that you have looked at all aspects of the problem, to keep your eyes open to new developments, and to avoid jumping to conclusions before all the facts are in.

CHAPTER 14

MODELS OF COGNITIVE
DEVELOPMENT: PIAGET AND PERRY

We will focus on the two theories of development which have been the most influential in the education of scientists and engineers: Piaget's theories of childhood development and Perry's theory of development of college students. To some extent they are complementary as both focus on different aspects of development, and since both Piaget and Perry discuss how students learn, this material ties in with Chapter 15.

These theories are important since they speak to what we can teach students and to where we want students to be when they graduate. Both theories postulate that students cannot learn material if they have not reached a particular level of development. Attempts to teach them material which they are unable to learn leads to frustration and memorization. As engineering students become more heterogeneous, the levels of student development in classrooms will also become more heterogeneous. Thus, it is becoming increasingly important to understand the levels at which different students function.

14.1. PIAGET'S THEORY

Jean Piaget was a Swiss psychologist whose research on the development of children has profoundly affected psychological theories of development and of the teaching of children. His theory has also been widely studied for its application to the teaching of science in grade school, high school, and college. Unfortunately, Piaget's writings tend to be somewhat obscure. We will present a significantly edited version focusing on those aspects of his theory which affect engineering education. Further information is available in Flavell (1963), Gage and Berliner (1984), Goodson (1981), Inhelder and Piaget (1958), Phillips (1981), Piaget (1950, 1957), and Pavelich (1984).

14.1.1. Intellectual Development

Piaget's theory conceives of intellectual development as occurring in four distinct periods or stages. Intellectual development is continuous, but the intellectual operations in the different periods are distinctly different. Children progress through the four periods in the same order, but at very different rates. The stages do not end abruptly but tend to trail off. A child may be in two different stages in different areas.

The sensimotor period, which is only of indirect interest to our concerns, extends from birth to about two years of age. In this period a child learns about his or her relationship to various objects. This period includes learning a variety of fundamental movements and perceptual activities. Knowledge involves the ability to manipulate objects such as holding a bottle. In the later part of this period the child starts to think about events which are not immediately present. In Piaget's terms the child is developing meaning for symbols.

The preoperational period lasts from roughly two to seven years of age. Piaget has divided this stage into the preoperational phase and the intuitive phase. In the preoperational phase children use language and try to make sense of the world but have a much less sophisticated mode of thought than adults. They need to test thoughts with reality on a daily basis and do not appear to be able to learn from generalizations made by adults. For example, to a child riding a tricycle the admonition "Slow down, you are going too fast" probably has no effect until the child falls over. This continual testing with reality helps the child to understand the meaning of "too fast." Compared to adults, the thinking of a child in the preoperational phase is very concrete and self-centered. The child's reasoning is often very crude, and he or she is unable to make very simple logical extensions. For example, the son of one of the authors was astounded when he heard that his baby sister would be a girl when she got older!

In the intuitive phase the child slowly moves away from drawing conclusions based solely on concrete experiences with objects. However, the conclusions drawn are based on rather vague impressions and perceptual judgments. At first, the conclusions are not put into words and are often erroneous (and amusing to adults). Children are perception-bound and often very rigid in their conclusions. Rational explanations have no effect on them because they are unable to think in a cause-and-effect manner. During this phase children start to respond to verbal commands and to override what they see. It becomes possible to carry on a conversation with a child. Children develop the ability to classify objects on the basis of different criteria, learn to count and use the concept of numbers (and may be fascinated by counting), and start to see relationships if they have extensive experience with the world. Unaware of the processes and categories that they are using, children are still preoperational. Introspection and metathought are still impossible.

At around age seven (or later if the environment has been limited) the child starts to enter the concrete operational stage. In this stage a person can do mental operations but only with real (concrete) objects, events, or situations. He or she can do mental experiments and can correctly classify different objects (apples and sticks, for example) by some category such as size. The child understands conservation of amounts. This can be illustrated with the results of one of Piaget's experiments (Pavelich, 1984). Two identical balls of clay are shown to a child who agrees they have the same amount of clay. While the child watches, one ball is flattened. When asked which ball has less clay, the preoperational child answers that the flattened ball has less clay. The concrete operational child is able to correctly answer this question. He or she becomes adept at addition and subtraction but can do other mathematics only by rote. In the concrete operational stage children also become less self-centered in their perceptions of the universe. Logical reasons are understood. For example, a concrete operational person can understand the need to go to bed early when it is necessary to rise early the next morning. A preoperational child, on the other hand, does not understand this logic and substitutes the psychological reason, "I want to stay up."

Piaget thought that the concrete operational stage ended at age eleven or twelve. There is now considerable evidence that these ages are the earliest that this stage ends and that many adults remain in this stage throughout their lives. Most current estimates are that from 30 to 60 percent of adults are in the concrete operational stage (Pintrich, 1990). Thus, many college freshmen are concrete operational thinkers; however, the number in engineering is small and is probably less than 10 percent (Pavelich, 1984). For reasons which will become clear shortly, concrete operational thinkers will have difficulty in an engineering curriculum. However, these people can be fully functioning adults. Piaget's theories at the concrete and formal operational stages measure abilities only in a very limited scientific, logical, algebraic sense. His theories do not address ethical or moral development. Thus a person may be a successful hard worker, a good, loving parent and spouse, and a good citizen, but be limited to concrete operational thought.

The final stage in Piaget's theory is the formal operational stage, which may start as early as age eleven or twelve, but often later. A formal operational thinker can do abstract thinking and starts to enjoy abstract thought. This person becomes inventive with ideas and starts to delight in such thinking. He or she can formulate hypotheses without actually manipulating concrete objects, and when more adept can test the hypotheses mentally (Phillips, 1981). This testing of logical alternatives does not require recourse to real objects. The formal operational thinker can generalize from one kind of real object to another and to an abstract notion. In the experiment with the balls of clay, for example, the formal thinker can generalize this to sand or water and then to a general statement of conservation of matter. This person is capable of learning higher mathematics and then applying this mathematics to solve new problems. When faced with college algebra or calculus the concrete operational thinker is forced to learn the material by memorization but then is unable to use this material to solve unusual problems. The formal operational thinker is able to think ahead to plan the solution path (see Chapter 5 for a further discussion of problem solving) and do combinatorial thinking and generate many possibilities. Finally, the formal operational person is capable of metacognition, that is, thinking about thinking.

14.1.2. Application of Piaget's Model to Engineering Education

The importance of the formal operational stage to engineering education is that engineering education requires formal operational thought. Many of the 30 to 60 percent of the adult population who have some trouble with formal operational thought appear to be in a *transitional* phase where they can correctly use formal operational thought some of the time but not all of the time. Engineering students in transition appear to be able to master engineering material (Pavelich, 1984). This probably occurs because they have learned that formal operational thought processes must be used in their engineering courses, but they have not generalized these processes to all areas of their life. This domain specificity of many students is one of the major criticisms of Piaget's theory (Pintrich, 1990).

The relatively small number of engineering students who are in the concrete operational stage will have difficulties in engineering. These students may make it through the curriculum by rote learning, partial credit, doing well in lab, repeating courses, and so forth. Concrete operational students can be identified by repeated administration of tests with novel problems on the same material (Wankat, 1983). On the first few tests students may be unable to work the problem either because of lack of knowledge or because of an inability to solve abstract problems. On the basis of a single test it is difficult to tell if lack of knowledge or poor problem-solving ability has caused the difficulties. Students who can use formal operational thinking learn from their mistakes, learn the missing knowledge, and fairly rapidly become able to solve difficult new problems. Students who are in the concrete operational stage do not appear to be able to learn from their mistakes on problems requiring formal operations. Thus, they make the same mistakes over and over. The solutions of these students do not appear to follow any logical pattern since they often just try something (anything) to see if it works and to see if they get any partial credit. These students have great difficulty in evaluating their solutions. In engineering, concrete operational students are likely to be quite frustrated and frustrating to work with.

The suggestion has been made repeatedly that freshmen-sophomore courses in engineering should be made available for nonengineering students (e.g., Bordogna, 1989). If this were done, the much higher percentage of concrete operational students in the general student population would likely cause problems in the course unless some type of screening or self-selection takes place.

14.1.3. Piaget's Theory of Learning

The presence of some concrete operational students in engineering leads us naturally to the question of how a student moves from one stage to another. This is another aspect of Piaget's theories. Piaget postulates that there are *mental structures* that determine how data and new information are perceived. If the new data make sense to the existing mental structure, then the new information is incorporated into the structure (*accommodation* in Piaget's terms). Note that the new data do not have to exactly match the existing structure to be incorporated into the structure. The process of accommodation allows for minor changes (figuratively, stretching, bending and twisting, but not breaking) in the structure to incorporate the new data. If the data are very different from the existing mental structure, it does not make any sense to incorporate them into the structure. The new information is either rejected or the information is *assimilated* or *transformed* so that it will fit into the structure. A concrete person will probably reject a concept requiring formal thought. If forced to do something with the data he or she will memorize even though the meaning is not understood. This is similar to memorizing a passage in a foreign language that one cannot speak. An example of transformation is a person's response to seeing a pink stoplight. Everyone "knows" that stoplights are red, and thus the pink stoplight will probably be registered as being red since red stoplights fit one's mental structure.

How does one develop mentally? How does one make the quantum leap from concrete to formal thinking? Mental development occurs because the organism has a natural desire to operate in a state of equilibrium. When information is received from the outside world which is too far away from the mental structure to be accommodated but makes enough sense that rejecting it is difficult, then the person is in a state of *disequilibrium*. The desire for equilibration is a very strong motivator to either change the structure or reject the data. If the new information requires formal thinking and the person is otherwise ready, then a first formal operational structure may be formed. This formal operational structure is at first specific for learning in one area and is slowly generalized (the person is in a transitional phase). The more often the person receives input which requires some formal logic, the more likely he or she is to make the jump to formal operational thought. Since this input takes place in a specific area, the transition to formal operations often occurs first in this one area. Also, a person with a less rigid personality structure and tolerance for ambiguity is probably more likely to make the transition. We emphasize that the transition to formal operations may not be easy.

Piaget developed a variety of experiments to test what stage children were in and to help them learn to make the transition to the next stage. Unfortunately, the experiments work well for testing the stage but not for moving people to the next stage. A method called the scientific *learning cycle* has been developed to help students in their mental development (Renner and Lawson, 1973; Lawson et al., 1989). In the scientific learning cycle the students are given first-hand experience, such as in a laboratory with an attempt to cause some disequilibration. The instructor then leads discussions either with individuals or in groups to introduce terms and to help accommodate the data and thus aid equilibration. Finally, students make further investigations or calculations to help the changed mental structure fit in with the other mental structures (organization). The scientific learning cycle is successful at helping people move to higher stages, but progress is very slow. Since concrete operational students may try hard but still have great difficulty in understanding abstract logic, the use of words like "obviously," "clearly," or "it is easy to show" by the professor is frustrating and demotivating to them. The scientific learning cycle is also useful for working with students who are already in the formal operational stage since these students also learn by being in a state of disequilibrium and using accommodation. The scientific learning cycle is discussed in more detail in Chapter 15.

Piaget's theory has partially withstood the test of time and partially been modified (Kurfiss, 1988). It is now generally agreed that individuals actively construct meaning. This has led to a theory called constructivism, which is discussed in more detail in Chapter 15. Piaget's general outline of how people learn and the need for disequilibrium has been validated. Disagreements with Piaget focus on the role of knowledge in learning. More recent researchers have found that both specific knowledge and general problem-solving skills are required to solve problems, while Piaget did not recognize the importance of specific knowledge.

14.2. PERRY'S THEORY OF DEVELOPMENT OF COLLEGE STUDENTS

William G. Perry, Jr., studied the development of students at Harvard University through their four years at the university. His team used open-ended interviews as the technique of measurement. Over a period of years a pattern of development could be distinguished among all the varied responses of the students. Perry then used this pattern of development to rate another group of students. This replication showed that the scheme was reproducible at least for the men at Harvard University. Since publication of the results in 1970 (Perry, 1970), interest in Perry's theory of development during the college years has grown until now his book is being called "the most influential book of the past twenty years" on how college students respond to their college education (Eble, 1988). Perry's study has been criticized since the group studied was quite homogeneous and consisted mainly of young men from privileged backgrounds. Additional studies since 1970 have essentially duplicated Perry's results and shown that his scheme has fairly general validity except that extensive modifications need to be made for the development of women (Belenky et al., 1986). See Kurfiss (1988) or Moore (1989) for references.

Although Perry's model has become quite influential in higher education in general, engineering education has lagged behind. The model appears to have been introduced in engineering education by Culver and his coworkers. Culver and Hackos (1982) presented an overview of Perry's scheme and discussed implications for engineering education. Fitch and Culver (1984) and Culver and Fitch (1988) presented data on the positions in Perry's model of engineering students, and discussed educational activities to encourage student development. Culver (1985a) described a workshop on Perry's model and discussed a developmental instructional model based on Perry's work. Culver (1985b) considered values in engineering education and specifically related them to Perry's model. Hackos (1985) discussed using writing to improve problem-solving skills and to enhance intellectual development. The next year Culver (1986) continued his series by discussing how Perry's model was useful in explaining the effects of motivation exercises. Culver (1987a) described applications of Perry's model in encouraging students to learn on their own and presented a workshop (Culver, 1987b) which was an overview of Perry's model and of applications to engineering education. Pavelich and Fitch (1988) measured engineering students' progress through Perry's positions and concluded that it is slow. Culver et al. (1990) discussed the redesign of design courses and curricula to aid the progress of students on Perry's model. [Note that in engineering education earlier efforts were made to tackle some of the problems clearly posed by Perry, but Perry's complete scheme was not used.]

It would be convenient if Perry's scheme started where Piaget's theory stops. Chronologically, the two theories do fit this way, but in other more important ways the theories are *not* a match. Perry does use Piaget's ideas of how students learn. That is, a certain amount of disequilibration is necessary for accommodation to occur. However, Perry's theory is not concerned with problem solving and the applications of logic as are the concrete and formal operational stages of Piaget's theory. Briefly stated, Perry's model is concerned first with how students move from a *dualistic* (right versus wrong) view of the universe to a more *relativistic* view, and second, how students develop commitments within this relativistic world. There is a strong learning connotation in Perry's model since students cannot understand or answer questions which are in a developmental sense too far above them.

14.2.1. Positions in Perry's Model

From his interviews and by extrapolation Perry (1970) postulated nine *positions* as shown in Figure 14-1. These positions and the movement from position to position represent the major contribution of Perry's model.

Position 1: Basic Duality. The person sees the world dualistically, right versus wrong. There are no alternatives. Authorities know all the answers. Men appear to identify with the authority figure while women do not (Belenky et al., 1986). The teacher as an authority is supposed to teach the correct answers to the students. Failure to do so means that the teacher is a bad teacher. Hard work and obedience will be rewarded. Authority is so all-knowing that all deviations from authority are lumped together with error and evil. Perry (1970) notes that this position is basically naive since there is no alternative or vantage point which allows the person to observe her- or himself.

Perry (1970) talked to freshmen after one year at Harvard. He did not talk to anyone in position 1 but inferred this position from student reports about what they had been like when they entered Harvard. Perry notes that this position's assumptions are incompatible with the culture of pluralistic universities and thus students will be unable to maintain this position if they stay at the university. Much of the confrontation with pluralism occurs in residence halls, which may be a good reason to strongly encourage freshmen to live in residence halls. Many other studies (e.g., Moffatt, 1989) have reaffirmed the importance of residence halls in the development of students. Students may start in this position because of a culturally homogeneous or narrow environment, but they will quickly lose their innocence at a university.

Confrontations with their basic dualistic position both in class and in residence halls cause disequilibration. The student tries to accommodate the new ideas of multiplicity. This can be done by moving to position 2 or, at least temporarily, by modifying position 1. The modified position 1 assumes that absolute truths exist, but that authorities may not know what these

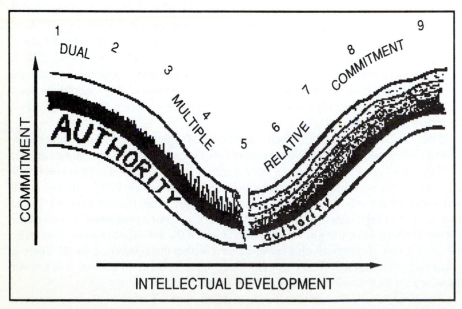

FIGURE 14-1   PERRY'S MODEL OF INTELLECTUAL DEVELOPMENT (Culver and Hackos, 1982)
(© 1982, American Society for Engineering Education)

truths are. Thus conflicts are explained since authority doesn't know the truth, but if one searches hard enough there is an absolute truth. This modified position itself leads to position 2 since the modified position admits that authorities can make errors. Unfortunately, there is another possible outcome to the stress induced by confronting multiplicity at the university. The student may leave.

In their study of the development of women, Belenky et al. (1986) included individuals from many social classes. By talking to women in social service agencies, they detected the presence of a position before (or below) position 1 which they called "silence." These women were from very deprived or abusive backgrounds. "Silent" women were unable to understand the words of others and were unable to articulate their own thoughts and feelings. With the steady increase in older students returning to college, some women who have once been in this position will become engineering students.

Position 1 is also the home of intolerance and bigotry. It appears to us that this is the basic position taken by some cults. Although engineering educators tend to shy away from moral arguments, there seem to be clear moral reasons to help students move out of position 1 into position 2.

Position 2: Dualism: Multiplicity Prelegitimate. In position 2 the student can perceive that multiplicity exists but still has a basic dualistic view of the world. There is a right and a wrong. Multiple views or indications that there are "gray" areas are either wrong or interpreted as authority playing games. Since it is possible for authority to be wrong, the absolutes are separate from authority. Thus, some authorities are smarter than others. This position may lead to the feeling that "I am right and authority is needlessly confused." The person may hold the view that there is one answer, but authority shows multiple answers as a game to make students learn how to find the one right answer.

An engineering student in position 2 can successfully solve problems, particularly closed-end problems, with a single right answer. These are the types of problems students in position 2 expect, and these students prefer engineering classes to humanities classes because the problems fit their dualistic mode of thought. In design classes, where problems have multiple answers, these students have difficulties, and they protest against open-ended problems. A student in position 2 wants the teacher to be the source of correct knowledge and to deliver that knowledge without confusing the issues. In this student's view a good teacher presents a logical, structured lecture and gives students chances to practice their skills. The student can then demonstrate that he or she has the right knowledge. From the student's viewpoint a fair test should be very similar to the homework.

Perry notes that students are bewildered and protest as they move from position 1 to position 2. The move from position 1 to position 2 may appear to be small; however, the student has made a major concession by allowing for some complexity and some groping into uncertainty.

In the two dualistic positions men and women use language differently. In general, men tend to talk and women listen. Since listening to authorities is the primary focus of women in the dualistic positions, Belenky et al. (1986) call these positions "received knowledge."

Position 3: Multiplicity Subordinate or Early Multiplicity. In position 3 multiplicity has become unavoidable even in hard sciences and engineering. There is still one right answer, but it may be unknown by authority. Thus the gap between authority and the one truth has been widened. The student realizes that in some areas the knowledge is "fuzzy."

This position has some built-in procedural conflicts. If authority does not yet know the answer, how can the professor evaluate the student's work? This is a considerable change from position 2 where honest hard work would presumably lead to the correct answers. Now, in position 3 honest hard work is no longer guaranteed to produce correct answers, and thus good grades seem to be based more on "good expression." The big question students ask is "What do *they* want?" The methods for evaluation become a very important issue and students want the amount of effort put into something to count. From the students' perspective a good professor clearly explains the methods used for determining the right answer even if he or she does not (temporarily) know the right answer, and the good professor presents very clearly defined criteria for evaluation.

For men education appears to play a significant role in the shift to multiplicity. From a developmental sense, one problem with engineering education is that there are few challenges at the lower levels to move the student into position 3 or 4. In class the challenges of multiplicity usually come in senior design classes and in graduate school. The lower-level classes are usually taught as if everything is known. This can lead to severe stress for students in a design course where multiple answers are expected and they are suddenly expected to function in a world with multiple answers. Students survive design courses but often do so without changing a great deal. This survival may occur because design is an isolated class which lasts only for one semester, and the legitimacy of multiplicity may not be reinforced in other classes or in the rest of the student's life. In addition, students who are academically very good can often hide from the challenges of multiplicity through competence (their design is likely to have fewer technical errors and they receive good grades). Beginning graduate students often become very frustrated as they try to determine what they are supposed to do. With less structure, fewer supports, a longer-term reward compared to seniors, and more pressure to adjust to a world of multiplicity, the graduate student's frustration is understandable. Graduate work in engineering and the physical sciences is similar to undergraduate work in the humanities in the respect that both confront the student with multiplicity and uncertainty.

For women formal education is relatively unimportant for the shift into "subjectivism" [the term used by Belenky et al. (1986) for multiplicity]. Women appeared to shift into subjectivism "after some crisis of trust in male authority in their daily lives, coupled with some confirmatory experience that they, too, could know something for sure" (Belenky et al., 1986, p. 58).

Position 4: Complex Dualism and Advanced Multiplicity.   The student tries to retain a dualistic right-versus-wrong position but realizes that there are areas of legitimate uncertainty and diversity of opinion. Students react to position 4 in one of two ways. They may conform to what authority seems to want and learn the forms of independent intellectual thought. These students learn that *independent-like* thought will earn them good grades. Genuinely independent thought has not yet been achieved or even considered as an issue. Most of the students Perry studied took this route. However, learning the forms is not enough, and these students may be tempted to *escape*.

The second reaction is that the student may oppose what authority wants in areas where multiplicity is important. The student may raise this multiplicity of opinions to a pervasive viewpoint that "anyone has a right to their own opinion." This raises areas of multiplicity and uncertainty to equal status with areas of dualism. "Everyone has a right to their own opinion" is obviously a wonderful position from which to fight authority. The danger of this position is that a bland "anything goes" attitude may prevail. The student may refuse to think since he or she believes everything can be solved by intuition. Men in this position fight authority openly, while women fight authority internally as "hidden multiplists" (Belenky et al., 1986). These women may be silently alienated from college. Since engineering does not affect their interior life, engineering may appear irrelevant and they may quit engineering even though they can do the work. This position was taken by fewer students and is probably rare for engineering students.

An engineer in position 4 can solve problems cleverly and creatively. The task of solving the problems becomes a game. Unfortunately, he or she cannot see that some problems are much more important than others. This person lacks vision and may solve problems considered unimportant or even immoral by others. Many engineering graduates with both baccalaureate and advanced degrees seem to be in positions 3 and 4.

Position 5: Relativism. In position 5 a person sees everything as relative, not because authority wants it that way but because that is the way he or she sees the world. There is a revolutionary switch from position 4 to position 5. In position 5 relativism becomes the common characteristic of everything and absolutes are a special case. One must then determine if complexity is *not* necessary. In position 4 the situation was the reverse: Dualism was the general principle, and relativism was a special case useful for certain classes of problems. Perry noted that this is often an extremely quiet revolution and that students hardly notice that it has occurred. The relativistic thought process becomes habitual without being noticed. For very focused students in engineering or science, position 5 may come as a shock when they realize that everything is relative in advanced classes.

Perry saw this position as occurring in three subpositions. First, the person divides the world into a relativistic area and into a dualistic area where authority still has answers. Then, the whole world is seen as relativistic, but this position alternates with a dualistic position. Finally, the whole world is seen as relativistic.

The relativistic position can be a very powerful one. There is room for detachment and objectivity. One can think previously forbidden thoughts. This ability to stand outside the situation and think objectively may, in Perry's words, "rank with language as the distinctive triumph of the human mind." The person in the relativistic position can get beyond the statement "all opinions are equal" by using the laws of evidence to develop positions which are more likely.

Belenky et al. (1986) noted that men and women may use different logical procedures in position 5, which they called "procedural knowledge." Most men and some women use the traditional logical approach with objective analysis and argument to form opinions. This *separate knowledge* or *objective knowledge* (Palmer, 1983) purposefully removes the person's personal experiences and feelings from the logical analysis. Separate knowledge emphasizes doubting and argument. It is the method one would expect from thinking types on the MBTI (see Chapter 13). This corresponds to the usual engineering approach. Arguments are supposed to be between positions, but many women have difficulty separating positions from people. Some women and men use an approach called *connected knowledge* which is an empathic treatment of divergent views. Connected knowledge personalizes knowledge and attempts to understand the reasons for another's way of thinking. Belief, not doubt, that the other is right from his or her viewpoint is the key stance of connected knowledge. Both feelings and thought are important. Connected knowledge would be expected of Myers-Briggs feeling types who are in position 5. In the same way that everyone has both feeling and thinking capabilities, everyone has the potential to learn both separate and connected knowledge. Individuals who strongly prefer to use connected knowledge as a way of understanding may find the environment at an engineering college somewhat hostile—since only separate knowledge is taught. Because these individuals can make major contributions as engineers (e.g., in life cycle design or conflict resolution), it is important to accommodate them in the educational system. For women the presence of a benign and encouraging authority appears to facilitate movement into position 5.

From the student's viewpoint in position 5 a good instructor acts as a source of expertise, but does not know all the answers since many answers are unknowable. This professor helps students become adept at forming rules to develop reasonable and likely solutions or solution paths. It is important for the professor to show that good opinions are supported by reasons. The student has become much more comfortable with being evaluated in a relativistic world and realizes that the evaluation is of her or his work and not of her or him.

There are problems in position 5. The world is full of possibilities, and there does not appear to be a clear way to choose. Decisions which were made earlier are now called into doubt. The student wonders whether engineering really is the right choice. Did he or she really marry the right person? And so on. Position 5 then represents both a period of strength and possibilities and a period of doubt and loneliness. Assuming that a person is eventually going to move into position 5, it is probably better to do so early while many important career and life decisions have yet to be made. Position 5 can also appear "cold" and even sinister to others because of the focus on method and the dissociation between means and values.

Position 6: Relativism: Commitment Foreseen. The way out of the uncertainty of relativism is *commitment*. In position 6 the student can see the need for commitment but has not yet made the commitment. This need for commitment may be seen as a logical necessity (this is likely for people who are a *T* on the MBTI or may be felt (people who are an *F*). Commitment may be looked forward to with eagerness, or the person may fight commitment. People who fight commitment may stay uncomfortably in position 5, or they may escape or retreat (these are discussed later).

Many students think they have already made firm commitments. Perry uses Commitment (with a capital C) to have a special meaning. Commitment is a mature decision made after one has accepted that the world can be viewed as relativistic and has seen all the possibilities. Previous decisions have been called into doubt and looked at objectively from a detached viewpoint. The new Commitment may be the same decision made previously, but the Commitment is deeper. Commitments can be made in a variety of areas such as career, religion, marriage, politics, values, and so forth. The Commitments one makes help set the person's identity and style. At this point one makes an objective decision on how much of the past to reject and how much to retain. This shedding of parts of the past is clearly different from adolescent rebellion which tends to be mindless.

In position 6 the person can see this need for Commitment, but the Commitment has not yet been made. People who move forward into position 7 often do so in one area of their lives at a time. They remain in position 6 in other areas.

For an engineering student who has invested a great deal of time in studying engineering, going through position 5 can be very unsettling. Position 6 can be something of a relief since the student sees that it is all right to commit to engineering if objectively that is a good decision. However, a major Commitment is not to be rushed, and the person may stay in position 6 for a while.

Positions 7 through 9: Levels of Commitment. Positions 7 through 9 are all levels of Commitment starting with initial Commitment in position 7. These positions represent degrees of development and depth of Commitment and are not as clearly defined as are the other positions. The person moves from position 6 into position 7 in one area by making a Commitment of his or her own free will. For some this is risky and may be done tentatively in relatively safe areas. As the person becomes more comfortable with making Commitments, he or she makes them in areas that are not as safe, eventually finding not only that a series of finite, discrete decisions have been made, but that a way of life has been developed.

Perry sees the student in position 7 first taking responsibility for who he is or will be in some major area of his life ("I'll stay in engineering"). In position 8, stylistic issues of Commitment become important. "If I am going to be an electrical engineer, how will I do it?" "What will my specialties be?" "What degrees should I get?" And so forth. Position 9 is a postulated position of maturity where the person has developed a sense of self in both Commitments and style. Perry postulated that this is a position reached some time after graduation. Women also make a commitment but it is to a life rather than the single Commitments men often make (Belenky et al., 1986).

Belenky et al. (1986) added insight into the thought processes of the Commitment positions. The thought process uses *constructed knowledge* where procedural knowledge gained from others is integrated with personal or "inner" subjective knowledge based on personal experience and introspection. This constructed knowledge allows the individual to integrate thought and feeling and avoid the compartmentalization which Belenky et al. (1986) perceive as a shortcoming of objective knowledge. At the levels of Commitment a good instructor needs to provide freedom so that students can learn what they need to learn (e.g., Rogers, 1969). The instructor also needs to forge linkages within the class (Palmer, 1983).

Perry's model is a staged model which tends to ignore the situational specificity of behavior and knowledge. Real people in real situations have the annoying tendency to be complex. They don't fit into one stage, but depending upon the situation may be in several different stages. Despite this difficulty, Perry's model is a very useful model for conceptualizing the development of college students. A more recent model which builds on Perry's model is the reflective judgment model (Kitchener, 1986).

14.2.2. Alternatives to Growth

Perry hypothesizes that natural growth is from position 1 toward position 9. At Harvard he saw many students graduate in positions 7 and 8. However, he notes that growth is not inevitable. In engineering it is likely that many students leave in positions 3 and 4. The three alternatives to growth are *temporizing, retreat,* and *escape.* Note that these names incorporate Perry's hypothesis that movement from position 1 toward position 9 is growth and thus is desirable.

Temporizing.   Growth does not occur linearly. Instead, periods of intense growth are commonly followed by pauses or plateaus. Perry defined *temporizing* as a pause in growth over a full academic year. All students go through plateau periods. Temporizing is just a rather long plateau and by itself is not bad. It may be a period in which the student gathers strength for the growth which lies ahead. In this case the student often seems aware that he or she is waiting for the correct combination of energy and will to move on. In an alternate mood of temporizing the student waits for fate to decide what will happen and may drift into escape.

Retreat.   Retreat is regression to earlier positions. The most dramatic such retreat is movement back to position 3 or 2 when the complexities of relativism and multiplicity become overwhelming. (Retreat into position 1 is also possible, but in Perry's study these students presumably dropped out of Harvard.) Retreat into dualism requires an enemy. The student must be on her or his guard against the pluralistic university. Students seem to be most susceptible to retreating to dualism when they rely on authoritarian structures for emotional control. Retreat also occurs from higher levels but is not as dramatic. For example, a student may retreat from position 6 or 5 to position 4 where he or she can hide in the concept that "everyone has the right to his or her own opinion."

Escape.   In escape the student avoids Commitment by exploiting the detachment afforded by positions 4 and 5. Perry's team noted two paths of escape both of which started from temporizing. In *dissociation* the student drifts into a passive delegation of responsibility to fate. She or he ends up in position 4. The alternate path is *encapsulation* which may be a favorite of engineering students. In encapsulation one avoids relativism by sheer competence in one's field. The student becomes very good at engineering but avoids any questions of deeper meaning or value. Engineers can use encapsulation to stay in position 4 or 5 for years. Escape need not be permanent, and people find different ways to resume growth.

14.2.3. Implications for Engineering Education

Perry's model has both value-free and value-laden implications for engineering education. Since the subject is less controversial, we will start with the implications which are relatively value-free. The major inescapable conclusion from Perry's model is that different students require different learning environments. This is no surprise since all models of learning come to the same conclusion. Students are not capable of understanding knowledge or questions which are too far above them as far as Perry's positions are concerned. If pushed to try to understand this material, they will become frustrated. How far above is too far? Perry does not address this issue. From our experience, questions which are one position above the student's position can, perhaps with considerable difficulty, be answered. Questions or knowledge two positions above the student's current position cause frustration. Students are capable of answering questions in positions below them although they may find these questions easy or may read too much into them. Appropriate teacher responses at each position were discussed with the descriptions of each position. How does the teacher provide an optimum learning environment for a heterogeneous class with students at a variety of levels? This is the key challenge of individualizing instruction, and there is no clear-cut answer. Some possible approaches were discussed in Chapters 5 through 10.

Most of the applications of Perry's model to engineering education involve the value judgment that growth on Perry's scale is desirable (at least up to some level) and should be fostered. Perry considered this question and decided that growth was both natural and desirable. However, his sample contained no engineering students and in many ways was quite narrow. The faculty at each school need to face the question of whether or not to encourage growth on Perry's scale. Failure to encourage growth is equivalent to a negative answer. Currently, engineering students show little progress toward higher Perry levels and may actually regress slightly during their engineering studies (Fitch and Culver, 1984; Pavelich and Fitch, 1988). Thus, if the faculty decide that growth is desirable, engineering education must be changed.

As noted previously, we feel that there are clear moral grounds for strongly encouraging students in position 1 to grow into position 2. Students and practicing engineers in positions 1 and 2 will have significant difficulty practicing engineering in our multiplistic society. Fortunately, the samples reported by Fitch and Culver (1984) and Pavelich and Fitch (1988) showed very few students who were clearly in position 2 (and none in position 1). A large number of students were in transition between positions 2 and 3, and the mean position for all engineering students was about 2.8. Students in transition between positions 2 and 3 can see and accept multiplicity in some areas, and they accept that authority does not have all the answers. This transition region appears to be the minimum region in which a student can successfully study and practice engineering. These engineers cannot see the big picture, and without further growth they are unlikely to advance significantly in their careers. Fitch and Culver (1984) also reported many students in position 3 and a few in the transition between position 3 and 4. No undergraduate engineering students in positions 4 or higher were observed. Pavelich and Fitch (1988) found that the written test used to measure students' developmental levels (Measure of Intellectual Development, the MID) was quite conservative. Interviews showed students who were at levels 4 and 5. This should be contrasted to Perry's sample of liberal arts students at Harvard where the average entering level was approximately position 4 and 75 percent were judged to be in position 7 or 8 at graduation. [Note: Perry appears to have had an unusual sample. Other studies have consistently found more students at lower levels (Kurfiss, 1988).]

The reasons for moving students to at least the transition between positions 2 and 3 are clear. Below this level they will have difficulty functioning as engineering students. Graduate students in thesis masters and Ph.D. programs will have trouble functioning below level 3 since they will not be able to answer the question "What do they want?" Research in graduate school seems to be structured to encourage the transition to position 3 if the student is not already there. Continued graduate study often moves the graduate student into position 4. Thus, engineering schools have implicitly made the decisions that undergraduates should reach at least the 2-to-3 transition and that graduate students should reach level 3 or 4 before graduation.

Is this sufficient? Probably most faculty will answer no. They want graduate students to operate at least at the level of the better students (position 4), and they want undergraduates to approach this level (say the 3-to-4 transition). In this regard Perry (1970) offers an interesting quote: "Fifty years ago [1920], our researches suggest, a college senior might achieve a world view such as that of Position 3 or Position 4 on our scheme and count himself a mature man."

Superficially, it is easy to conclude that engineering education must change and take students past position 4. However, there are many dangers to this.

1 Taking the student to position 5 will fill the student with doubts about engineering as a profession. If a school purposely takes a student to position 5, the school must ethically help her or him to at least position 6. Some of these students will decide to make a Commitment to another profession.

2 It is difficult to take engineering students to position 5 even if we decide we want to. Engineering education at the undergraduate level reinforces positions 2 and 3, and at the graduate level does little to push students to position 5. Engineering students are very adept at escaping into competence once they reach position 3 or 4.

3 Many employers are happy with the current graduates at both undergraduate and graduate levels. This includes engineering schools as employers of Ph.D.s.

4 A consensus of engineering professors does not exist.

5 Absolute standards in physical laws are a useful mental construct despite the Heisenberg uncertainty principle. Perry's relativism can undermine this absolute standard (Graff et al., 1991.)

6 Some professors feel that there *should* be absolute standards in engineering ethics (Graff et al., 1991). This is a different value judgment than Perry's.

There are reasons for encouraging students to move beyond their current positions.

1 Growth appears to be natural and in this sense is "good."

2 Growth into positions 7 through 9 appears to be necessary to function well in important positions such as vice-president, dean, president, and CEO (e.g., see Florman, 1987, p. 178). In a technological society we need more engineers in these positions.

3 "The main trouble with engineers has not been their lack of morality. It has been their failure to recognize that life is complex" (Florman, 1976, p. 27).

4 For women, movement to higher-level positions is empowering and helps them act as equals with men.

What types of activities and teaching encourage growth? Fitch and Culver (1984), Culver (1985a,b), and Culver and Fitch (1988) make the following suggestions based on the work of Lee Knefelkamp. First, since highly structured courses reinforce the lower levels, the curriculum should be restructured so that courses become progressively less structured. Second, a diversity of learning tasks is required, which means that the use of a single textbook in a course probably is not enough. Third, students need concrete learning experiences such as case studies, team projects, industrial experience, and so forth. These experiences should be designed to reinforce diversity. Fourth, a learning environment which supports risk taking needs to be developed in engineering classes and in the university as a whole. Additional suggestions can be added from the research of Belenky et al. (1986). First, the student needs assurance that she is capable, and this support is needed from the beginning. Successful programs for women in engineering always include a significant component of support. Unfortunately, many women distrust praise from male professors. "The women worried that professors who praised their minds really desired their bodies" (Belenky et al., 1986, p. 197). Second, separating evaluation from instruction is valuable for many students. It is difficult for many professors to be supportive when they know they will have to evaluate later. Evaluation and instruction can be separated by using separate competency examinations scored by outsiders or by having separate help classes taught by instructors not involved in the graded class. Third, professors need to think out loud instead of presenting prepackaged thoughts as finished solutions (see Chapter 5 for further discussion of this). Finally, it is particularly important for professors of engineering and science "to avoid the appearance of omniscience" (Belenky et al., 1986, p. 216).

One learning environment designed to encourage intellectual growth is the *practice-theory-practice* model developed by Lee Knefelkamp which has been applied in engineering education by Culver (1985a, 1986, 1987a). In this model a concrete experience (practice) is used to introduce the concept. Then theory is developed to explain the experience. Finally, further practice is used to reinforce the theory and to provide an extension to other material. This type of cycle appears to be particularly important for women who found concepts useful in understanding their experience but balked at an abstract approach devoid of experience (Belenky et al., 1986). To be effective for producing intellectual development, the experiences and theory must be understandable at the stage of development of the student, but the experience must also challenge the student. Activities appropriate for graduate students are probably inappropriate for freshmen. Learning cycles which encourage intellectual growth are discussed in detail in Chapter 15.

Perry found that dormitory living was very important for moving students out of the lower levels. He also found that liberal arts courses were very valuable in helping students grow. Can liberal arts courses help engineering students grow on Perry's scale? The answer appears to be of the "yes, but" variety. Florman (1987) is strongly in favor of liberal arts for engineers, yet he notes (p. 173), "One need not be a broadly educated scholar in order to be a topnotch engineer." Liberal arts courses can be useful, but some restructuring is probably needed. Certain courses such as beginning language courses and beginning economics courses have little effect. In other liberal arts courses engineers need to be mixed in with students from other areas. Putting all engineers into the same class defeats much of the purpose of achieving diversity. Since engineering students often see liberal arts courses as unimportant, the engineering *faculty* has to work hard to change this opinion. And since the students do not take a critical mass of liberal arts courses, ideas of multiplicity and relativism need to be reinforced in engineering classes. The liberal arts courses should be selected to challenge the student successfully no matter what his or her level. Dissonance, which is necessary for change, can be generated by writing or discussion but not by multiple-choice tests. Thus, courses must have significant writing or discussion.

One additional implication of Perry's model is somewhat disturbing. Mature students may find beginning engineering and science courses intellectually unchallenging and perhaps even stultifying. These students may find liberal arts or the social sciences more intellectually fulfilling and drop engineering. Evidence for this is contained in the study of Tobias (1990) exploring why some students drop science and engineering. A paradoxical result of this is that with current course levels it may be advantageous to delay intellectual growth until students have completed the lower-level courses. Obviously, there are many views on how much engineering education should do to move students on Perry's scale, but there is absolutely no correct view.

14.3. CHAPTER COMMENTS

Piaget's theories have had a major impact on the teaching of science, but little impact in engineering. One possible reason for this difference is that engineers generally teach only engineers, whereas scientists teach everyone in the university. Thus, engineering classes have a small percentage of students who are in the concrete operational stage. Since there are few of them and most of them do not survive in engineering, it has been easy for engineering professors to ignore them.

Teaching Perry's model to graduate students can be an interesting experience for an engineering professor. We have encountered strong resistance to Perry's value judgment that growth on his model is positive. This resistance came from strongly religious graduate students. The most palatable presentation of Perry's model for these students clearly separated the observed behavior (the positions) from the value judgment. In addition, these students preferred to consider relativism as a way in which people *can* look at the world instead of Perry's formulation that this *is* the way the world is.

There is a strong tendency to demand equal rights and opportunities for women without looking at the real differences between men and women which were discovered in the research by Belenky et al. (1986). In our opinion equal opportunity does not imply an education that is exactly the same for men and women. Since many current educational practices are more tuned to the ways men learn and develop, women have less opportunity to benefit. Ideally, an equal opportunity education would involve many educational opportunities which are helpful for students with very different developmental needs and methods of knowing. Belenky et al. (1986) noted that there are different methods of knowing then those identified by Perry (1970). The different paths appear to be the result of socialization, not hardwired gender differences. Once we admit to more than one approach to knowledge, it is logical while studying multiplicity to expect multiple paths caused by different socialization procedures in different societies. What other paths might there be? Palmer (1983) discusses a spiritual path. Another path is through dreams. The Senoi tribe in Malaysia clearly uses dreams as a path to knowledge (Garfield, 1974). Although it is difficult to conceive of teaching engineering in Western society through dreams, they have often played a role in the solution of technical problems. Other paths to knowing exist and have been explored by anthropologists.

14.4. SUMMARY AND OBJECTIVES

After reading this chapter, you should be able to:

• Describe Piaget's theory to another engineering professor and discuss its implications for engineering education.

• Describe Perry's theory to another engineering professor and discuss the specific value judgments in this theory. Do you agree or disagree with these values? Outline how Perry's theory as modified by your value judgments impacts engineering education.

• Discuss how the development of men and women may differ. Outline the consequences of this in engineering education.

HOMEWORK

1 Several students come to you to complain bitterly that Prof. Whatastar gives tests that are very different from the reading and homework assignments; he also requires solving problems the students have never seen before.

a Use Piaget's model and Perry's model to explain what is going on.

b What can the students do to prepare better for Prof. Whatastar's exams?

2 Explore the differences and similarities between a good sophomore engineering course and a good graduate-level engineering course. Bolster your discussion with evidence from the theories discussed in this chapter.

3 Students who earn B.A. degrees and then return to take engineering courses often do not do well. Although there may be a variety of reasons for this phenomenon, discuss possible reasons based on Piaget's and Perry's theories.

4 Should you share with students their level on Perry's scheme? Discuss the pros and cons of sharing this information.

5 Individuals in the concrete operational stage often become very frustrated by science and engineering courses in college. What course of action would you take? Explain why. Do you believe universities should try to help concrete operational students grow into the formal operational stage?

6 Should engineering education be reorganized to produce larger gains on Perry's scale? Note: Do all three parts.

a Assume the answer is yes and argue in favor of changing the system.

b For part a, discuss what changes you would recommend.

c Assume the answer is no and argue against changing the system.

7 Classify yourself on Perry's model or on the modified model for women.

8 It is often noted that Piaget was really an epistemologist and not a psychologist. Do you agree? Does it really matter?

9 Do the third objective in Section 14.4.

REFERENCES

Belenky, M. F., Clenchy, B. M., Goldberger, N. R., and Torule, J. M., *Women's Ways of Knowing: The Development of Self, Voice and Mind*, New York, Basic Books, 1986. (An interesting key source, but the jargon can be difficult.)

Bordogna, J., "Entering the '90s: A national vision for engineering education," *Eng. Educ.*, 79 (7), 646 (Nov. 1989).

Culver R. S., "Applying the Perry Model of intellectual development to engineering education," *Proceedings ASEE/IEEE Frontiers in Education Conference*, IEEE, New York, 95—99, 1985a.

Culver, R. S., "Values development in engineering education," *Proceedings ASEE/IEEE Frontiers in Education Conference*, IEEE, New York, 199–205, 1985b.

Culver, R. S., "Motivation for continuing education," *Proceedings ASEE/IEEE Frontiers in Education Conference*, IEEE, New York, 105–111, 1986.

Culver, R. S., "Who's in charge here? Stimulating self-managed learning," *Eng. Educ.*, 297 (Feb. 1987a).

Culver, R. S., "Workshop: Rational curriculum design," *Proceedings ASEE/IEEE Frontiers in Education Conference*, IEEE, New York, 391—396, 1987b.

Culver, R. S. and Fitch, P., "Workshop: Rational Curriculum Design," *Proceedings ASEE Annual Conference*, ASEE, Washington, DC, 563–568 (1988).

Culver, R. S. and Hackos, J. T., "Perry's model of intellectual development," *Eng. Educ.*, 73, 221 (Dec. 1982).

Culver, R. S., Woods, D., and Fitch, P., "Gaining professional expertise through design activities," *Eng. Educ.*, 533 (July/Aug. 1990).

Eble, K. E., *The Craft of Teaching*, 2nd ed., Jossey-Bass, San Francisco, 1988.

Fitch, P. and Culver, R. S., "Educational activities to stimulate intellectual development in Perry's scheme," *Proceedings ASEE Annual Conference*, ASEE Washington, DC, 712—717, 1984.

Flavell, J. H., *The Development Psychology of Jean Piaget*, D. Van Nostrand, New York, 1963.

Florman, S. C., *The Existential Pleasures of Engineering*, St. Martin's Press, New York, 1976.

Florman, S. C., *The Civilized Engineer*, St. Martin's Press, New York, 1987.

Gage, N. L. and Berliner, D. C., *Educational Psychology*, 3rd ed., Houghton-Mifflin, Boston, 1984.

Garfield, P., *Creative Dreaming*, Simon and Schuster, New York, 1974.

Goodson, C. E., "An approach to the development of abstract thinking," *Proceedings ASEE Annual Conference, American Society for Engineering Education*, ASEE, Washington, DC, 187–193, 1981.

Graff, R. W., Leiffer, P. R., and Helmer, W., "The influence of the Perry model in teaching engineering ethics," *Proceedings ASEE Annual Conference*, ASEE, Washington, DC, 902, 1991.

Hackos, J. T., "Using writing to improve problem solving skills and enhance intellectual development," *Proceedings ASEE/IEEE Frontiers in Education Conference*, IEEE, New York, 186–190, 1985.

Inhelder, B. and Piaget, J., *The Growth of Logical Thinking from Childhood to Adolescence*, Basic Books, New York, 1958.

Kitchener, K. S., "The reflective judgment model: Characteristics, evidence, and measurement," in Mines, R. A. and Kitchener, K. S. (Eds.), *Adult Cognitive Development: Methods and Models*, Praeger, New York, 76–91, 1986.

Kurfiss, J. G., *Critical Thinking: Theory, Research, Practice, and Possibilities*, ASHE-ERIC Higher Education Report No. 2, Association for the Study of Higher Education, Washington, DC, 1988.

Lawson, A. E., Abraham, M. R., and Renner, J. W., *A Theory of Instruction: Using the Learning Cycle to Teach Science Concepts and Thinking Skills,* Monograph 1, National Association for Research in Science Teaching, Cincinnati, OH, 1989.

Moffatt, M., *Coming of Age in New Jersey: College and American Culture,* Rutgers University Press, New Brunswick, NJ, 1989.

Moore, W. S., "The learning environment preferences: Exploring the construct validity of an objective measure of the Perry scheme of intellectual development," *J. Coll. Stud. Develop.*, 30, 504 (Nov. 1989).

Palmer, P. J., *To Know as We Are Known: A Spirituality of Education*, Harper Collins, San Francisco, 1983.

Pavelich, M. J., "Integrating Piaget's principles of intellectual growth into the engineering classroom," *Proceedings ASEE Annual Conference*, ASEE Washington, DC, 719—722,1984.

Pavelich, M. and Fitch, P., "Measuring student's development using the Perry model," *Proceedings ASEE Annual Conference*, ASEE, Washington, DC, 668—672, 1988.

Perry, W. G., Jr., *Forms of Intellectual and Ethical Development in the College Years: A Scheme,* Holt, Rinehart and Winston, New York, 1970. (This is considered a key source on the development of college students.)

Phillips, J. L., Jr., *Piaget's Theory: A Primer*, W.H. Freeman, San Francisco, 1981.

Piaget, J., *The Psychology of Intelligence*, Harcourt and Brace, New York, 1950.

Piaget, J., *Logic and Psychology,* Basic Books, New York, 1957.

Pintrich, P. R., "Implications of psychological research on student learning and college teaching for teacher education," in W. R. Houston, M. Haberman, , and J. Sikula (Eds.), *Handbook of Research on Teacher Education*, MacMillan, New York, 926–857, 1990.

Renner, J. W. and Lawson, A. E., *The Physics Teacher*, 165–169 (March 1973) and 273–276 (May 1973).

Rogers, C. R., *Freedom to Learn*, Merrill, Columbus, OH, 1969.

Tobias, S., *They're Not Dumb, They're Different*, Research Corporation, Tucson, AZ, 85710–2815, 1990. [Free copies of this book can be obtained directly from the Research Corporation, 6840 East Broadway Boulevard, Tucson, AZ.]

Wankat, P. C., "Analysis of student mistakes and improvement of problem solving on McCabe-Thiele Binary distillation tests," *AIChE Symp. Ser.*, 79 (228), 33 (1983).

CHAPTER 15

LEARNING THEORIES

In Chapter 14 we discussed Piaget's dictum that individuals construct their own knowledge structures. By continually testing these knowledge structures against the external world and then adapting them to fit that world, most individuals acquire a knowledge structure which "works" reasonably well in their world. For most individuals a "working" structure or model must be socially acceptable. This is true even of scientific concepts. The resulting structure may not be "true" in any absolute sense. For example, many engineering students start freshmen physics with the belief that a constant force must be applied to keep an object moving at constant speed. This belief results from years of pushing wagons, riding bikes, and driving cars. For these purposes this "knowledge" is adequate. In first-year physics, Newton's laws and friction are introduced, and the knowledge structure has be reconstructed. Such a reconstruction may be difficult (this is discussed later), but once developed it is adequate for most engineering and physics courses. In relativistic physics, students find that the newtonian model is not adequate, and a new model must be incorporated into their knowledge structure. This more complicated knowledge structure includes newtonian physics and driving a car as special cases.

15.1. CONSTRUCTIVISM AND THE SCIENTIFIC LEARNING CYCLE

What makes students go through the agony of such reconstructing? The answer appears to be the disequilibrium caused by new data which cannot be explained by the old model, and the inability to solve required problems. Bodner (1986) notes that many students find mathematical arguments and lectures with little discussion insufficient reason to discard the pre-newtonian model. Experiments with an almost frictionless system (such as a dry ice puck) are required to make students revise their model of the world. The inconsistencies between a student's model of the world and these new data should be forcefully pointed out. The second step is the availability of a plausible and understandable new concept or model which can eliminate the disequilibrium by explaining the new data. The student will restructure or assimilate new data only if accommodation fails and he or she is motivated to reconcile anomalies and reduce inconsistencies.

This example illustrates several important points about the constructivistic theory. Since the pre-newtonian model has been reinforced by years of practice where it worked, this knowledge structure is securely lodged in the brain. Removing any entrenched knowledge structure will be difficult. Thus, an extended period of time focused on Newton's laws is required both in and out of class, which helps to explain why learning new material is often slow. Frequent and timely feedback on mistakes helps to strengthen the necessary but not sufficient disequilibrium. Since forming new knowledge structures is difficult, students must be motivated. Direct contact with faculty can have a very positive effect on reorganization of the knowledge structure, particularly for students who identify with authority figures. The reorganization is aided by presenting information in hierarchical form with explicitly stated rules for generating hierarchies (Kurfiss, 1988). Learning new material in a form which is easy to recall from memory is aided if students are given objectives which help them key on important material and if the material is presented in a well-organized fashion (Kiewra, 1987).

The usual lecture-homework sequence requires formal operations. Students still in the concrete operational stage in physics have difficulty revising their knowledge structures. For those in this stage, the concrete operations of the laboratory can be instrumental in helping them accept the new organization of knowledge. The laboratory exercise has other advantages as well. In the laboratory the student must be active, unlike in a lecture where a passive approach is allowed and often encouraged. Reconstruction requires active mental effort by the student. The laboratory is also often a group activity which encourages students to discuss their understanding of physics actively, and the experience provides support from the group. Finally, this example helps to explain why beginning physics is widely considered to be the most difficult first-year course (Tobias, 1990). Many students are overwhelmed by the need to use formal reasoning to revise well-entrenched commonsense knowledge structures quickly and totally in a large class which often appears unfriendly.

It is interesting to compare the constructivist view of learning with the traditional view of knowledge which is implicitly assumed by many professors. In the traditional view knowledge exists independent of the individual. The mind is a *tabula rasa,* a blank tablet, upon which a picture of reality can be painted. If the student is attentive, learning occurs when the teacher unloads his or her almost perfect picture of reality through well-designed and well-presented lectures. Most experienced professors can attest that this model does not work for most students. Unfortunately, the traditional model focuses on the delivery system and not on the learner. Or, in computer language, the focus is on output devices and not input devices. The minds of the learners are not blank tablets upon which the teacher can write at will. The constructivist theory says the tablets are not initially blank and only the individual can do the writing. The traditional delivery system, the noninteractive lecture, satisfies the conditions of the traditional theory, but not the conditions of the constructivist theory. Fortunately, lectures can be modified so that the conditions necessary for learning are satisfied. These conditions are discussed in the remainder of this chapter, and specific modifications of the lecture method were given in Chapter 6. Following constructivist theory, the professor will become a facilitator of learning instead of a purveyor of knowledge. At times this facilitation is aided by lecturing, and at times it is not.

There are exercises and homework assignments professors can use to help students develop a knowledge structure. One useful assignment for every book chapter or section of the course is the development of a *key relations chart* (Mettes et al., 1981). A key relations chart lists and diagrams the key ideas, equations, relations, definitions, and so forth, on a single page. The instructor can first illustrate this procedure by handing out his or her own chart for a chapter; then students can be required to do the same for homework. The chart can be evaluated for accuracy, completeness, and conciseness. Finally, the assignment is no longer made, but students are urged to continue developing the charts. Some professors allow students to consult key relation charts during tests. Since the preparation of such a chart is a useful exercise, this is an interesting alternative to open book tests.

A related exercise is to have small groups of students develop a *memory board* (Woods et al., 1975), which is similar to a key relations chart but is significantly more complete and is prepared as a group exercise. It can include more equations, rules, interrelationships, and problem-solving hints. Construction of a memory board is a group activity, which makes it useful for support and motivation, particularly for the extroverts in the class. Working in groups also provides social pressure for students to change constructs which appear to be incorrect.

A third related exercise is to have individual students or groups of students develop *concept maps* or *networks* (Smith et al., 1985; Smith, 1987). A concept map or network visually represents the relationship between concepts, usually two-dimensionally. Both the hierarchical relationships and the key cross-links between concepts are shown. Concept maps are complementary to key relations charts and memory boards since the concept map does not give equations, definitions, or ideas. It shows the relations between concepts without full explanation of the concept. Since it is a visual representation, a concept map is often fairly easy to remember (see Section 15.2.2). Students need to be taught how to construct concept maps and then encouraged to develop them on their own. Smith et al. (1985) illustrate a scoring model for evaluating concept maps. See Figure 5-1 for an example of a concept map.

Constructivism can help to explain how individuals solve problems. Problem solving appears to require both a general problem-solving strategy and specific knowledge (Kurfiss, 1988; Pintrich, 1990). For routine problems, the specific knowledge structure is probably sufficient since it includes a pattern for solving routine problems. When confronted with unusual problems, the solver finds that no pattern exists for solving them. General problem-solving heuristics help one to start reconstructing the knowledge structure to solve the problem. Without specific content knowledge the general procedures are insufficient. Thus, engineering professors need to teach content and procedures. Teaching problem solving is the topic of Chapter 5.

Piaget's ideas and constructivism have led to a theory of how to teach science which is known as the *scientific learning cycle* (see Figure 15-1). (In the literature this is simply called the *learning cycle*. We have added the word "scientific" to differentiate it from Kolb's learning cycle.) This method was independently developed by Robert Karplus in physics and Chester Lawson in biology [see Lawson et al. (1989) for a historical perspective and complete references]. It has been extensively used and tested in science education at a variety of school

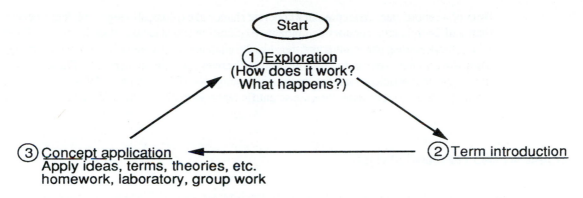

FIGURE 15-1 SCIENTIFIC LEARNING CYCLE

levels. There is considerable experimental evidence that the scientific learning cycle is more effective in teaching science than are the more traditional methods.

In the exploration phase, students explore new phenomena with minimal guidance; for example, given a new mechanical linkage or a new circuit, their assignment can be to determine how it works. In this phase they discover for themselves some of the patterns and concepts involved. The exploration can be done individually or in groups.

In the second phase, called term introduction, invention, conceptual invention, or concept introduction, the professor introduces terms and definitions. Students are encouraged to use these new terms to describe the patterns as completely as possible. The professor then fills in the missing parts of the pattern to give a complete scientific picture. This phase can be accomplished through lecture, readings, video, guided discussion, and so forth.

In the third phase, concept application, concept expansion, or idea expansion, students apply the new ideas, terms, and patterns to new examples. For instance, if the exploration phase involves development of a new physical law, then the law can be applied in new ways. This phase can involve homework, group discussions, or laboratory.

Although developed originally for use with laboratory manipulations in the exploration phase, the scientific learning cycle can be modified for other types of experiences. For example, the exploration phase can involve a computer simulation game which allows students to explore the simulated properties of some process or device. Alternatively, students can explore through video, slides, or even a lecture-question format. The key is to have students discover concepts on their own instead of being "spoon-fed."

The scientific learning cycle follows the ideas of constructivism. The exploration phase uses experiences (often concrete) to present data which cannot be explained by the students' existing knowledge structures. Students are encouraged to develop new knowledge structures by assimilation or accommodation, and the teacher ensures that this information is encoded with the correct terms. The concept application phase helps to organize the new knowledge structures.

The scientific learning cycle can easily be adapted to engineering education if appropriate laboratory equipment or computer simulation games are available. Adoption of the learning cycle to lecture-style classes is more problematic but is certainly possible. Demonstrations in front of an entire class can represent a concrete chance to explore, although with less freedom than with individual laboratory equipment. Exploration can also take place in lectures if the instructor describes phenomena and then has the students "experiment" by asking questions. The instructor has to be careful to allow them to discover concepts on their own. This approach may seem less efficient than the traditional lecture, but if efficiency is defined as student learning per amount of time, then the scientific learning cycle is more efficient.

15.2. LEARNING AND TEACHING STYLES

Individual preferences for learning and teaching are varied. Since mismatches can cause problems, professors should understand these styles. We have already explored learning styles in some depth, particularly in Chapters 13 and 14. These previous discussions on learning and teaching styles will not be repeated, but connections will be noted.

15.2.1. Dichotomous Styles

Many investigators have described dichotomies in learning styles. The Meyers-Briggs scheme includes the sensing-intuition dichotomy, while Belenky et al. (1986) introduce the dichotomy between separate and connected knowing into Perry's scheme. In addition, both Piaget and Perry note the dichotomy between rote memorization and true learning. Other ways of looking at dichotomous learning styles are briefly discussed below.

Reflection versus impulsivity (Claxton and Murrell, 1987) measures the tendency either to reflect over possible answers or to impulsively select a solution. This appears to be a relatively stable trait, but individuals can be taught either to slow down or to speed up. Students who lean toward impulsivity need to be taught to slow down so that they at least read all the possible answers. Students who reflect for such a length of time that they either become immobilized or take an excessively long time on tests can become a bit more impulsive. When people live or work together for a long period, they tend to approach each other on this dichotomy (that is, some learning occurs).

Information processing can be either deep or shallow (Claxton and Murrell, 1987; Schmeck, 1981). Deep processors learn the meaning and connections of ideas, whereas shallow processors tend to learn in terms of symbols and by memorization. For example, a deep processor learns the meaning of an equation and is able to use the equation if the symbols are changed. A shallow processor learns the equation in terms of symbols. If the meaning of symbols is changed, the shallow processor may have considerable difficulty in using the equation. Most students are capable of both types of processing. The professor, through homework and tests, exerts considerable control over which type they use. If the homework and tests emphasize rote learning, then shallow processing is reinforced. This is probably a good reason for not requiring the memorization of a large number of equations. Students in the concrete operational stage of development or on the dualistic levels of Perry's model may not be able to do deep processing, since deep processing skills appear fairly late in the developmental process.

Another learning style dichotomy involves deductive versus inductive learners (Felder and Silverman, 1988). Deductive reasoning starts with general principles and then deduces consequences from these general principles. For example, a variety of specific equations can be deduced from very general equations such as Maxwell's equations or the Navier-Stokes equations. Inductive reasoning starts with specifics and then proceeds to induce generalities. Inductive reasoning may appear to be a slower way to present new material, but it is the natural learning style. The inductive reasoning process is the natural way to construct a knowledge structure in a new area and is the style used in the scientific learning cycle. Inductive reasoning can be used by individuals at any level of development, whereas deductive reasoning requires that the individual be in the formal operational stage. When students are seeing the material for the second time, deductive reasoning is a very effective presentation style. Since a preliminary knowledge structure exists in this case, they have something on which to build their deductions. The apparent success of deductive reasoning in these cases has seduced many professors into employing deductive reasoning at all times. Introductory textbooks are much easier for students to understand if they are written in an inductive style, starting with fairly specific simple cases and building to generalities. A deductive style may be advantageous for advanced textbooks where students are seeing the material for the second or third time. At Arizona State University Anderson (1991) found that engineering students preferred an inductive style, while professors preferred to teach deductively. Clearly, there is a mismatch.

Field-independent versus field-sensitive learning represents another useful dichotomy for understanding the dynamics of teaching and learning (Claxton and Murrell, 1987; Robinson and Heinen, 1975). Field-independent individuals are less cognizant of the surroundings or field when they are working on a given task. For instance, these individuals can study effectively in a crowded, noisy college union. Field-independent individuals are more likely to be autonomous, and they often self-select into analytical fields such as engineering, mathematics, and science. Field-sensitive individuals are strongly influenced by authority figures and peer groups. They tend to be more people-oriented and are often good at working with others because they are aware of subtle messages. Achievement in a course does not appear to correlate with this dichotomy, but attitude and survival in a curriculum probably do. Groups which are underrepresented in engineering, women and some minorities, have a large percentage of field-sensitive individuals. Teaching methods such as collaborative learning which are attractive to field-sensitive individuals will probably help retain individuals in engineering (see Chapter 11).

People appear to process information either serially (sequentially) or globally (holistically) (Claxton and Murrell, 1987; Felder and Silverman, 1988). Serialists take information in logical sequence and build their knowledge structures step by step. They can function quite well without seeing the big picture and they learn best in well-defined, logical classrooms. Since most elementary and high school classrooms follow a sequential procedure, serialists often do quite well in school. Holistic learners are driven early in the process to create a knowledge structure which shows the big picture even though most of the details are missing. As they learn, holistic learners fill in the details. Serialists tend to be better at details, and holists are better at overviews or seeing how everything fits together. Obviously, skill at both tasks is useful. Advance organizers are extremely useful for holists and are probably ignored by most serialists. Since globalists often struggle, particularly in introductory courses, it is important for professors to provide some aid and encouragement. In advanced classes globalists may have an advantage since they can see connections and do syntheses which are difficult for serialists. At Arizona State University sequential learning was the preferred learning mode for engineering students and the preferred teaching style of professors (Anderson, 1991).

The final dichotomy to be discussed involves active and reflective processing of information (Kolb, 1984; Stice, 1987; Claxton and Murrell, 1987; Felder and Silverman, 1988). This dichotomy is part of the Kolb learning cycle which is discussed in Section 15.3. Active experimenters want to do something with the information in the external world. For example, they want to discuss, teach, solve, or make something. They want to try the activity and learn by doing. This dimension is closely related to extroversion. Reflective individuals want to process the information internally (introversion). They want to ponder it. However, a noninteractive lecture is optimum for neither style of learner. As in the case of all the dichotomies discussed, individuals can learn to learn better if they can use both techniques when appropriate. Anderson (1991) found that engineering students prefer active processing, while the preferred teaching style is reflective.

Whether these dichotomies are independent constructs appears to be doubtful. Claxton and Murrell (1987) report that Kirby (1979) hypothesized that there may be only two fundamental groups which he calls "splitter" and "lumper" types and which overlap with left-brain and right-brain analyses. The splitters include field-independents, serialists, abstract, separate-knowledge individuals, whereas the lumpers include field-sensitive, holistic, concrete, connected-knowledge individuals. If this is true, then the dichotomies are not independent, but each dichotomy adds to the picture of how people learn. However, individuals are complex and have the disturbing habit of not fitting into any theory.

15.2.2. Auditory, Kinesthetic, and Visual Modes

People use three different modes for perceiving the world: auditory, kinesthetic, and visual. Everyone without a major physical handicap has the ability to use all three modes. For example, at a feast you can first enjoy the sight of the food and the table. Then you can enjoy the smell, taste, and feel (all kinesthetic) of the food and drink. Finally, after the meal you can sit back and enjoy the feast again by talking about how wonderful it was. As in other aspects of learning, most of us have developed a favorite mode of perception for learning about the world. This favorite mode affects how we learn in different situations (Felder and Silverman, 1988; Murr, 1988; Waldron, 1986).

Kinesthetic learning includes taste, touch, smell, and feelings. Kinesthetic learning is important for chefs, athletes, therapists, artists, skilled craftspersons, and others. Kinesthetic learning occurs in engineering education when students work in laboratories and handle real components such as circuit boards, valves, and machine tools. Passing objects around during a lecture not only spices up the class but also incorporates kinesthetic learning. Touch can be useful to understand the smoothness of objects or the heat generated when a bearing is binding. The sense of smell can be used as part of the learning process for food process engineers, chemical engineers, and environmental engineers. Smell can help tell if a process is operating correctly or incorrectly. Feelings or affective aspects of learning are always present. Success and praise can help engender a positive attitude (feelings) toward the course, while failure and criticism do the reverse. Although criticism is often necessary, professors should never try to humiliate or belittle students. Writing about something is a good way to learn, partly because it involves both kinesthetic and auditory learning.

Visual learners prefer to process information in pictures, and they prefer to learn from pictures, charts, diagrams, figures, actual equipment, photographs, graphic images, and so forth. This appears to be the preferred mode of learning for most people (Barbe and Milone, 1981) and was the preferred mode for engineering students (Anderson, 1991). The phrase, "A picture is worth a thousand words," is a common-sense way of saying that most people prefer visual learning. Visual information appears to be easier to understand and place into memory than words (Kiewra, 1987). Visual learning can be incorporated into engineering education in a variety of ways. Plotting equations to show their shape makes them much more real for many students. This can be done conveniently with calculators with plotting screens. Graphical solution methods are easier for many students to understand than solving equations analytically. Showing that the intersection of two curves is the simultaneous solution of two equations helps students understand what this means. Graphical solutions to more complex problems such as a McCabe-Thiele diagram in distillation or a Bode plot in control, help many students understand the solution procedure. Showing graphical integration procedures and comparing these to Simpson's rule or other integration procedures helps clarify for the student the meaning of the integration procedure. Correlations of data should be shown both in a figure with the scatter of data and as an equation with the correlation coefficient. Equipment sketches and diagrams should be insisted on for the solution of all problems. Computer-aided three-dimensional diagrams can help to clarify complex concepts in mechanics and other areas. Field trips or at least professionally produced videos of plant sites help students see the "real thing." For many students this one-time exposure to real equipment makes an entire semester of equations and problem solving much more understandable. Students in co-op programs also benefit from this aspect of visual education.

Auditory teaching methods are most commonly used in Western education systems. This includes lectures and print material. Reading in Western cultures is a visual representation of auditory processing techniques. In contrast, Chinese ideograms are closer to visual processing, and Eastern education has a more visual character (Murr, 1988). Writing words or equations on the blackboard is also a visual representation of an auditory method. Few people prefer to use auditory learning if given a choice; however, the Western educational system does not usually provide for a choice. Successful students have adjusted to auditory teaching styles before they reach college. One of the basic tenets of learning theory is that learning is more thorough and is retained better if multiple modes are used to input and process the information. Stice (1987) reports on some early data from the Socony-Vacuum Oil Company which supports this contention. For reading alone, the learner's retention was 10 percent; for hearing alone, 26 percent; and for seeing, 30 percent. If the learner both saw and heard, retention was 50 percent; if the learner said something, retention was 70 percent; and if the learner said and did something, the retention was 90 percent. Thus, auditory styles of teaching should be heavily supplemented with visual and, to a lesser extent, kinesthetic learning opportunities. Opportunities for the student to speak, write, and solve problems should be incorporated in the course. With a little creativity this can often be done without major changes in the course format or coverage. Since visual learning is the preferred style for most students, it is also useful to consider if the entire course can be presented in a mainly visual style. This revision would probably require major changes in the course.

15.3. KOLB'S LEARNING CYCLE

Kolb (1984, 1985) developed a two-dimensional circular or three-dimensional spiral model of how people learn (see also Atkinson and Murrell, 1988; Claxton and Murrell, 1987; Felder and Silverman, 1988; McCarthy, 1987; Stice, 1987; Svinicki and Dixon, 1987). Kolb's model starts with two dichotomies which are considered to be orthogonal to each other. The first of these, active experimentation (AE) versus reflective observation (RO), was discussed briefly in Section 15.2.1. This dichotomy refers to how individuals prefer to transform experience into knowledge. Individuals who favor active experimentation like to get things done and see results. Reflective observers prefer to examine ideas from several angles and to delay action.

The second dimension in Kolb's theory is the dichotomy between abstract conceptualization (AC) and concrete experience (CE). This dimension distinguishes between how an individual grasps or takes in information. Abstract conceptualizers prefer logical analysis, abstract thinking, and systematic planning. Individuals who favor concrete experience want specific experiences and personal involvement, particularly with people, and tend to be nonsystematic.

Kolb considers each of these four areas to be steps in learning. McCarthy (1987) modified and extended Kolb's model to apply it to teaching. The complete learning cycle shown in Figure 15-2 requires all four steps; thus, a proficient learner is able to complete all steps in the cycle although he or she prefers certain modes of operation. The cycle can be entered at any of the four steps, but usually starts with the concrete experience method of grasping information. This information is then transformed or internalized by reflective observation (RO). For complete learning the individual should continue around the circle and use abstract conceptualization to perceive the information that has now been changed by reflection. Next the learner processes the information actively and does something with it. For complex information the circle is traversed several times in a spiral cycle. The spiral may extend through several courses and on into professional practice as the individual learns the material in more and more depth.

Kolb's learning cycle is a theory describing the steps required for complete learning. Unfortunately, students often take short-cuts and employ only one or two stages in the cycle, which results in significantly less learning. A study of the retention of knowledge showed 20 percent retention when only AC was used, 50 percent with RO and AC, 70 percent with CE, RO, and AC, and 90 percent when all four stages were employed (Stice, 1987). Most college education is geared to abstract conceptualization, but retention (hence long-term learning) is

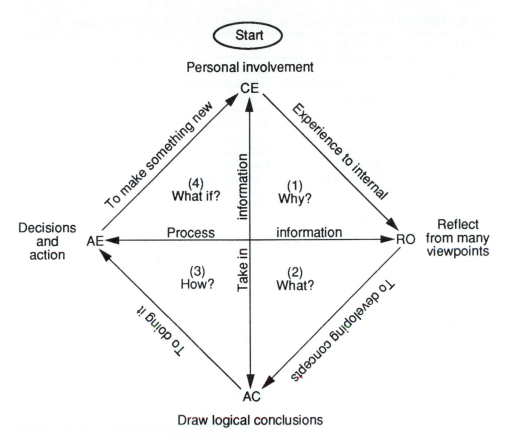

Start

Personal involvement
CE

Experience to internal

To make something new

(4)
What if?

(1)
Why?

information

Decisions
and
action
AE ◄──────── Process   information ──────► RO

Reflect
from many
viewpoints

Take in

(3)
How?

(2)
What?

To doing it

To developing concepts

AC

Draw logical conclusions

FIGURE 15-2 MODIFIED KOLB'S LEARNING CYCLE.
Learning styles: quadrant 1, divergers; quadrant 2, assimilators; quadrant 3, convergers; quadrant 4, accommodators.
Based on Kolb (1984), McCarthy (1987), Harb et al. (1991), and Svinicki and Dixon (1987).

enhanced by use of other stages in addition. Requiring more active involvement by students increases learning because additional stages in the learning cycle are used. Cooperative education and summer jobs aid learning because they involve the student in doing and in concrete experience.

Kolb's learning cycle is useful for conceptualizing how people learn and for developing courses and training programs (Claxton and Murrell, 1987; McCarthy, 1987). Stice (1987) first discussed applications in engineering education. A lecture (RO) can be followed by requiring students to think about the ideas (AC), do homework (AE), and observe demonstrations or do laboratory experiments (CE). Retention should be significantly better than in a course requiring only regurgitation of lecture (RO) and homework (AE).

McCarthy (1987) showed that Kolb's theory is similar to many other theories of learning. She extensively modified Kolb's theory and applied it to teaching a variety of topics at all levels. McCarthy's 4MAT system has been applied to engineering classes by Harb et al. (1991), Terry et al. (1991), and Todd (1991). We will discuss the modified and extended Kolb learning cycle or 4MAT system in more detail.

TABLE 15-1   TEACHING AND LEARNING ACTIVITIES (Harb et al., 1991; McCarthy, 1987; Svinicki and Dixon, 1987)

Diverger (1)	Assimilator (2)	Converger (3)	Accommodator (4)
Motivation	Information and facts	Try it	Do it themselves
"War" stories	Lecture	Homework problems	Self-select projects
Brainstorming	Reading	Laboratory	Design
Observations:	Instructor or	Simulations	Open-ended problems
Field trips	TV demonstration	CAI	Write problems
"On street"	Patterns	Problem solving	Field trips
Logs	Organizing	Short answer	Work experience
Journals	Analyzing	Reports	Simulations
Role playing	Objective tests	Demonstrations	Teach yourself
Discussion	Library Work	Experiment	Teach someone else
Questioning	Problem-solving examples	Tinker	Think tank
Visualization	Seminars	Record	
		Make things work	

This teaching and learning system starts each instructional unit with concrete experience (CE) and leads to reflective observation (RO). The student learns why the material is important in the first quadrant of Figure 15-2. This is the motivation step which professors often skip. McCarthy (1987) suggests performing first a right-brain-mode activity and second a left-brain-mode activity to create reasons for learning material. The right-brain-mode activity can be experimental such as going out "on the street" and seeing and feeling the need for a bridge at a specific location. The left-brain-mode activity can then reflect on the need for the bridge. McCarthy (1987) suggests breaking down the learning activities in all four quadrants into both right and left activities. Possible teaching and learning activities are listed in Table 15-1.

In the second quadrant students move from reflective observation (RO) to abstract conceptualization (AC). They think and learn concepts. The key question is what? What are the facts? What body of knowledge are the students supposed to learn? For students studying bridge building various aspects of bridge design are covered in class. The teacher's role is to teach. This quadrant is normally the major part of typical engineering courses, and activities are listed in Table 15-1.

In the third quadrant students move from thinking to doing. They want to answer the question How does it work? This is where homework assignments, laboratory sessions, and fieldwork fit into engineering education. In the example on bridge building, students can do homework on bridges and test model bridges in the lab. The professor coaches them and facilitates their efforts but lets them do it themselves. Engineering and technology programs include at least some courses where the third quadrant is heavily used. Activities are listed in Table 15-1.

In the fourth quadrant students remain active and move from active experimentation to concrete experience. This completes the cycle, but the students return to concrete experience with a very different understanding of the knowledge. In this fourth quadrant they can teach themselves and others, ask what-if questions, and do something with the knowledge. They can create their own experiment or construct a model of their design. For example, for the class on bridges students can choose from a variety of projects such as designing a new bridge, building a model, producing a portfolio of bridge photographs, and so forth. Other activities are listed in Table 15-1.

The usual college education uses what McCarthy (1987) calls a "pendulum style" of teaching. That is, it oscillates between quadrants 2 and 3. This style never goes around the entire cycle. Thus students are seldom motivated and seldom have the opportunity to do it themselves unless they have co-op or summer jobs. The pendulum style reduces retention and, as we shall see shortly, does not satisfy the favorite learning style of many students.

Kolb also developed a theory of learning styles (Kolb, 1984, 1985; McCarthy, 1987). A short psychological test which provides numerical scores for the grid is available (Kolb, 1985). The four styles are illustrated in Figure 15-2. Convergers prefer abstract conceptualization (AC) and active experimentation (AE) (quadrant 3). They enjoy logic, practical application of ideas and theories to solve problems and are often quite focused. They tend to use deductive reasoning and are good at solving problems with a single answer. Many engineers, technologists, computer scientists, and physical scientists are convergers. The favorite learning style of convergers is in quadrant 3 where they can do experiments and design equipment. If too convergent, these individuals may tend to act without reflection and to think without feeling. As a result, they may be perceived as being arbitrary and cold. Since convergers need to relate theory to practical applications, case studies, laboratory, field trips, and work experience are a very helpful part of their education.

Assimilators prefer abstract conceptualization and reflective observation (Quadrant 2). They are excellent at understanding information and developing logical forms, prefer inductive reasoning, and are good at creating theoretical models. They can be contrasted with convergers since they do not worry about practical aspects. They do share the AC aspect with convergers and are often more interested in ideas than in people. Many teachers, writers, lawyers, mathematicians, scientists, and engineers with a scientific bent are assimilators. Assimilators often do well in lecture classes, and their favorite learning style is in quadrant 2. Assimilators are systematic planners, but they may ignore the human aspect.

Accommodators prefer active experimentation and concrete experience (Quadrant 4). They are similar to convergers in that they like to act and to get things done. They differ from convergers in that they are less logical and are more people-oriented. If the theory does not fit the experiments, they will often discard the theory and go with what works. They enjoy new experiences and are often willing to take risks. Accommodators are often found in business or large organizations where they enjoy marketing, sales, managing, politics and public relations. They do well in hands-on group activities in class or group laboratory assignments. They prefer quadrant-4 activities. Accommodators may be seen as pushy and nontheoretical (a no-no in engineering education), and they rely heavily on trial and error.

Divergers are the opposite of convergers, preferring concrete experience and reflective observation (Quadrant 1). Often imaginative, emotional, and good at seeing the global picture, they tend to do well in working with people, recognizing problems, and generating many alternatives. Unfortunately, if too divergent, they may not make decisions and will not get things done. Divergers often become artists, actors, personnel managers, counselors, and social workers. In a classroom, divergers do well in quadrant-1 activities such as group exercises, particularly brainstorming-type activities.

TABLE 15-2 DISTRIBUTION OF PREFERRED LEARNING STYLES (McCarthy, 1987)

Diverger (1)	Female (%)	Male (%)	Total (%)
Learning styles:			
Diverger (1)	25.0	19.4	23.0
Assimilator (2)	27.5	37.5	31.1
Converger (3)	14.8	23.5	17.5
Accommodator (4)	32.7	19.6	28.5
Dimensions:			
Concrete (1 plus 4)	57.7	39.0	51.5
Abstract (2 plus 3)	42.2	61.0	48.5
Reflective (1 plus 2)	52.5	56.9	54.1
Active (3 plus 4)	47.5	43.1	45.9

It is important to note that these are *preferred* styles, but that everyone has the capability to use and the need to develop all four styles. Working through Kolb's entire cycle automatically has students use all styles. In addition, every student has an opportunity to shine when the learning activity is in her or his favorite quadrant. The distribution of preferred learning styles for teachers and administrators was determined by McCarthy (1987) and is given in Table 15-2. It is interesting to note that higher percentages of men than of women are assimilators and convergers, which are the typical engineers, scientists, and technologists. Men tend to prefer abstract methods for taking in information, while women prefer more concrete approaches. These style preferences are not cast in stone. Students who are in a program which heavily emphasizes a given learning style tend to shift their preferences toward that style (if they survive). Also, as people get older they tend to process information more reflectively and less actively.

Individuals who prefer any of the four learning styles can find a niche where they will be successful engineers. After school, accommodators tend to move toward management, sales, and marketing; divergers move toward personnel and creative positions. Convergers tend toward hard-core engineering jobs such as plant operations, design, and construction. Assimilators gravitate toward research, development, and planning. Since technically trained people are needed in all these jobs, it is important to design educational programs to retain students with each of these styles. In school, convergers and assimilators are likely to find more kindred spirits among both teachers and their peers. Thus, it is the accommodators and the divergers who are most at risk in engineering education.

Teachers also have styles. If these styles differ from those of their students, the mismatch can cause problems. For example, assimilators emphasize logic, abstract theories, and ideas without applying them to practical problems. Convergers in the class do not consider the class to be practical and may not see the practical applications of the material. All students may have problems applying the material if later classes are taught in a convergent fashion. This mismatch often explains why engineering students are unable to use the mathematics they studied earlier. The teacher can help all students by including all aspects of Kolb's learning cycle. This provides some activities that are appropriate for each student, and helps each student broaden his or her repertoire of skills.

15.4. MOTIVATION

Regardless of the student's learning style and basic intelligence, he or she will not learn if not motivated. Unfortunately, "nobody can't teach nobody nothing" (Kolstoe, 1975, p. 61). Thus, student motivation is crucial to learning. Although much of this motivation is beyond the teacher's control, he or she can do a great deal either to motivate or demotivate students.

Motivation is usually considered either intrinsic or extrinsic. Intrinsic motivation is internal. It often satisfies basic human needs which include physiological needs, as well as the need for safety, belongingness, love, esteem, and, finally, self-actualization (Maslow, 1970). Extrinsic motivation is externally controlled and includes many things that the instructor can do, including grading, providing encouragement and friendship, and so forth. The differences between intrinsic and extrinsic motivation are not always sharp. For example, a high salary might be considered to be an extrinsic motivator, but it can also enhance an individual's self-esteem. Both intrinsic and extrinsic motivation will be discussed in terms of Maslow's theory of human needs and motivation.

15.4.1. Student Motivational Problems

Students can have a variety of motivational problems. Since the "cure" often depends upon the problem, it will be helpful to list some of the problems briefly.

1 The student does not want to study engineering or even to be in college. A surprising number of students are in engineering because of parental pressure. Failure is one way the student can prove that the parents are wrong. Research clearly shows that students who do not believe in the importance of education have lower success in school (*What Works*, 1986).

2 The student is not under pressure to be in engineering but is uncertain if engineering is the best choice. Since many outstanding engineers were once in this category, a major motivational effort may be appropriate. Since students need to see meaning in their studies, the motivation effort can focus on this. Once purpose is instilled, these students can become outstanding engineers.

3 The work ethic is absent. Many students coast through high school and find engineering painfully hard work. Installing a work ethic at this late date may be difficult, but it is important for success in engineering.

4 The background in prerequisites is inadequate. Success is very motivating, but with an inadequate background students may be unable to be successful in a specific course or in the entire curriculum.

5 The student feels isolated and perhaps discriminated against. This can particularly be a problem for women and minorities who are traditionally underrepresented in engineering. It can also be a problem for international students.

6 The student finds engineering classes or classes in general distasteful. If the student's learning styles are very different from the professors' teaching styles, the student may find classes unrewarding even if they are not difficult. Some students find engineering classes too competitive or feel they never get rewarded for their efforts.

7 External problems are overwhelming. A death in the family, health problems, financial difficulties, relationship problems, and so forth, can prevent students from being motivated in their studies.

8 The student becomes overly anxious during tests or while doing homework. The discomfort caused by excessive anxiety can reduce motivation. High stress on tests is detrimental to all students but hits women harder than it does men (McKeachie, 1983). Anxiety and stress can be controlled by desensitization procedures (such as giving more tests), by relaxation methods (see Section 2.7), and by giving the student more control of the grade he or she will earn.

9 The student wants only a grade or a degree and does not care about learning the material. Although the professor may think that the student is motivated for the wrong reason, these motivations can be used to get the student to learn.

10 The student is not intelligent enough. We placed this reason last since, contrary to the opinion of many professors, the lack of intellectual ability is seldom the major reason for a lack of motivation, although it may contribute, particularly for concrete operational students. A significant body of research shows that "accomplishment in a particular activity is often more dependent upon hard work and self-discipline than on innate ability" (*What Works*, 1986).

15.4.2. Maslow's Hierarchy of Needs

According to Maslow's (1970) theory of motivation, which has become widely accepted, individuals have a hierarchy of needs (Figure 15-3). When a need is unfulfilled, the individual is very motivated to fulfill that need. Once needs at the lower levels are satisfied, higher-level needs become important and the individual becomes motivated to satisfy these needs. If one of the lower-level needs is suddenly not satisfied, then this need becomes the most important need until it is again satisfied. For example, a Ph.D. in engineering who is lost in the woods and starving thinks only about food and rescue, not about abstract theory. Maslow noted that the hierarchy is not invariably followed by all individuals.

Western society tries to satisfy the physiological and safety needs for everyone, although not always successfully. Since professors and most students have these needs satisfied, we tend to ignore their importance. Professors need to remember that for some of their poorer students these needs may be very important. It is difficult to focus on studying if one is wondering where money for food or rent will come from. This type of external problem needs to be solved with financial aid, not by exhortations to study. A student who is terrified to walk back to a dorm after dark will not benefit from help sessions or the availability of a computer laboratory. These safety needs must be met by proper campus lighting, police patrols, and an escort service before the student can focus on studying.

When students leave home to go to college, they often find that the needs for belonging and love are no longer satisfied. Parents and friends several hundred miles away may be insufficient to satisfy these needs. Part of the adjustment process for freshmen, transfer

FIGURE 15-3   MASLOW'S HIERARCHY OF NEEDS

students, and graduate students involves satisfying the belongingness needs in a strange location. The adjustment process tends to be worse for freshmen because they have less experience in satisfying these needs on their own. The school can help by encouraging students (and for freshmen, their parents also) to visit before registration. Mixers and other get-togethers are useful in helping new students meet others. Living in a residence hall is particularly helpful to freshmen and also helps their development on Perry's scale (see Chapter 14).

Professors have an important role to play in helping to satisfy belongingness needs. Retention of students is significantly enhanced when students are integrated into the university both socially and academically (Smith, 1989). Academic integration includes contact with faculty and staff, involvement in the curriculum, and academic performance. Students who have made significant contact with a faculty member during the first six weeks of the semester are more likely to become academically integrated and remain at the university. To make contact with students the professor must at a minimum learn everyone's name. A more active approach such as inviting small groups of students to his or her house or for coffee at the student lounge can have a positive impact. It is interesting that significant contact almost always occurs for new engineering graduate students, but at large universities is often absent for freshmen. Students who do not want to be in engineering or who are unsure about engineering have more difficulty achieving academic integration. Counseling, support, and encouragement can help these students. The ability of engineering to satisfy other needs may help them become academically integrated. Thus, spending some time in introductory classes talking about the many joys and advantages of being an engineer helps some students get past a difficult period. Strong negative feedback attacks both the need for belonging and esteem. Unfortunately, the sting of negative feedback lasts much longer than the glow from positive feedback (Boschman, 1987). Professors need to be creative in finding ways to use positive instead of negative feedback.

Students with very different learning styles often do not feel that they belong in engineering. A relatively small amount of course modification to include other learning styles can help these students feel they belong. These modifications were discussed in Section 15.2. A particularly important change for many students is to make learning more cooperative and less competitive (Smith, 1989). Cooperative group exercises and grading which does not pit students against each other can help convince them that the true adversary is ignorance, not the professor or each other. The need to belong can have a negative impact on the student's desire to study since some groups may exclude students who do too well in class. This can be combated by developing groups such as honor societies, study groups, or professional organizations where academic excellence is appreciated.

A major need that can be fulfilled in class is that for esteem. Grades are often the most important motivating device (McKeachie, 1986) because they directly relate to the esteem needs, and grades are under the professor's control. Achievement, reputation, and self-respect can all be enhanced by good grades. The perception that one is doing well is very motivating. Excusing students from the final because of good grades during the semester can be an excellent motivator for the better students. Yet grades won't motivate if students believe that high grades will interfere with their belonging, and the belongingness needs are unfulfilled. When unfulfilled, the lower-level needs are more important. Good grades must also be seen to be achievable. Students with poor academic backgrounds and poor study habits quickly learn that they cannot achieve good grades. For them, grades are a demotivator. Remedial help and tutoring can help these students succeed. Another modification which involves considerable effort, but is extremely valuable for some students, is to use a flexible time frame and allow the students to spend more time learning. This can be done in mastery or self-paced classes (see Chapter 7). Since every student can achieve if given sufficient time and encouragement, these classes can be very motivating.

Needs for esteem and belongingness are also met by respect from faculty and by positive feedback. Eble (1988) states that respecting students as human beings without requiring them to prove themselves is one of the most important things a teacher can do to help them grow. Feedback should be immediate, and if at all possible should contain some positive aspects. Effort should be praised even if it is somewhat misplaced. Professors can learn from successful coaches in this respect. For example, in basketball when a player fouls, the coach may praise the player for a good hustle and then correct him or her for the foul. Negative feedback should be avoided if at all possible, but if necessary it should be focused entirely on the performance and not on the person. Unfortunately, negative reinforcement may result in unexpected and undesired behavior changes such as avoiding class entirely to avoid being yelled at. Criticizing a student as lazy is an attack on the person. In the long run, it is usually more productive to point out that the performance is not up to the student's ability and is not satisfactory. Smiles, nods, and encouragement for responses are all positive reinforcement. Greeting a student by name with a smile in the hall or in your office is also positive reinforcement which can help to meet the student's esteem needs. This reinforcement is unexpected and intermittent and thus is very powerful. Many students who leave engineering cite discouragement and the lack of support as major reasons (Hewitt and Seymour, 1992).

Assignments and tests motivate students to keep up with the class since they tap into the need to be successful and avoid failure. Motivation for doing tests and assignments appears to be highest when there is a fair but not certain chance for success (McKeachie, 1986). The professor should introduce assignments and tests with positive expectations for student performance. These positive expectations are in themselves motivating (Peters and Waterman, 1982; *What Works*, 1986). Success is motivating. It is worthwhile to ensure that there is some aspect of an assignment or course at which each student can be successful. The workload should be reasonable since excessive work is demotivating and reduces the chance of success.

The prospect of a good salary upon graduation is often considered to be a crass extrinsic motivator. Based on Maslow's theory, there are often good reasons why the promise of salary is a strong motivator. If the student experiences periods when physiological or safety needs are not met, then the salary can be a way of ensuring this does not happen again. Engineering should promote itself as a way up and out of poverty. Parental pressure to go into engineering may arise from the parents' desire to have a son or daughter earn a good salary. If satisfying parents helps meet belongingness and love needs, then the student may be positively motivated. For many students the salary helps to satisfy the need for esteem. Since salary after graduation is a long way off for a freshman or sophomore, the more immediate reinforcement of a summer or a co-op job may be a better motivator.

The chance to present a paper at a meeting and to be a coauthor on a published paper can help meet a student's need for esteem and reputation. This can be a tremendous motivator for graduate and undergraduate students. Students work harder on research when they have a self-imposed deadline (paper presentation or the desire to graduate) than when pushed by the professor.

The highest level in Maslow's hierarchy, self-actualization, is the need for individuals to reach their potential. The need to self-actualize is what causes individuals to write poetry at 2 A.M. when they have to report to a respectable, well-paying job at 8 A.M. Cooking gourmet meals when something simpler would suffice may represent the need to self-actualize. Creativity and the need to create can be considered part of the need to self-actualize. Maslow notes that for extremely creative individuals the need to create may be more important than the lower needs. People require time to learn how to satisfy their needs. Thus self-actualization occurs in mature individuals and based on Maslow's studies is uncommon. Self-actualized students are more likely to be encountered in graduate or continuing education classes.

Self-actualized individuals have a need to guide their own destiny. In class they appreciate the chance to do individual projects and delve into a topic of their choice at considerable depth. Bonus problems and other methods which give them some control over what they do are appreciated. In research they want to guide their own projects. The professor's job is to step back and serve as a resource person when asked.

Maslow notes that cognitive needs are present throughout the five stages. There is joy in learning and creating which can be used to motivate. However, professors must make an effort to remove barriers that prevent students from achieving the joy of learning. The professor's enthusiasm and joy in learning the subject can be contagious. Sleeping students are not learning. Lecturing with energy, excitement, and some humor at least keeps students awake. And students enjoy classes more and learn more when the professor performs (see Section 6.3).

The force of curiosity is most evident in young children and in self-actualized individuals. Professors can use curiosity as a positive motivator in the classroom. For example, in a lecture questions can be asked and not be answered. We have found that questions which ask the students to use their engineering knowledge to explain nature often pique their interest. Why does a car window frost over at night when the window on an adjacent building does not? What is wind chill? Or, have the student estimate how long it will take for a person to respond on a very long-distance telephone call. Other variations of the socratic approach can be used. The important point is to ask questions which are thought-provoking for a group of students. This use of curiosity, like all motivating techniques, will work for only a portion of the class.

At all levels of Maslow's hierarchy the locus of control is important. People who believe they have some control over their work life are more strongly motivated (Peters and Waterman, 1982). Students can be provided with a modicum of control with grade contracts, a choice of projects, a choice of problems on a test, or a vote on the test date. Graduate students, in particular, can be given significant control over their projects and often respond with extraordinary energy.

All writers on motivation in college teaching (e.g., Eble, 1988; Ericksen, 1974; and McKeachie, 1986) note that teachers need to be creative in developing motivational techniques. With a creative effort the professor can often find just the right thing to do to motivate a particular student. For example, we have seen graduate students become very motivated when given the opportunity to present a paper at a meeting or to tutor students. The chance to coauthor a research paper has sparked some undergraduates. Having a piece of equipment actually constructed and used while on a co-op assignment has turned students on to engineering. Taking a mastery class and being able to succeed academically for the first time in college has been a tremendous motivator for some students. One student obtained the help he needed once a professor took the time to sit and talk with him about the potential career consequences of his inability to communicate. Informal parties at a professor's house have helped many students feel at home at the university and thus have satisfied their belongingness needs. Often it is the attention and not the actual action which increases the students' motivation. This is the famous "Hawthorne effect" (e.g., see Peters and Waterman, 1982). A professor can motivate classes by continually creating the Hawthorne effect by always experimenting. Professors control motivation in a class by their actions. If they give lip service to creative problem solving but always emphasize drill on homework and tests, the students will do drills. To obtain creative solutions there must be a focus on the activity. Many other examples could be cited.

15.5. CHAPTER COMMENTS

This chapter is not a complete picture of how individuals learn because that complete picture is not yet known or even sketched out. Individuals who prefer a global learning style may find this fragmentation disconcerting. However, enough is known and well documented by research that we have been able to make some firm recommendations about what is known to work. Many of the suggestions can be tried piecemeal with little effort. In the space available we have been unable to cover all the theories which can be used to understand learning and improve engineering education. In particular, the research on right- and left-brain functioning and the research on expert systems has not been included. The interested reader might start with Edwards (1989), Gazzaniga (1970), McCarthy (1987), and Springer and Deutsch (1989) for right-left brain research, and Smith (1987) for expert systems and artificial intelligence applications in engineering education.

Our experience in teaching this chapter is that some students become extremely excited about Kolb's theory. They read his and McCarthy's books, do a project using his theory, and plan on incorporating his theory into their classes.

15.6. SUMMARY AND OBJECTIVES

After reading this chapter, you should be able to:

• Extend Piaget's theory to the constructivism theory. Explain how constructivism and the scientific learning cycle can be used to improve engineering education.

• List and discuss the dichotomous learning and teaching styles. Type yourself on these styles. Discuss what you could do to improve your teaching.

• Delineate how auditory, kinesthetic, and visual styles affect learning and how they can be incorporated in engineering education.

• Explain Kolb's learning cycle and the implications of this theory in engineering education.

• Explain Maslow's theory of needs and discuss applications in engineering education.

HOMEWORK

1 Develop a key relations chart for this chapter.
2 Develop a concept map for this chapter.
3 Pick a topic in one of your engineering classes.
   a Determine how to teach it using the scientific learning cycle.
   b Determine how to teach it using Kolb's learning cycle.
   c Compare parts a and b.
4 Do the second objective in Section 15.6 (list dichotomous learning/teaching styles).
5 Do the third objective in Section 15.6 for a specific engineering class.
6 Choose a student whom you know well and who is not strongly motivated. Analyze this student by Maslow's theory. Determine some interventions which might help motivate this individual. Try one or two of the interventions.
7 Analyze the scientific learning cycle in terms of Kolb's learning cycle. Note which steps in the scientific learning cycle match quadrants in Kolb's cycle. Compare the order of steps. Both methods have been shown to work. Comment on why both approaches work. Which would you prefer to use? Why?

REFERENCES

Anderson, M. R., "Characterizations of the graduate career change woman in engineering: Recruitment and retention," *Proceedings ASEE/IEEE Frontiers in Education Conference*, IEEE, New York, 248, 1991.

Atkinson, G., Jr., and Murrell, P. H., "Kolb's experiential learning theory: A meta-model for career exploration," *J. Couns. Develop.* 66, 374, 1988.

Barbe, W. B., and Milone, M. N., "What we know about modality strengths," *Educ. Leadership*, 378 (Feb. 1981).

Belenky, M. F., Clenchy, B. M., Goldberger, N.R., and Torule, J.M., *Women's Ways of Knowing: The Development of Self, Voice and Mind*, Basic Books, New York, 1986.

Bodner, G. M., "Constructivism: A theory of knowledge," *J. Chem.Educ.*, 63, 873 (1986).

Boschman, E., *Ten Teaching Tools. Ten Secrets to Total Teaching Success*, Kendall/Hunt, Dubuque, IA, 1987.

Cashin, W. E., "Motivating students," Idea Paper No. 1, Center for Faculty Evaluation and Development, Kansas State University, Manhattan, KS, 1979.

Claxton, C. S., and Murrell, P. H., *Learning Styles: Implications for Improving Education Practices*, ASHE-EPIC Higher Education Report No. 4, Association for the Study of Higher Education, Washington, DC, 1987.

Dansereau, D. F., "Technical learning strategies," *Proceedings ASEE/IEEE Frontiers in Education Conference*, IEEE, New York, 165, 1986.

Eble, K. E., *The Craft of Teaching*, 2nd ed., Jossey-Bass, San Francisco, 1988.

Edwards, B., *Drawing on the Right Side of the Brain*, rev. ed., Jeremy P. Tarcher, Los Angeles, 1989.

Erickson, S. C., *Motivation for Learning*, University of Michigan Press, Ann Arbor, MI, 1974.

Felder, R. M., and L. K. Silverman, "Learning and teaching styles in engineering education," *Eng. Educ.*, 674 (April 1988).

Flammer, G. H., "Applied motivation—A missing role in teaching," *Eng. Educ.*, 519 (March 1972).

Gazzaniga, M., *The Bisected Brain*, Apple-Century-Crofts, New York, 1970.

Harb, J. N., Durrant, S. O., and Terry, R. E., "Use of the 4MAT system in engineering education," *Proceedings ASEE/IEEE Frontiers in Education Conference*, IEEE, New York, 612, 1991.

Hewitt, N. M., and Seymour, E., "A long discouraging climb," *ASEE Prism*, 1(6) 24 (Feb. 1992).

Kiewra, K. A., Memory-compatible instruction," *Eng. Educ.*, 285 (Feb. 1987).

Kirby, P., *Cognitive Style, Learning Style, and Transfer Skill Acquisition*, Information Series No. 195, Ohio State University, National Center for Research in Vocational Education, Columbus, OH, 1979.

Kolb, D. A., *Experiential Learning: Experience as the source of learning and development*, Prentice-Hall, Englewood-Cliffs, NJ, 1984.

Kolb, D. A., *Learning Style Inventory*, McBer & Co., Boston, 1985.

Kolstoe, O. P., *College Professoring: Or, Through Academia with Gun and Camera*, Southern Illinois University Press, Carbondale, IL, 1975.

Kurfiss, J. G., *Critical Thinking: Theory, Research, Practice, and Possibilities*, ASHE-ERIC Higher Education Report No. 2, Association for the Study of Higher Education, Washington, DC, 1988.

Lawson, A. E., Abraham, M. R., and Renner, J. W., *A Theory of Instruction: Using the Learning Cycle to Teach Science Concepts and Thinking Skills*, Monograph *1*, National Association for Research in Science Teaching, Cincinnati, OH, 1989.

Lowman, J., *Mastering the Techniques of Teaching*, Jossey-Bass, San Francisco, 1985.

Maslow, A., *Motivation and Personality*, 2nd ed., Harper and Row, New York, 1970.

McCarthy, B., *The 4MAT System. Teaching to Learning Styles with Right/Left Mode Techniques*,

EXCEL, Barrington, IL, 1987. (The format of this book may seen strange to engineers, but the book rapidly gets into practical and theoretical details.)

McKeachie, W. J., "Student anxiety, learning and achievement," *Eng. Educ.,* 724 (April 1983).

McKeachie, W. J., *Teaching Tips*, 8th ed., D.C. Heath, Lexington, MA, 1986.

Mettes, C. T. C. W., Pilot, A., Roosink, H. J., Kramers-Pals, H., "Teaching and learning problem solving in science. Part II: Learning problem solving in a thermodynamics course," *J. Chem. Educ.,* 58(1), 51 (Jan. 1981).

Murr, L. E., "Engineering education in the visual culture," *Eng. Educ.,* 170 (Dec. 1988).

Peters, T. J. and Waterman, R. H., Jr., *In Search of Excellence. Lessons from America's Best-Run Companies*, Harper and Row, New York, 1982. (Although about business, many of the ideas are directly applicable to education.)

Pintrich, P. R., "Implications of psychological research on student learning and college teaching for teacher education," in W. R. Houston, M. Haberman, and J. Sikula (Eds.), *Handbook of Research on Teacher Education*, MacMillan, New York, 926—857, 1990.

Robinson, J. E., and Heinen, J. R. K., "Some implications of cognitive styles for the teaching-learning process," *Educ. Res. Methods* 7(4), 87 (Summer 1975).

Schmeck, R., "Improving learning by improving thinking," *Educ. Leadership,* 38, 384 (1981).

Smith, D. G., *The Challenge of Diversity: Involvement or Alienation in the Academy?* Report No. 5, School of Education and Human Development, The George Washington University, Washington, DC, 1989.

Smith, K. A., "Educational engineering: Heuristics for improving learning effectiveness and efficiency," *Eng. Educ.,* 274 (Feb. 1987).

Smith, K. A., Stanfield, A. M., and Macneal, R., "Constructing knowledge bases: A methodology for learning to synthesize," *Proceedings ASEE/IEEE Frontiers in Education Conference*, IEEE, New York, 374—381, 1985.

Springer, S. P., and Deutsch, G., *Left Brain, Right Brain*, 3rd ed., W. H. Freeman, New York, 1989.

Stice, J. E., "Using Kolb's learning cycle to improve student learning," *Eng. Educ.,* 291 (Feb. 1987).

Svinicki, M. D. and Dixon, N. M., "The Kolb model modified for classroom activities," *Coll. Teach.,* 35(4), 141 (1987).

Terry, R. E., Durrant, S. O., Hurt, P. K., and Williamson, K., "Implementation of the Kolb learning style theory in a faculty development program at Brigham Young University," *Proceedings ASEE Annual Conference*, ASEE, Washington, DC, 54, 1991.

Tobias, S., *They're Not Dumb, They're Different*, Research Corporation, Tucson, AZ, 1990. [Copies of this book can be obtained directly from the Research Corporation, 6840 East Broadway Boulevard, Tucson, AZ 85710-2815.

Todd, R. H., "The how and why of teaching an introductory course in manufacturing processes," *Proceedings ASEE/IEEE Frontiers in Education Conference*, IEEE, New York, 460, 1991.

Waldron, M. B., "Modeling of visual thinking and creativity," *Proceedings ASEE Annual Conference*, ASEE, Washington, DC, 488, 1986.

What Works, U.S. Department of Education, Washington, DC, 1986. (Copies can be obtained by writing to *What Works*, Pueblo, CO 81009.)

Woods, D. R., Wright, J. D., Hoffman, T. W., Swartman, R. K., and Doig, I. D., "Teaching problem solving skills," *Eng. Educ.,* 238 (Dec. 1975).

EVALUATION OF TEACHING

It is natural to want to know how well one has done on a given task. In its simplest form, evaluation of teaching allows an instructor to obtain this feedback. Once collected, the data can be used to help the instructor improve the course, compare instructors, reward or punish the instructor, or inform potential students. Since improvement of teaching without this feedback is unlikely, we are in favor of this use of teaching evaluations. Unfortunately, the evaluation of teaching has become embroiled in controversy, partially because of the other uses of evaluations.

In this chapter we will start with a discussion of formative and summative evaluations and the objectives of each; then we will consider the validity of student evaluations, correlations with other methods, and extraneous variables which affect student evaluations. Since student evaluations are only one of many procedures which have been used for evaluating teaching, we next discuss the various other methods.

Many professors in psychology and education have devoted their careers to studying the evaluation of teaching. Although many questions remain, there is a large body of scientifically valid knowledge about the subject. We intend to tap into this knowledge so that the reader can intelligently discuss the issues surrounding the evaluation of teaching. This background information will give the reader a distinct advantage over most engineering professors who discuss these issues on the basis of ancedotal evidence and biases.

16.1. FORMATIVE AND SUMMATIVE EVALUATIONS

Essentially, a course can be evaluated at any time during or after the semester or term. Evaluations made during the course, called formative evaluations, are aimed at eliciting comments from students so that the professor can make in-course corrections. These evaluations can be as simple as passing out comment cards and asking the students to respond anonymously to two questions such as:

What do you like about this course?
What about this course could be changed to improve your learning?

They are useful if the professor changes things that are not working. If, for example, the comments reveal that the TA is not available during office hours, the professor can take steps to correct this problem early in the semester. The evaluations can also allow the professor to do something he or she wants to do, but which might not go over well without the empowerment of student comments. For example, if one or two students are monopolizing the professor's time during questions and discussion, other students will likely complain on the comment cards. The professor can then say in a positive sense that he or she has been asked to involve more students in the discussion or questions.

There are other types of formative evaluation. Chatting with students informally during the semester often points out what is or is not working. Formal weekly meetings with a group of students representing the class is another way of obtaining useful feedback during the semester. Chatting with the TA can also be illuminating since TAs often have a good idea of what is or is not working. Critically evaluating the results of quizzes or tests may show that certain critical concepts have not been learned. The professor may want to adjust the syllabus to provide more time for these concepts. Watching the students' nonverbal behavior and asking them if they understand is also a type of formative evaluation which can be used in every class period.

Summative evaluations, which are done at the end of the course or well after the course is over, are used for a variety of purposes, some of which are controversial (see Sections 16.2 and 16.4). Of course, summative evaluations provide feedback to the professor. Since professorial self-evaluations are often very high (Centra, 1980), student evaluations can provide a salutary dose of reality. When the professor has done a good job, the feedback is a welcome pat on the back. Summative follow-up evaluations by alumni can also provide feedback as to what course material has proven to be particularly useful in industry (see Section 16.4).

Summative student evaluations can also be helpful in instructor and course improvement. The more specific the comments, the more useful they are for course improvement. Answers to very general rating questions such as "This is one of the best courses I have ever taken" are not useful for course improvement. Questions on the textbook, handouts, availability of help, homework, tests, lectures, and so forth, can provide the professor with specific areas needing improvement. Based on dissonance theory (when the person's self-evaluation and the feedback received from others differ, dissonance is generated and the person reacts to reduce this dissonance), professors should act to improve their teaching based on student ratings. Unfortunately, many studies have shown little or modest improvement in teaching resulting from the use of course evaluations *alone* (Aubrecht, 1979, 1981; Centra, 1980; Lowman, 1985). A meta-analysis by Cohen (1980) shows that there is improvement, but it is modest.

What does work to improve teaching? For a start, specific questions on student ratings coupled with consultation with another professor (Aubrecht, 1981; Eble, 1988; McKeachie, 1986, 1990). Without a consultant most professors either rationalize the ratings or "just try harder." The consultant helps the professor focus on an action plan to solve the problems pointed out in the ratings. This person can make specific suggestions of what to try and can also be supportive. A specific development plan with informal follow-ups can be developed for the remainder of the semester or for the next semester. Since professors are busy and have many obligations in addition to teaching a specific course, we recommend that the consultant be an interested professor in his or her own department. Then the consultant will understand the constraints the professor is acting under and will not make recommendations which are impossible. McKeachie (1986) suggests that there is no reason to wait until the end of the semester to administer the evaluation form. The student evaluation can be useful for course improvement in the current semester if it is administered from the third to the fifth week of the semester.

Student evaluations, whether formative or summative, are useful because they improve student morale. The chance to register an opinion is helpful even if no one pays any attention. Of course, if it is clear that someone is paying attention and the instructor responds to the comments and improves the course, then student morale will improve even further (Abbott et al., 1990). Although it would be manipulative to give students an opportunity to evaluate courses merely to increase student morale, the increase in student morale when evaluations are used for other purposes is obviously a side benefit.

Administrative use of student evaluations tends to be quite controversial (Eble, 1988; Johnson, 1988; Lowman, 1985), especially when salary, promotion, and tenure decisions are involved. The first problem is that student evaluations are often not well administered. It is not unheard of for professors to hand out the evaluations and then to throw away poor evaluations before turning them in for scoring. A uniform administration procedure must be used to avoid this or other abuses (see Section 16.2.2). One possible solution is to use a separate rating form for administrative purposes and to administer it in a senior seminar course (Milligan, 1982). Second, many professors do not trust the reliability or validity of student evaluations. This issue can be partly put to rest with scientific data (see Section 16.3). Unfortunately, if the administrator using the evaluations does not understand the effect of extraneous variables, the evaluations can be misused. For example, evaluations of professors in classes with less than fifteen students tend to be quite high. This needs to be taken into account when professors are compared. A related problem is that the specific questions which are so useful for course improvement are not useful for overall administrative evaluations (Centra, 1980). Only the overall course and instructor ratings are useful for this purpose since the overall ratings have the highest correlations with student learning. To avoid inadvertent abuses, only the overall ratings should be sent to administrators and promotion committees. The alternative of a separate rating form for administrative use only would also solve this problem. A fourth problem is that few professors are uniformly excellent or uniformly poor in all types of courses (Murray et al., 1990). Poor ratings may only represent poor casting of the professor in a course. What types of courses a professor can teach well is obviously useful information, but using student ratings in a single course is not a fair procedure for setting raises or deciding on promotions. Ratings over a long time period for a large number of courses are needed.

Evaluation of teaching for administrative use by faculty or chair visits to the classroom are even more controversial than the use of student ratings. Since ratings based on visits by professors not trained in the evaluation of teaching tend to be much less reliable than student ratings, this practice should not be used for administrative purposes. (Faculty visits can be useful for course improvement; see Section 16.4.)

A final use of student ratings is as information for other students who are potential consumers of the courses (Canelos and Elliott, 1985; Marsh, 1984). Some universities have a long tradition of student-run evaluations which are then published in student guides. There is no doubt that these guides do have an effect on the elective courses which students sign up for. The aim of informing the consumer of what an instructor and course will be like is probably laudable. Unfortunately, student-run ratings and guides may be poorly controlled (and in effect uncontrollable). It is not unusual for some of the guides to be extremely biased, particularly during periods of political upheaval. Engineering courses are usually not heavily represented in these guides since few engineering students join these student groups and since few engineering courses are electives.

16.2. METHODS FOR DOING STUDENT EVALUATIONS

Since student evaluations are now the most common method for evaluating instruction, we will focus on them in this section and in Section 16.3. However, student evaluations by themselves cannot completely evaluate instruction; thus, they should be used in conjunction with other evaluation methods (see Section 16.4).

16.2.1. Types of Student Evaluations

If the purpose of the course evaluation is entirely feedback to the instructor for the purpose of course improvement, then informal evaluation procedures can be used. Both formative and summative evaluations can be made with comment cards, either with or without cues to the students on what to focus on. If there are specific questions of interest, the professor can generate a student questionnaire (Cook, 1975). But for administrative use or for research purposes, professor-generated questionnaires and comment cards are not suitable.

For administrative purposes, global questions on teaching effectiveness should be used since they have the highest correlations with student achievement (Centra, 1980). A simple alternative is to have all the seniors rate the professors on a scale from 1 to 5. Milligan (1982) suggests doing this for each professor regardless of the number of different courses the student has taken from that professor. Since most professors cannot teach all courses with equal skill (see Section 16.3.3), it is probably better to do this evaluation course by course for each teacher. There is an advantage to separating the course improvement and administrative functions of student evaluations, since professors are more likely to use formalized course evaluations if they know they will not be used by the administration.

TABLE 16-1   COMMERCIALLY AVAILABLE FORMS FOR STUDENT EVALUATION OF FACULTY
(JOHNSON, 1988)

Form	Comments	Source
Cafeteria	Allows instructor to choose 40 from 200 items. Five (5) overall core items are automatically added. Room for comments. Five (5) point scale.	Center for Instructional Services Purdue University STEW B14 West Lafayette, Indiana 47907
Course Instructor Evaluation Questionnaire (CIEQ)	Up to 63 items instructor generated with 7 on student backgrounds. Has open-ended items. Four (4) point scale.	Division Education Foundation and Admin., College of Education University of Arizona P.O. Box 302 Tucson, Arizona 85721
Instructional Assessment Form (IAS)	General evaluation, diagnostic feedback, information about students and course. Instructor can add items. Can include open-ended items. Six (6) point scale.	Educational Assessment Ctr. University of Washington 453 Schmitz Hall P.O. Box 30 1400 N.E. Campus Parkway Seattle, Washington 98195
Instructional Development and Effectiveness Assessment System (IDEA)	Variable number items. Can be instructor-generated. Instructor and course characteristics, evaluate progress towards course objectives, self-rating by students. Can use open-ended questions. Five (5) point scale.	Center for Fac. Evaluation and Development Kansas State University 1627 Anderson Avenue Box 3000 Manhattan, Kansas 66502
Student Instructional Report (SIR	Thirty-nine (39) items, can use instructor-generated, no open-ended. Five (5) student background items. Instructor and course characteristics, course and instructor variables. Four (4) point scale.	Educational Testing Serv. (ETS) ETS College and University Prog. Princeton, New Jersey 08541-0001

Many universities use formalized course evaluation procedures which are often administered by either a separate learning center or a student organization. The forms used are usually machine-scorable, multiple-choice questionnaires with space available for student comments. The students usually rank a variety of questions on 4- to 7-point scales. Usually both specific items such as "The textbook was well written and understandable" and global ranking items such as "Overall, this course ranks highly" are included in the questionnaire. The forms may allow for instructor selection of items from a large pool, and it may be possible for the instructor to add additional items. Marsh (1984) notes that since good instruction can have many dimensions, the forms must be multidimensional; that is, many different aspects of instructional ability need to be considered.

A large number of course evaluation forms have been developed and are available for a nominal fee. Some of them are listed in Table 16-1. Johnson (1988) gives two samples of the available forms. Jakubowski (1982) shows a form generated following the comments of a student panel. Janners and Tampas (1986) discuss the development of a form locally so that the faculty will accept its use. They present their final result. Fowler (1978) shows both a multiple-choice form and a form with open-ended questions. A detailed questionnaire for formative evaluations has been developed by Davis and Alexander (1976). Marsh (1984) notes that if student evaluations are to be used for research purposes, the form needs to be carefully designed. Many of the commercially available forms are adequate, and no form has been shown to be superior, which is why many different forms are in use.

16.2.2. Administration of Student Evaluations

Several studies have shown that the way student evaluation forms are administered can affect student ratings (Aubrecht, 1979; Centra, 1980; Marsh, 1984). Professors who make verbal comments requesting high rankings because of their importance in promotion and tenure decisions may well get higher rankings, particularly if the comments are subtle instead of blatant. There is also a built-in bias if the professor is present when the students fill out the evaluation forms. In addition, professors, like students, are subject to human frailty, and both have been known to cheat occasionally.

To avoid these problems a uniform procedure for administering student evaluations should be used throughout the department. The professor should not be present when students are filling out the forms. A trustworthy administrative assistant, TA, or even the department chair should administer the evaluations. A standard procedure such as the following should be followed by this person:

1 Bring in the forms and pencils needed.
2 Announce to the class why he or she is there and state that it is departmental policy that the professor not be present.
3 Describe the purpose of the forms, state what they will be used for, and note that evaluations are important and need to be done carefully.
4 Pass out the forms and pencils.
5 Give simple instructions. Be sure to note that 1 is high (or low).
6 When all the students are finished, collect the forms and put them into an envelope. Seal the envelope.
7 Deliver the envelope to the agency which scores the forms.

What should be done with the results of the evaluations once they have been scored is somewhat controversial. Certainly they should be provided to the professor for course improvement. Professors should be encouraged but not forced to discuss their evaluations with another professor or an instructional consultant. They should also be encouraged to discuss the ratings and an improvement strategy with the class since this increases the students' satisfaction (Abbott et al., 1990). The use of voluntary evaluations for administrative purposes can cause problems if norms are reported. Since those who volunteer are mainly professors who are most interested in teaching and who are good at teaching, the norms are skewed to high rankings. For administrative uses a required rating of all the professors in the department is preferable.

16.3. STUDENT EVALUATIONS: RELIABILITY, VALIDITY, AND EXTRANEOUS VARIABLES

Many faculty members complain that student evaluations do not mean anything, arguing that they are not reliable, that students can be bought with grades, that the ratings are not valid, that alumni, not students, should do the rating, and so forth. Unfortunately, engineering professors who would never dream of doing an engineering design without data are willing to complain about student evaluations with no data. In this section a sampling of the available scientific data which allows one to discuss these complaints rationally will be presented. Before discussing the questions of reliability, validity, and extraneous variables in detail, we will note that the complaints are often somewhat misplaced. Students are generous evaluators. For example, in a study only 11 percent of 852 engineering classes were rated as below average (Centra, 1980).

16.3.1. Reliability of Student Ratings

Reliability of student ratings means that they are consistent for whatever it is they are measuring. The internal consistency of student ratings is quite good and becomes excellent as the number of students doing the rating increases. For the IDEA rating system Aubrecht (1979) reports the following correlation coefficients:

$r = 0.69$ (ten students)
$r = 0.81$ (twenty students)
$r = 0.89$ (forty students)

For the SEEQ rating system Marsh (1984) reports correlation coefficients that are slightly higher:

$r = 0.6$ (five students)
$r = 0.74$ (ten students)
$r = 0.90$ (twenty-five students)
$r = 0.95$ (fifty students)

Thus, the internal consistency (the agreement of students in the same class) is quite high.

A second measure of reliability is stability. Are the raters stable over time and are the professors stable over time? The correlation coefficient for students in 100 classes when they were asked to rate the class after it was over and at least one year later was $r = 0.83$ (Aubrecht, 1981; Marsh, 1984). When the same instructor was evaluated for the same course but in different years (which means different student raters), the correlation coefficients varied from $r = 0.62$ to $r = 0.89$ with a mean value of $r_{mean} = 0.74$ (Marsh, 1984; Murray et al., 1990). Thus, we can conclude that both student raters and professors teaching the same course are stable.

There is no reason to believe that professors will be equally proficient at teaching all courses. When the same instructor was rated in the same year in different courses, the correlation coefficients varied from $r = 0.33$ to $r = 0.55$ with a mean value of $r_{mean} = 0.42$ (Marsh, 1984; Murray et al., 1990). These correlation coefficients are significantly lower than those obtained for the same instructor teaching the same course. This result is discussed in more detail at the end of Section 16.3.3.

16.3.2. Validity of Student Ratings

Validity means that student ratings are measuring what they are supposed to be measuring. Do student ratings actually measure teaching quality? This is a much harder question to answer than questions of reliability, but sufficient research reports are available to give an affirmative answer.

Critics of student ratings often claim that student achievement is the outcome that we should study. Do student ratings correlate with student achievement? There is broad agreement in the literature that a reasonably strong positive correlation exists between student achievement and student ratings (Aubrecht, 1979; Centra, 1980; Cohen, 1981; Greenwood and Ramagli, 1980; McKeachie, 1986, 1990; Marsh, 1984). The conclusive study was the meta-analysis of Cohen (1981) who looked at all available studies relating student achievement and student ratings in courses with multiple sections taught by different instructors. The global ratings were highly correlated with the final examination scores. Cohen (1981) found correlation coefficients of $r = 0.50$ based on questions about instructor skill, $r = 0.47$ based on questions about the global rating of the course, and $r = 0.43$ based on questions about the global rating of the instructor. Thus, sections where students learned more rated the instructor and the course higher than sections where students learned less.

Student ratings also have modest positive ratings when compared to other methods of evaluating instruction. The correlation coefficients between student ratings and ratings by professors ranged from $r = 0.60$ to $r = 0.70$ if the professor had not visited the classroom (Aubrecht, 1979; Marsh, 1984). The correlation between student ratings and administrator ratings where the administrator had not visited the classroom was $r = 0.47$ (Aubrecht, 1979). If the colleague had visited the classroom before rating the instructor, then the correlation coefficient with student ratings was $r = 0.20$ (Aubrecht, 1979; Marsh, 1984). This number is low partially because the reliability of ratings based on colleague visits is low (see Section 16.4). When professors did not visit a colleague's classroom, they apparently based at least part of their ratings on discussions with students. Thus, these ratings correlate significantly higher than those based on visits.

The correlations of professors' self-rating of their teaching ability with student ratings has been extensively studied. Correlation coefficients between student ratings and a general instructor self-rating are about $r = 0.19$ (Greenwood and Ramagli, 1980). When the instructors do a self-rating for a specific course, then the correlations with student ratings are significantly higher, $r = 0.45$ to $r = 0.49$ (Marsh, 1984). Most professors rate themselves higher than the students do, and about 30 percent of the time significantly higher.

Factor analysis has been used to determine what students are rating. The results of these studies show that students do not just give a single global rating but include several factors. Aubrecht (1979) states that a typical breakdown of factors with the most important factor first is:

1 **Skill.** Interesting presentation, intellectual stimulation, clarity.
2 **Rapport.** Concern for students, classroom interaction.
3 **Structure.** Organization, course preparation.
4 **Difficulty.** Amount of work demanded.

A similar but more detailed list of seven factors is given by Marsh (1984):

1 **Learning and value.** Challenge, subject interest, amount of material learned.
2 **Enthusiasm.** Interest, humor.
3 **Organization.** Objectives, clear explanation.
4 **Group interaction.**
5 **Individual rapport.** Provides help and answers questions.
6 **Breadth of coverage.**
7 **Examinations and grading.**

Wilson's (1972) list includes the first five factors given by Marsh.

Higgins et al. (1991) prepared a list of the top ten characteristics of instruction by asking their engineering students to generate such a list. Their list in order of importance is:

1 Organized and prepared.
2 Simple, straightforward instruction with complete examples.
3 Good communication and pronunciation.
4 Real-life applications and analogies.
5 Sound knowledge of subject matter.
6 Open to questions during and after class.
7 Goals clearly stated at beginning.
8 Interested in subject.
9 Lots of examples.
10 Logical order and avoids tangents.

The procedure used to create this list is significantly different from factor analysis. In addition, the list was generated from a rather small number of engineering students instead of a large number of students from many areas as was done for the factor analysis. The engineers were somewhat more pragmatic and wanted examples, but other comments were similar to the factor analysis lists.

These analyses are further proof of the validity of student ratings. Students have rated by reasonable criteria for good teaching. These ratings also agree with the two-dimensional model of good teaching which was presented in Chapter 1.

16.3.3. Effects of Extraneous Variables on Student Ratings

Critics attack the validity of student ratings because of the effect of extraneous variables. They state that ratings are affected by the time at which the class is taught, who is taught, the grades given, the class size, the type of course, the age and gender of the professor, and so forth. This attack is partially correct since extraneous variables do affect student ratings, but the effect is usually quite small and is not enough to make a good teacher look poor, or vice versa (McKeachie, 1990). In this section we will explore what Marsh (1984, p. 730) calls "the witch hunt for potential biases in students' evaluations."

Initial Student Motivation and Expectations. Students who expect a course to be good often find this to be true (McKeachie, 1986). The correlation between the student's initial liking for the subject and the student's rating of the course at the end of the semester range from $r = 0.42$ to $r = 0.49$ which are quite high (Aubrecht, 1979, 1981). Student enthusiasm and prior interest account for much of the background or extraneous variable effect (Marsh, 1984). Initial student motivation is such an important variable that in the IDEA system for teacher evaluation, initial student motivation is used in combination with class size to establish norm groups for comparison purposes (Aubrecht, 1979).

Class Size. The second most important extraneous variable is class size, but the correlation coefficients are significantly less than for initial student motivation. When there are fewer than fifteen students in a class, the ratings are significantly higher than they are otherwise. Students enjoy the close personal contact with the professor and with other students, which is almost automatic in classes this small (Centra, 1980). As the class gets larger the ratings decrease and the correlation coefficients obtained are generally from $r = -0.10$ to $r = -0.30$ (the correlation is negative since ratings are smaller for larger classes) (Aubrecht, 1979; Koushki and Kuhn, 1982). For very large classes (more than 200 students), several studies show that ratings go back up (Marsh, 1984; Koushki and Kuhn, 1982). This may occur because departments assign their best teachers to large classes. Note that some studies have found no effect of class size in engineering courses (Canelos and Elliott, 1985).

Academic Field. There are small but significant effects based on the academic discipline when all other variables are controlled. Aubrecht (1979) reported that humanities, fine arts, and language had slightly higher rankings than social or physical sciences, mathematics, and engineering. Koushki and Kuhn (1982) found that at Clarkson University the arts and sciences and industrial distribution had slightly higher ratings than either engineering or management.

Although there is not complete agreement between these studies, they do agree that engineering students give ratings at the low end of the spectrum. Thus, cross-field comparisons are somewhat difficult.

Course Type. Engineering professors commonly believe that laboratory courses receive low ratings. Kuriger (1978) found that this was indeed true and that these courses had much lower ratings than classes dispensing theory. Kuriger (1978) also found that engineering elective courses had better ratings than required courses in the engineering discipline, which had higher rankings than core engineering classes. Seniors and graduate students rated classes slightly higher than other students even when the type of class was the same (Kuriger, 1978). Koushki and Kuhn (1982) also found that electives and courses in the discipline had higher ratings than core courses, but they did not observe a difference between elective and required courses in the discipline. The hours that the class meets also makes a small difference, with classes meeting at the convenient times of midmorning and midafternoon receiving the highest rankings (Koushki and Kuhn, 1982).

Grades and Course Workload. A common criticism is that professors can buy ratings by requiring very little work and by easy grading. The first hypothesis is clearly wrong. Studies show that students rate courses with high workloads higher than courses with low workloads (Marsh, 1984). The effect of grades is much more complex. Grades earned previously from the same instructor do not affect ratings in the course (Canelos and Elliott, 1985). Although Kuriger (1978) found a negligible correlation between grades and ratings in engineering courses, pooled studies over many classes show correlation coefficients between expected grade and ratings ranging from $r = 0.1$ to $r = 0.3$ (Aubrecht, 1979). However, one needs to be careful not to confuse correlation with causation. When the studies are controlled for prior interest in the subject and for the effect of workload in the course, most of the correlation disappears (Aubrecht, 1981; Marsh, 1984). What remains is mainly from students who are receiving A's. Marsh (1984) discusses three possible hypotheses for the remaining slight effect of expected grade on course rating. These hypotheses are:

1 Grading leniency. The students rate the course higher because they expect a grade higher than they have really earned. There is no empirical evidence for this hypothesis.

2 Validity. Better learning in the course as illustrated by a higher grade leads to a better rating of the course.

3 Student characteristics. Students who earn better grades have some characteristic which leads them to rate the course higher. This is correlation without causation.

Professors. A large number of professor effects have been studied. First, Kuriger (1978) found that professors who had won teaching awards received significantly higher rankings than professors who had not. This is no surprise and represents another sign of the reliability of student ratings. Kuriger (1978) also found that professors received better ratings than American TAs who had higher ratings than foreign TAs. Presumably, the professors are more experienced. However, younger professors do better than older professors (Canelos and Elliott, 1985). By a slight amount, associate professors received the highest ratings, but this disappeared when only electives were considered (Kuriger, 1978). Students do react positively to very expressive teachers, and these teachers may get overly generous ratings (McKeachie, 1990). However, one of the items that students consider a constituent of good teaching is enthusiasm, and expressiveness is interpreted as enthusiasm. Neither the gender of the instructor nor the knowledge of the subject matter affects the ratings (McKeachie, 1990). The question of how the professor's research affects teaching ability and ratings has been extensively studied and has proven to be complex. Research either has no effect or a slight positive effect on ratings. This issue is discussed in detail in Section 17.3.

Instructor Personality. Murray et al. (1990) did a very interesting study on the interactions between the professor's personality and the type of course. The most important conclusion from their study is that few professors are good teachers in all types of courses and few teachers are poor teachers in all types of courses. Casting the professor in the appropriate type of course is important. The authors suggest that professors should determine what type of course they do well in and then stay with that type of course as much as possible. The three general categories of courses which were clearly different were introductory and general courses which were held in large lecture halls, junior- and senior-level electives which were much smaller discussion classes, and methodology courses which were very work-intensive. Professors who were extroverts, yet compulsive enough to handle the details of large classes, received high ratings in the large lecture classes. Professors who were extroverted, friendly, and supportive yet flexible received high ratings in the discussion classes. Ambitious, competent, hardworking, and confident professors did well in the methodology courses. The only personality trait which correlated with high ratings in all categories of courses was leadership, which they defined as taking initiative and getting things done. Note that this study involved psychology courses and may not generalize to other fields. In a separate study Sherman and Blackburn (1975) found that instructor pragmatism was positively related to ratings in natural science courses but not in courses in humanities or social sciences. Instructor amicability was related to ratings in humanities but not in natural or social sciences. From this one could hypothesize that pragmatic instructors would receive higher ratings in engineering courses.

16.3.4. Can a Professor "Buy" Student Ratings?

Yes, a professor can "buy" student ratings by two different methods. First the professor can load all the extraneous variables in her or his favor. Thus, the professor could arrange to teach a small, nonlaboratory, elective class to seniors and graduate students. The course would be scheduled at a convenient time, and the TA would be from the United States. If possible, the students would be initially interested in the material. The professor would give A's to all the students on the A-B border. This set of conditions can buy a slightly higher rating, but it cannot turn a poor teacher into a good one.

The second approach is to present material clearly and communicate with the students. Organize the material and give clear objectives. Follow a logical presentation scheme with a minimum of tangents. Present many examples and real-life applications. Cultivate a pragmatic, let's-get-things-done attitude. Show enthusiasm, interest, and a love for the subject.

Stimulate the students intellectually and have a significant breadth of coverage. Be available for questions both in and out of class. Have a sense of humor. Use a good textbook which is integrated into the course. Arrange matters so that the workload is high, but not unreasonably so. Have fair examinations and a clearly defined grading system. Encourage group interactions both within and outside the class. Develop a team concept with the students—a team whose job it is to learn the material. Keep the students active and incorporate a variety of modes of presentation. If all these things are done, then the professor will have done a good job and will have earned the high ratings he or she will receive.

16.4. OTHER EVALUATION PROCEDURES

Student evaluations, though useful, are neither the sole nor the best way to evaluate a course. They miss, for example, the richness of ideas which can be obtained with interview techniques, and students are also often not qualified to evaluate content. A combination of techniques can make up for the deficiencies of student ratings.

Student interviews can be a much richer source of information than student ratings, but they are time-consuming. Interviews should not be done by any of the professors who are being evaluated. If the department chair can arrange to interview all the graduating seniors, a significant amount of information can be obtained about the performance of professors, the curriculum, and miscellaneous items. Except in regard to courses the students have taken recently, the information is not likely to be specific enough to help professors improve courses. Thus, the interviews should supplement course evaluations. For valid information to be obtained on professors, a high percentage of the students need to be interviewed; otherwise, only students with complaints may come in. Although the main advantage of interviews is that students have the freedom to bring up whatever they want, some structure helps control the time and ensures important topics are covered. Setting a time limit in advance is useful since it helps the interviewer structure the interview and control time.

An alternative to individual student interviews which takes much less time is the Small Group Instructional Diagnosis (SGID) method (Abbott et al., 1990) in which a facilitator and the instructor first meet to discuss the course. The facilitator then meets with the class in the absence of the instructor and forms small groups which discuss the strengths of the class, areas requiring change, and recommendations for change. Each group reports to the class, and the facilitator collects and summarizes the reports for the class. He or she then clarifies the ideas until the class agrees that the summary is accurate. This class meeting can take place in a single fifty-minute period. The facilitator and the instructor then meet to discuss the students' concerns and recommendations. A strategy for improving teaching is developed. The instructor returns to class and extensively discusses the facilitator's report and the proposed improvement strategy. Of the methods tried, students preferred the group interview procedure to the use of standardized rating forms. With either the group interview or standardized rating forms, students were more satisfied when the instructor responded extensively to the student evaluations (Abbott, et al., 1990).

Self-ratings by instructors are useful for course improvement, although the correlations with student ratings are low. Since many faculty rate themselves high, with 30 percent significantly higher than the students' evaluations, self-ratings should be used as only one part of the course evaluation system. Instructors are more realistic in their self-ratings when they focus on a specific course. Use of some type of questionnaire such as the course evaluation guide developed by Lindenlaub and Oreovicz (1982) helps to ensure that the instructor has not missed any important areas. Course improvement is highest when the self-evaluation is discussed with a supportive but critical consultant. This is particularly true regarding the pace of the course and the workload. Natural science professors typically underestimate the pace and workload (Greenwood and Ramagli, 1980), and engineering professors probably do also.

Consultation can be used with any of the other techniques such as student ratings. Course improvement is much more likely if the ratings are shared with a consultant, probably because it is much harder to avoid the signals that some improvement is needed. An unstructured conversation—letting the person just talk about teaching—can be very useful in providing insights (Elbow, 1986). The unstructured conversation can also be pleasurable since many professors enjoy talking about teaching and do not do so as often as they would like. The consultation can also be structured around a student evaluation or a classroom visit by the consultant.

Visits in class can be a natural extension of consultation since they give the instructor and the consultant more to talk about. Unfortunately, most professors are not trained in classroom observation, and the correlation coefficient between the ratings done by different faculty raters after visits is $r = 0.26$ (Marsh, 1984), which is quite low. Despite this, peer visits are useful since the professor visiting the class is likely to provide some feedback, both positive and negative, that the students do not. Student evaluations are much more reliable than faculty evaluations, possibly because the students see the professor many more times than a professor visiting the classroom does. Although there are advantages to an unstructured procedure during visits (Elbow, 1986), correlation coefficients are likely to be higher if a structured procedure is followed. Acheson (1981) discusses one such procedure developed for college classrooms, while Andrews and Barnes (1990) discuss several of the highly structured instruments which are used for evaluating primary and secondary schools. If engineering colleges are ever forced to make assessments, the wealth of experience from primary and secondary schools should be used to show what does and does not work.

Administrative ratings are similar to peer ratings (Greenwood and Ramagli, 1980). An administrator often bases her or his ratings on informal information gathered from students. Administrators have one disadvantage compared to professors in visiting classrooms and in doing evaluations. Untenured professors in particular are likely to be intimidated by them. The advantage that department heads have is that it is part of their job to help young faculty improve their teaching, and many young faculty members report that such a person was the only professor with whom they had discussed teaching (Boice, 1991).

A systematic follow-up of alumni is quite appealing. Many professors argue that the alumni are older and hopefully wiser, have a feel for what is important in industry, and rate professors differently than students. Alumni follow-ups routinely result in very high agreement with ratings by current students (Canelos and Elliott, 1985; Centra, 1980). Since evaluations from current students are cheaper, are easier, and result in a higher rate of usable returns, the college of engineering at Pennsylvania State University stopped doing alumni ratings of professors and switched to student ratings (Canelos and Elliott, 1985).

Discussions with engineers in industry are useful as part of a content evaluation. Students are perhaps least able to evaluate content, which is probably best done by a team of professors and engineers from industry. One advantage of ABET visits is that content is evaluated, albeit by only one person. In general, new professors who do not have industrial experience are likely to err on the side of being too abstract. Professors heavily involved in research are likely to put too much of their research in courses. Older professors who are doing neither research nor consulting may be presenting obsolete material.

Videotape has some application in course evaluations, particularly in considering some of the performance aspects of teaching (Centra, 1980). However, the presence of a video camera in the classroom can inhibit both professor and students. The result is a somewhat artificial class which will not be completely representative. Elbow (1986), who tried videotaping classes, was not sure that it was particularly helpful, and he noted that if a videotape is shown to a consultant, the professor should pick one that the professor is satisfied with. Our conclusion is that videotaping is probably worth doing once so that the professor can watch for annoying mannerisms.

Many critics of student evaluations claim that what should be analyzed is student learning or student achievement. As noted in Section 16.3.2, there is a positive correlation between student ratings and test scores. Although direct measurement of student learning to evaluate courses may be preferable, it is difficult (Centra, 1980; Davis and Alexander, 1976; Greenwood and Ramagli, 1980). One major difficulty is that students vary tremendously both within a class and from year to year. Should the evaluation be considered positive if many students score well on the test even though they may not have learned much new material? In other words, should it be the increase in knowledge or the total knowledge that counts? Should the learning of the better or the more poorly prepared students be counted differently? The better prepared students will probably score higher on tests but may learn less new material than students with poorer preparation.

Another problem with direct measurement of learning is that some type of standardized test must be used. Instructor-prepared tests can easily be written to cover what the instructor thinks the students know. This biasing of the test may well be unintentional. An alternative problem is that the instructor may teach to the test if he or she knows what is covered on the exam.

Measures of learning must include all levels of the taxonomies which are important in the course objectives. In the cognitive domain it is easiest to test at the three lowest levels, but certainly analysis, synthesis and evaluation are important in engineering education. The affective domain also needs to be included. Most professors and students would agree that a course in which students learn the material but hate it is not a good course.

Despite these problems with the direct use of student learning for the evaluation of teaching, it should be used to supplement other evaluation methods. In particular, student learning should be used for course improvement. Tests should be analyzed first for discrimination (see Chapter 11.2.2) and then to see if there are topics which students are not learning. If there are, then extra time or a different teaching strategy is needed. Once the problem areas have been pinpointed, the problems and possible solutions should be discussed with another professor.

Often professors try to teach too much material, and the easiest way to increase student learning is to cover less material but do a better job with that material.

Classroom observations or "classroom research" can also be used to determine what the students in a classroom are learning (Cross, 1991). One assessment technique is "minute papers." Toward the end of the class ask the students: (1) What is the most important thing you learned today? or (2) What questions do you still have? Not only do minute papers require the students to be active and construct their own knowledge, but they also provide useful feedback to the instructor. A perusal of the students' responses may show where your message is not getting across.

16.5. CHAPTER COMMENTS

The style in this chapter differs from that of previous chapters in that we have tried to cite all our facts, and in some paragraphs almost every sentence has a reference citation. This was done because of the controversial nature of evaluations of teaching. We wanted to be sure that our facts were backed by the research literature on evaluating teaching, and that skeptical readers could check our sources.

We are in favor of student evaluations and other methods of evaluating teaching since we believe that they help improve the teaching of undergraduates. Naturally, all these methods could be improved. However, there does not seem to be a justification for not evaluating teaching just because improvements are needed. There is clearly enough empirical evidence to show that student evaluations can separate good teachers from poor teachers. On the other hand, there is also evidence that student ratings are not a fine instrument and, for example, we cannot say that someone who ranks third out of twenty faculty is necessarily better than someone who ranks fourth.

16.6. SUMMARY AND OBJECTIVES

After reading this chapter, you should be able to:

• Discuss the advantages and disadvantages of formative and summative evaluations.
• Explain the various uses of teacher evaluations and discuss the controversies surrounding them.
• Discuss the various types of student ratings and how they should be administered.
• Discuss the reliability of student ratings and contrast it to the reliability of other evaluation methods.
• Discuss the validity of student ratings. Defend a position pro or con that student ratings are valid.
• Delineate the extraneous variables which affect student ratings and outline a procedure

to minimize the effects of these variables.

• On the basis of your personality determine the type of courses in which you are most likely to do a good or a poor teaching job.

• Discuss other evaluation procedures and how they can complement student ratings to help improve teaching.

HOMEWORK

1 Informally discuss your teaching with a colleague.
2 Ask a master teacher if you can visit her or his class. Make arrangements to do so and then discuss teaching with the master teacher.
3 Develop a simple formative evaluation instrument to use in your classes.
4 Obtain a copy of your university's summative form for student evaluations. Evaluate the evaluation form. Is it adequate? If not, how could it be improved?

REFERENCES

Abbott, R. D., Wulff, D. H., Nyquist, J. D., Ropp, V. A., and Hess, C. W., "Satisfaction with processes of collecting student opinions about instruction: The student perspective," *J. Educ. Psychol.,* 82, 201 (1990).

Acheson, K. A., "Classroom observation techniques," Idea Paper No. 4, Center for Faculty Evaluation and Development, Kansas State University, Manhattan, KS, 1981.

Andrews, T. E. and Barnes, S., "Assessment of Teaching," in Houston, W.R., Haberman, M., and Sikula, J. (Eds.), *Handbook of Research on Teacher Education,* MacMillan, New York, chap. 32, 1990.

Aubrecht, J. D., "Are student ratings of teacher effectiveness valid?" Idea Paper No. 2, Center for Faculty Evaluation and Development, Kansas State University, Manhattan, KS, 1979.

Aubrecht, J. D., "Reliability, validity, and generalizability of student rantings of instruction," Idea Paper No. 6, Center for Faculty Evaluation and Development, Kansas State University, Manhattan, KS, 1981.

Boice, R., "New faculty as teachers," *J. Higher Educ.,* 62, 150 (March/April, 1991).

Canelos, J. J., and Elliott, C. A., "Further investigations of teaching and course effectiveness evaluation. An ongoing project at Penn State Engineering," *Proceedings ASEE/IEEE Frontiers in Education Conference*, IEEE, New York, 77, 1985.

Centra, J. A., "The how and why of evaluating teaching," *Eng. Educ.,* 205 (Dec. 1980).

Cohen, P. A., Effectiveness of student-rating feedback for improving college instruction: A meta-analysis of findings," *Res. Higher Educ,* 13, 321 (1980).

Cohen, P. A., "Student ratings of instruction and student achievement: A meta-analysis of multi-section validity studies," *Rev. Educ. Res.,* 51, 281 (1981).

Cook, D. I., "Write your own questionnaire," *Eng. Educ.,* 353 (Jan. 1975).

Cross, K. P., "Effective college teaching," *ASEE Prism,* 27 (Oct. 1991).

Davis, R. H. and Alexander, L. T., "Evaluating instruction," *Guides for the Improvement of Instruction in Higher Education, No. 3*, Michigan State University, East Lansing, MI, 1976.

Eble, K. E., *The Craft of Teaching*, 2nd ed., Jossey-Bass, San Francisco, 1988.

Elbow, P., *Embracing Contraries: Explorations in Learning and Teaching*, Oxford University Press, New York, 1986.

Fowler, W. T., "Improved feedback to the lecturer," *Eng. Educ.,* 250 (Dec. 1978).

Greenwood, G. E. and Ramagli, H. J., "Alternatives to student ratings of college teaching," *J. Higher Educ,* 51, 673 (Nov./Dec. 1980).

Higgins, R. C., Jenkins, D. L., and Lewis, R. P., Total quality management in the classroom: Listen to your customers," *Eng. Educ.,* 12 (Jan./Feb. 1991).

Jakubowski, G. S., "What students think about their teachers," *Eng. Educ.,* 372 (Feb. 1982).

Janners, M. Y., and Tampas, P. M., "Developing policies and procedures for evaluating instruction," *Eng. Educ.,* 675 (April 1986).

Johnson, G. R., *Taking Teaching Seriously: A Faculty Handbook*, Texas A&M University Center for Teaching Excellence, College Station, TX, 1988.

Koushki, P. A., and Kuhn, H. A. J., "How reliable are student evaluations of teachers?" *Eng. Educ.,* 362 (Feb. 1982).

Kuriger, W. L., "Some statistics regarding student-faculty evaluations," *Eng. Educ.,* 211 (Nov. 1978).

Lindenlaub, J. C., and Oreovicz, F. S., "A course evaluation guide to improve instruction," *Eng. Educ.,* 356 (Feb. 1982).

Lowman, J., *Mastering the Techniques of Teaching*, Jossey-Bass, San Francisco, 1985.

McKeachie, W. J., *Teaching Tips: A Guidebook for the Beginning College Teacher*, 8th ed., D.C. Heath, Lexington, MA, 1986.

McKeachie, W. J., "Research on college teaching: The historical background," *J. Educ Psychol.,* 82, 189 (Spring 1990).

March, H. W., "Student's evaluations of university teaching: Dimensionality, reliability, validity, potential biases, and utility," *J. Educ Psychol.,* 76, 707 (1984).

Milligan, M. W., "Evaluating teaching effectiveness for administrative decisions," *Eng. Educ.,* 374 (Feb. 1982).

Murray, H. G., Rushton, J. P. and Paunonen, S. V., "Teacher personality trait and student instructional ratings in six types of university courses," *J. Educ. Psychol.,* 82, 250 (1990).

Sherman, B. R. and Blackburn, R. T., "Personal characteristics and teaching effectiveness of college faculty," *J. Educ. Psychol.,* 67, 124 (1975).

Wilson, R. C., "Teaching effectiveness: Its measurement," *Eng. Educ.,* 550 (March 1972).

CHAPTER 17

PROFESSIONAL CONCERNS

Professors have a variety of professional concerns, from obtaining tenure to professional growth, which directly or indirectly affect their teaching. Matters of faculty development for a successful and enjoyable career confront them with responsibilities of professional ethics and the necessity that they be ethical professionals. The sections that follow will deal with these matters in turn.

17.1. PROMOTION AND TENURE

We will first consider the pros and cons of tenure and then discuss promotion procedures along with the widely perceived criteria for promotion. Finally, we'll consider appropriate actions for untenured professors desiring to be promoted.

17.1.1. Tenure

Tenure is essentially a lifetime guarantee of a job at a university as long as the university continues to teach the subject and as long as the professor is not found guilty of any heinous crime. Our discussion of tenure relies heavily on the well-researched article by Segal (1974). Tenure was invented to protect a faculty member's right to say things in her or his area of competence. This right is now called "academic freedom."

Prior to the development and widespread adoption of tenure it was not unusual for a professor to be "summarily dismissed" for saying something that the president or board of trustees of the institution disliked. Clearly, the American Association of University Professors (AAUP) was reacting to abuses when its 1915 Declaration of Principles was adopted. Amended in 1940, this declaration advocates:

1 Bestowing tenure on all associate and full professors.
2 A probationary period with a maximum length of seven years.
3 Explanation of the grounds for dismissal.
4 Written notification and a hearing before a faculty committee prior to dismissal.

Most universities use the AAUP guidelines as the basis for their individual variations of tenure. A typical variation is the separation of promotion to associate professor from tenure which others closely link.

Tenure does have several advantages. It has proven to be the best protection for academic freedom. There are numerous instances of abuses by institutions, but sanctions established by the AAUP are embarrassing to the institution and do force most institutions to use due process for tenured professors, a protection not enjoyed by untenured professors.

For some professors the granting of tenure serves to unleash a latent creative ability which can lead to major scholarly advances. The newly tenured professor may feel free to try risky research or to attack the scholarly establishment. Although this flowering does not always occur, the possibility that it might occur is a strong argument in favor of tenure. One additional advantage is that tenure forces the institution to make a carefully considered decision at a defined point in time. Otherwise, many institutions, like many individuals, would procrastinate and not make hard decisions. When the department chair needs to fill out the teaching roster, it would be quite easy to keep someone barely adequate in place.

Like any structure invented in response to abuses, tenure can be abused. First, the process of granting tenure often does not follow the AAUP ideal of faculty control. Even if administrators do not vote or have a limited vote, their presence on committees certainly has an effect on tenure decisions. Of course, the AAUP is an advocacy group, and their ideal may not be in the best interests of all universities.

A second abuse of tenure is by professors. Perhaps the major charge against tenure is that it inbreeds mediocrity (Segal, 1974). Once mediocre professors become promoted they may promote other mediocre professors and the entire faculty rapidly becomes mediocre. As the faculty slides downhill, the truly excellent professors may decamp for greener pastures. The danger in the tenure decision is that it is a guess at a fairly early stage about what a professor will do for the next thirty or so years. If too fine a cut is made, some excellent people may be let go, and they may well bloom elsewhere. If the cut is too easy, mediocre or lazy individuals may be retained.

Tenure often places untenured professors under enormous pressure, while tenured professors are under almost no pressure. This pressure on assistant professors pushes them to do research that is rapidly publishable but not necessarily important. The untenured professor is told to focus and not become a broadly educated scholar. Changing one's research area from one's Ph.D. subject may be the kiss of death even if the now older and wiser professor can see more productive research areas. The push for tenure can also severely limit the time an untenured professor spends on improving teaching (see Section 17.1.2).

The pressures of tenure also skew the institution's resources. Assistant professors are often given light or nonexistent teaching loads and committee assignments. This is done to let them devote time to research. In the best circumstances this strategy works well, although in the worst circumstances the assistant professor leaves before ever having produced anything. In addition, this procedure may reduce the teaching load below the critical mass necessary for the assistant professor to learn how to become an effective, efficient teacher.

Finally, the very idea of academic freedom can be abused by some professors. Academic freedom is meant to protect professors in their areas of competence. There are professors who wander outside their areas of competence and still expect to be protected by academic freedom. It is also not unheard of for professors to convert a class from one topic to another discipline and then argue that it is their academic freedom to do so. Since our colleagues in areas such as philosophy, political science, and religion really do need the protection of tenure for academic freedom, we are in favor of retaining tenure.

17.1.2. Structure of the Promotion Process

Promotion and tenure systems have significant differences from institution to institution, but the general pattern of the process is similar. We will describe a representative pattern. Untenured professors should determine both the written and the unwritten rules for tenure at their university.

Typically, the promotion process starts in the fall. The promotion document is prepared by the candidate's department, usually with considerable input from the candidate. The departmental primary committee, consisting of the full and sometimes the associate professors in the department, receives a copy of the document. The candidate is fully discussed at the primary committee meeting and a vote, usually by secret ballot, is taken. Support from the candidate's department and chair is necessary, but not sufficient, for promotion.

If the candidate is successful at the departmental level, the nomination is sent to the next level which is often the school (such as the school of engineering) level. The department head or a representative makes a presentation to this committee, and another vote is taken. If successful, the nomination is sent to the university level where yet another committee discusses and votes on it. Finally, the nomination is sent to the board of trustees for approval. The board has the legal right to vote no, but fortunately most boards are wise enough to leave promotion decisions to the faculty. By now, it is spring and candidates who are naturally nervous are reduced to quivering jelly.

The details of exactly when this all occurs, who votes, how many votes are required to pass, and so forth, vary greatly. Often the only way to find out is to ask.

17.1.3. Criteria for Promotion and Tenure

The criteria for promotion also vary greatly. Although often not written down, time in grade is usually included. Many schools adhere to the AAUP guidelines with promotion being considered during the sixth year so that unsuccessful candidates can be given the seventh year to find another position. Many schools have an unwritten but firm minimum number of years (four or five) required before the candidate will be considered. Since schools have both written and unwritten criteria, an untenured professor is advised to develop a relationship with a mentor (Balachandran and Udoka, 1990; Engelken, 1986). The written criteria at most schools include research, teaching, and service. These requirements should certainly be read carefully since they contain some useful information and some nuggets of truth. At research universities the actual criterion for promotion to associate professor and for receiving tenure has been

RESEARCH / RESEARCH / RESEARCH

which is usually translated into

PUBLISH / PUBLISH / PUBLISH

(Sisson, 1982; Addy and Dutton, 1987; Boyer, 1990). Reporting on a 1989 Carnegie Foundation survey of faculty, Boyer (1990) found that 83 percent of faculty at research universities agreed with the statement "In my department it is difficult for a person to achieve tenure if he or she does not publish." This number is up from 44 percent in 1969. Among engineering professors 63 percent strongly agreed with this statement. See Table 17-1 for additional information from this survey. In particular, the responses to Q2 further support the importance of publications. Recently, some evidence has appeared that many schools have revised the unwritten promotion criterion to

PUBLISH / PUBLISH / MONEY / ADEQUATE TEACHING

The addition of two requirements corresponds to a general tightening of the tenure requirements at most universities.

The importance of bringing in money is shown in Q3 in Table 17-1. The argument for the need for sponsored research is that professors cannot continue to do excellent research without support, and the peer review process measures quality. A small amount of institutional self-interest may also enter the picture. The importance of teaching is shown in Q4 in Table 17-1. The results in Q4 probably understate the importance of teaching since the requirement for adequate teaching seems to operate as a minimum condition which must be surpassed but then is not considered further. Since bad teachers continually cause the department and particularly the chair a great deal of grief, the requirement for adequate teaching is clearly in the best interests of the department. Obviously, one can argue with the values that only adequate teaching is necessary; our purpose here is to report what *is*, not what could or should be.

An untenured professor needs to know the details of what counts for how much in the various areas. This search will lead into many subjective areas (Watson, 1991). For instance, not all publications are equal. Ideally, the quality of all publications would be determined by careful scrutiny, but this is a difficult subjective judgment. Although attempts are made to determine quality, it is common to use other criteria as a substitute. For technical papers, refereed articles in a major journal are more important than refereed articles in a minor journal,

TABLE 17-1   TENURE QUESTIONS FROM 1989 NATIONAL SURVEY OF FACULTY (E.L. Boyer, *Scholarship Reconsidered: Priorities of the Professoriate*, (Princeton, NJ: The Carnegie Foundation for the Advancement of Teaching, 1990). © 1990 The Carnegie Foundation for the Advancement of Teaching. Reprinted with permission.)

Q1. In my department, it is difficult for a person to achieve tenure if he/she does not publish

	Strongly agree	Agree with reservation	Neutral	Disagree with reservation	Strongly disagree
Research Institution	83	12	1	3	2
Engineering	63	18	7	7	4

Q2. How important is the number of publications for granting tenure in your department?

	Very Important	Fairly Important	Fairly Unimp.	Very Unimp.	No Opinion
Research Institution	56	39	4	1	1
Engineering	43	40	10	5	3

Q3. How important are research grants received by the scholar for granting tenure in your department?

	Very Important	Fairly Important	Fairly Unimp.	Very Unimp.	No Opinion
Research Institution	40	36	16	6	2
Engineering	49	28	17	4	2

Q4. How important are student evaluations of couses taught for granting tenure in your department?

	Very Important	Fairly Important	Fairly Unimp.	Very Unimp.	No Opinion
Research Institution	10	41	30	16	12
Engineering	17	38	31	10	4

which are more important than referred notes, which are more important than articles in refereed proceedings, which are more important than nonrefereed articles. Nontechnical articles are less important than any of the above. Thus, the journal is used as a substitute for a direct measure of quality. Since there may be little difference in the time and energy required for publishing in prestigious journals, assistant professors are often advised to publish in these journals.

Presentations at conferences and universities also count, but in a different way. Many schools ask professors in the candidate's area to evaluate the candidate's research. It is easier for the professor to remember the candidate's research and to write a favorable letter if the professor knows the candidate. Excellent presentations and informal discussions at meetings help one to develop these personal connections.

Who the candidate writes publications with is also scrutinized. Papers from the candidate's thesis are expected and count positively only if they are of exceptional quality or quantity. Since the thesis papers are expected but really do not count, it is important to finish them as soon as possible. This is one advantage of having a postdoctoral appointment. Under no circumstance should an assistant professor start a job before he or she has completed the requirements for a Ph.D. Once these papers have been completed, the candidate needs to sever the umbilical cord to the adviser. This is particularly important for professors who stay at the school where they earn their Ph.D. Besides papers from the thesis, the candidate should have a mix of papers written by her- or himself, with colleagues and with students. If all papers are written with colleagues, members of the promotion committees will wonder if the candidate is independent, and if all papers are solos, the question will be whether the candidate can work with others.

Support for research is necessary to continue doing quality research and to support graduate students. As is the case with publications, not all research support is counted equally. At many research universities grants from certain government agencies such as NSF, NIH, and NASA are more valued than other grants. External support is always more highly valued than internal university support. The most weight is given to grants with the candidate as the principal investigator (PI). Grants for which the candidate is a co-principal investigator or investigator also count but not as much.

Fortunately, most schools do not expect that assistant professors will have graduated Ph.D.'s within the six-year probationary period. Assistant professors are expected to have graduate students who are conducting research. At least some papers should be coauthored with these students. However, because of the six-year time constraint, assistant professors should not expect the research of their students to be sufficient for promotion and tenure.

A final comment on research: Many full professors want to see a big, long-term research plan. What will the candidate be doing five and ten years from now? Develop a research plan to help guide your activities and to help impress the full professors.

Teaching counts, but not enough (Addy and Dutton, 1987). Since no one benefits from bad teaching, most departments want proof that teaching is at least adequate. Although the lack of a large number of student complaints may be sufficient proof, it is better to obtain positive proof by regularly obtaining student evaluations of the class. Unfortunately, at most research universities excellent teaching helps only in borderline cases. For example, if the promotion case looks to be a little early on the basis of research alone, excellent teaching may make the difference. Excellent teaching can be proven with teaching evaluations and teaching awards. In Chapter 16 we noted that teaching evaluations need to be used with care in promotion decisions. A uniform procedure for administration should be followed for distributing and collecting the forms. Items which ask for overall ratings should be used since they correlate more highly with student learning. Adjustments should be made for extraneous factors such as class size, time of day, or unpopularity of classes (such as laboratory courses). Finally, since different personalities do better in different types of courses, ratings should be collected for a variety of courses. The National Science Foundation has begun giving grants for curriculum and course development. These grants are quite competitive and will obviously count toward promotion, but it is too early to tell how much.

For promotion to associate professor and for receiving tenure, service has very little clout at most universities. One cannot totally ignore service, since failure to do one's share of committee work and other types of departmental service will be a negative factor. However, once a reasonable share has been done, more will not help. Professional society activities are also expected, but moderation is again the key. Once a professor has tenure, service does count and is rewarded more than teaching but less than research (Sisson, 1982).

A final unwritten area is general conduct and personality. Promotion is not a case where "nice guys finish last." All things being equal, it is easier to promote a personable individual and easier not to promote a nasty person than vice versa. A talented nasty person will be promoted, but a mediocre nasty person probably will not. If you act in a collegial fashion, do your share willingly, get things done on time, and have a generally positive outlook on life, then you will benefit if your promotion is not clear-cut. Part of the tenure process involves the decision that the candidate fits in with the institution (Watson, 1991). This paragraph may seem unfair, but remember that in industry the ability to get along and work with a team is even more highly prized than in academia.

Universities do change and the criteria for promotion and tenure change. We believe that publishing and research support will continue to be important, but that universities will be able to redefine scholarship to some extent so that a broader range of activities is rewarded. This follows the main conclusion of the Carnegie report (Boyer, 1990). There is a clear swing toward increasing the importance of teaching, though such a nationwide trend may not be followed by a given university. As with the weather, it is often easier to talk about rewarding good teaching than to actually do anything about it. Some of the unhappiest people we know are professors who were hired to do one thing (teaching) and then had the university change and ask them to do something else (research). Professors need to watch the trends at their university.

17.1.4. Actions for Untenured Professors

Many professors want to argue with the values universities use to set priorities for promotion and tenure. Doing this can lead to many spirited discussions with the participants leaving feeling morally superior. However, a professor ignores the established reward system at his or her peril. Our observation is that universities do not punish professors for excellent teaching and for spending time with students. What universities punish professors for (by denying tenure or promotion) is not doing what the university asked for (research and money). To survive with your moral esteem intact, determine how to do both what you want and what the university wants. Since the norm for new faculty is a work week of about fifty-five hours (Beaufait and Harris, 1989), there is enough time to get everything done if you work efficiently.

What can you do as an untenured professor to increase the odds that you will be promoted and receive tenure? The first is to find out as clearly as possible what the target is, especially since the requirements for promotion and tenure represent a moving target that is not clearly defined. Thus, the opinions of several professors are important. Once the target has been identified, develop a plan (see Chapter 2) that focuses first on activities and priorities and then on appropriate schedules and to-do lists. List those things which count for promotion at your school and list those that you want to do. Plan an effective way to combine teaching and research. This can be done by teaching classes in your research specialty. Discuss with your chair the teaching assignments for the next several years and see if you can get a commitment that you will teach one course several times in a row.

Develop a tentative schedule for doing and publishing research. This schedule needs to include plans for writing proposals, visiting funding agencies, training new graduate students, doing research, going to meetings, writing papers, and so forth. Since plans like these are usually overly optimistic, plan to get more done than will be needed to secure your promotion. Then if some of the plans are delayed, you will still have done enough.

Your plans should be developed for the entire untenured period at a sustainable pace. If you can do some research that will come to fruition quickly and some that will take more time to mature, you will have a steady stream of papers coming out. Since this is a five- to six-year period, not a one-month orgy of work, you need to include time to relax. Take one day off every week. Schedule an extra day to relax by flying to meetings on a Saturday to get the cheaper airfares. Schedule a week of vacation every year. In the long run these breaks will increase your efficiency, and you will get more done.

Professors need to keep a running record of things that they do (Beaufait, 1990) to ensure that all pertinent information is included in the curriculum vita. This is important in order to avoid selling oneself short in the promotion and tenure document. For instance, if you give three or four seminars every year at different universities, at the end of five years you will have accumulated between fifteen and twenty visits. If these are not written down, it is very easy to forget one or more of them. It is convenient to keep a running vita either in a computer file or on paper. Get into the habit of recording things right after you have done them.

The world does not end when tenure is denied. Most engineers who are denied tenure go into industry (Watson, 1991). Their salary and job satisfaction are often higher than in academia. If teaching was a positive part of the academic experience, there are many part-time teaching opportunities available.

17.2. FACULTY ENVIRONMENT

There is widespread grumbling in the professorial ranks (Beaufait and Harris, 1989; Boyer, 1990; Eisenberg and Galanti, 1982; Friel, 1985; Mooney, 1991), yet in many ways professors like their jobs (Boyer, 1990; Mooney, 1991). The reasons for these mixed messages are worth exploring.

Perhaps the best sources of information on the attitudes of faculty are the extensive faculty surveys done by the Carnegie Foundation for the Advancement of Teaching (Boyer, 1990) and by the Higher Education Research Institute at the University of California at Los Angeles (see Mooney, 1991). The signs of dissatisfaction are widespread and are reported in Table 17-2. From the responses to question Q1 in Table 17-2, one can see that 50 percent of the engineering professors are more interested in teaching than research. There is an obvious difference

TABLE 17-2 FACULTY SATISFACTION QUESTIONS FROM 1989 NATIONAL SURVEY OF FACULTY (E.L. Boyer, *Scholarship Reconsidered: Priorities of the Professoriate*, (Princeton, NJ: The Carnegie Foundation for the Advancement of Teaching, 1990). © 1990 The Carnegie Foundation for the Advancement of Teaching. Reprinted with permission.)

Q1. Do your interests lie primarily in research or teaching?

	Research	Lean to research	Lean to teaching	Teaching
Research Institution	18	48	24	9
Engineering	7	43	23	27

Q2. The pressure to publish reduces the quality of teaching at my university.

	Strongly agree	Agree with reservation	Neutral	Disagree with reservation	Strongly disagree
Research Institution	24	29	10	23	15
Engineering	24	29	13	19	15

Q3. During the past two or three years, financial support for work in my discipline has become harder to obtain.

	Strongly agree	Agree with reservation	Neutral	Disagree with reservation	Strongly disagree
Research Institution	38	25	21	13	3
Engineering	29	23	34	12	2

Q4. I hardly ever get to give a piece of work the attention it deserves.

	Strongly agree	Agree with reservation	Neutral	Disagree with reservation	Strongly disagree
Research Institution	13	33	12	30	13
Engineering	22	29	15	24	9

Q5. My job is the source of considerable personal strain.

	Strongly agree	Agree with reservation	Neutral	Disagree with reservation	Strongly disagree
Research Institution	15	32	12	24	16
Engineering	16	33	18	20	12

TABLE 17-2 (CONT.)

Q6. If I had it to do over again, I would not become a college teacher.

	Strongly agree	Agree with reservation	Neutral	Disagree with reservation	Strongly disagree
Research Institution	6	7	11	25	51
Engineering	8	5	11	21	54

Q7. I feel trapped in a profession with limited opportunity for advancement.

	Strongly agree	Agree with reservation	Neutral	Disagree with reservation	Strongly disagree
Research Institution	5	9	10	19	56
Engineering	6	10	13	16	56

Q8. This is a poor time for any young person to begin an academic career.

	Strongly agree	Agree with reservation	Neutral	Disagree with reservation	Strongly disagree
Research Institution	7	15	16	38	24
Engineering	11	17	15	32	25

Q9. Please indicate the degree to which your academic discipline is important to you.

	Very important	Fairly important	Fairly unimportant	Not at all important
Research Institution	77	21	2	0
Engineering	75	23	2	0

Q10. Please indicate the degree to which your department is important to you.

	Very important	Fairly important	Fairly unimportant	Not at all important
Research Institution	48	39	11	2
Engineering	52	42	6	0

Q11. Please indicate the degree to which your college or university is important to you.

	Very important	Fairly important	Fairly unimportant	Not at all important
Research Institution	30	50	17	4
Engineering	41	43	16	1

between this and the perceived requirements for tenure that are reported in Table 17-1. Another source of dissatisfaction is the perception that publication pressures reduce teaching quality (see Q2 in Table 17-2). More than half of the professors at research institutions and more than half of the engineering professors agree with this statement. The interaction of teaching and research will be discussed in more detail later. There is also substantial agreement that it has become more difficult to obtain financial support (Q3, Table 17-2). Professors also report that it is difficult to put sufficient time into any project (Q4, Table 17-2).

These sources of dissatisfaction add up to considerable strain on faculty (Q5, Table 17-2). Approximately half of faculty members report considerable strain. The Higher Education Research Institute survey (Mooney, 1991) reported that the following were major sources of stress:

1 Time pressures (reported by 83.5 percent of professors surveyed).
2 Lack of personal time (79.8 percent).
3 Teaching load (65 percent).
4 Managing household responsibilities (63.7 percent).
5 Committee work (57.5 percent).
6 Colleagues (54.2 percent).
7 Students (50.4 percent).
8 Research or publishing demands (50.4 percent).
9 Faculty meetings (49.6 percent).

Boyer (1990) reports that when the data are looked at on the basis of age, the youngest faculty members report considerably more strain than any other age group. Clearly, there is a price to pay for trying to earn promotion and tenure. This is strongly supported by anecdotal evidence (e.g., Howard, 1980).

Table 17-2 also lists several questions which show that in some ways college professors are satisfied with their jobs. Q6 shows that most professors would become college professors again despite everything they now know. In addition, Q7 shows that most professors do not feel trapped, and Q8 shows that most think that now is a good time to start an academic career. Clearly, there is something satisfying about being a professor when it is compared to the alternatives. Q9 to Q11 show that the academic discipline, department, and university are all-important to professors but that the discipline has the highest level of allegiance.

What does all this mean? There appear to be some major satisfactions to being a college professor. But there are some demotivating factors at work, some of which have increased in recent years. These factors include pressure on faculty, red tape, too many administrative responsibilities, too many courses to teach, inadequate staff support, lack of modern equipment, excessive workload, lack of influence, tenure requirements, lack of collegiality, a poor administration, and the low value placed on teaching (Beaufait and Harris, 1989; Boyer, 1990; Eisenberg and Galanti, 1982; Engelken, 1986; Friel, 1985; Mooney, 1991; Yao and Michael, 1987). Interestingly, salary and fringe benefits are no longer the major problems they once were.

It is easy to complain and not present possible solutions. In the remainder of this section and in the next section on faculty development, we will discuss what can be done to improve the environment for college professors. Obviously, more money would help but is probably not forthcoming. What can be done with no or modest amounts of money? Boyer (1990) strongly urges universities to find new ways to define scholarship and to develop new methods for the evaluation of teaching. Both of these actions would reduce some of the demotivating stress and ease the strain, particularly on untenured faculty.

Collegiality is a caring about one's colleagues. It involves both informal and formal sharing of the load required for an excellent department. It involves cooperation instead of competition. In a collegial atmosphere everyone is glad when one professor wins an award since the whole department has won. Working and playing together lead to collegiality. In a collegial atmosphere everyone works within the system and tries to change things without being disruptive. Like good will, collegiality is a fragile resource easily lost and difficult to regain. Unfortunately, the competitive atmosphere of research universities causes collegiality to suffer (Astin, 1985). Malacious gossip, vendettas, paranoia and false accusations, temper tantrums, pettiness, and bickering all lead to a poisonous atmosphere. One way to start to regain collegiality is to reinstitute TGIF with other faculty and graduate students. Another start is the development of ad hoc faculty groups to learn about new developments in mathematics, science, or engineering. Since young faculty members in particular complain about a lack of collegiality (Boice, 1991), an organized luncheon series to discuss teaching methods can be very helpful.

As noted in Q2 in Table 17-2, there is widespread belief that research can decrease the quality of teaching. This belief is only partially supported by the data on teaching evaluations. From a review of the literature, Canelos and Elliott (1985), Eble (1988), Feldman (1987), and McKeachie (1986) state that studies show little correlation between effective research and effective teaching, but these studies were not confined to engineering. On the other hand, Kuriger (1978) found that the teaching ratings of engineering professors who did no research were considerably lower than those of professors who did research. The ratings of professors doing a moderate amount of research were slightly better than those of faculty with a large amount of research. If only elective courses were considered, then teachers doing a large amount of research did slightly better than those doing a moderate amount. Bresler's (1968) study of scientists and engineers at Tufts University agreed with Kuriger's study, except that Bresler found that professors who did extensive research received higher ratings in all courses.

The disagreement between studies is an indication that the relationship between teaching and research is complex. Murray et al. (1990) found that few teachers are either good or poor in all courses. Professors who are ambitious, competent, hardworking, and confident tend to receive high student ratings in methodology courses which are very work-oriented. These same personality traits are highly correlated with research productivity. Thus, for this one type of course one might expect a correlation between student ratings and research. However, correlation does not imply causation. There is also a possibility that the pressure to do research, obtain funding, and publish has become worse, and that research interferes more with teaching than it did in 1968 or 1978.

The widespread belief that research interferes with teaching probably comes from ancedotal evidence and the self-knowledge that one could do better if more time were available. In addition, as Rugarcia (1991) argues, the direct link between engineering research and the teaching of undergraduates is rather weak. Ideally, research or other scholarly activity reinforces teaching and both the teaching and the research improve. In engineering this is most likely to happen in elective courses since the professor has more freedom to discuss research. The advantages of doing research include developing faculty who are vital and enthusiastic (Roberds, 1988), and the faculty in some sense remain learners themselves. Research may also harm teaching if fewer faculty are teaching, the students are neglected, curriculum development is neglected, or the uncertainty of being on "soft" money lowers faculty morale (Roberds, 1988; Cavin and O'Neal, 1991). A balance of research and teaching is required for each individual faculty member (Turns, 1991; Cavin and O'Neal, 1991) and for the faculty as a group (Rugarcia, 1991).

One problem which may adversely affect the faculty environment in the future is a shortage of engineering teachers. Although this may drive salaries up, a shortage would also increase workloads and the sense that there is never enough time to do anything right. Such a shortage might also cause salary compression and competition for professors, so that the easiest way to obtain a large raise would be to change jobs. The competition for new professors would also probably drive up start-up costs and reduce the money available in the department for other projects.

The data on doctoral recipients are not encouraging. In 1989, there were 4536 Ph.D. degrees awarded in engineering and only 40.9 percent of them were to U.S. citizens (Anonymous, 1991). Only 8.2 percent of the degrees were awarded to women, 1.4 percent to Blacks, and 2.1 percent to Hispanics. After receiving the degree, only 23.1 percent of the recipients planned employment in an educational institution.

There are several possible solutions to the impending shortage of qualified faculty. One could increase the pool by increasing the number of B.S. engineers and by increasing the percentage that go on to graduate school. It would be particularly advantageous to increase the number and percentages of women and minorities in engineering. This requires action from grade school through high school up to the undergraduate years (Lowman, 1991). We can encourage more students to go to graduate school by stopping the current "burnout process" (Barber et al., 1989), explaining the advantages of graduate school, increasing the stipend, providing teaching (Newton and Scholz, 1987) and research opportunities to undergraduates, pointing out the long-term economic return of graduate school (Kauffman, 1985), developing one-day workshops for undergraduates on graduate education (Blackmond, 1986), and selling students early on the joys of being a professor (Barber et al., 1989; Landis, 1989).

Another solution is to increase the percentage of Ph.D. engineers who become professors. Since salaries are competitive, other aspects of a professor's job need to be made more appealing. Innovative plans to lessen the sting of the probationary period for tenure may help. The employment guidelines of at least one engineering society now call on employers to expand opportunities for minorities and women, to encourage professional development, to provide employees assistance with dependent care, and to be flexible in hours and duties (AIChE, 1990). Innovative maternity and paternity leaves and plans to handle "the two-career problem" could attract well qualified engineers into teaching. Tickton (1982) lists a series of approaches used by different universities to attract and retain qualified professors. Matier's (1991) study shows that the reputation of the school is the major factor in assistant professors' choices of a first academic position. Other important factors over which the department has more control are teaching and research loads, teaching assignments, research opportunities, congeniality of associates, and rapport with departmental leaders.

Another approach is to change the definition of qualified. Wouldn't engineers with many years of industrial design experience be qualified and probably more qualified to teach design, laboratory, and possibly other courses than professors with no industrial experience? Perhaps innovative contracts will be needed to hire these engineers at the right level without typical tenure and publication concerns. Could more use be made of "loan" engineers or engineers from industry on sabbatical? Engineering departments should use their creative problem-solving abilities to solve the faculty shortage problem.

17.3. FACULTY DEVELOPMENT

The real quality of a university is not the facilities but the faculty and staff. Universities need to make a long-term commitment to faculty development or they will risk having older, tenured faculty memberswho are both obsolete and burned out. It is essential that engineering faculty remain current with technological advances and industrial practice. One argument in favor of having engineering faculty do research is that research keeps them current. This is true, but often only in the professor's narrow specialty. Only very large departments can afford the luxury of having professors teach only in their special area. Most professors teach some courses that are not their specialty, but if they do not make an effort to stay current, the course will soon become somewhat stale. For the purpose of teaching undergraduate courses, other methods for staying current such as writing a textbook, consulting, writing review papers, and attending workshops may be more effective than research.

A second reason faculty development is needed is that the roles of a professor change during his or her career (Graham, 1986; Sloan, 1989). The first three years are spent learning how to teach and starting on research. During this period new professors usually receive less help and mentoring than they want (Boice, 1991; Sloan, 1989). For the second two or three years, assistant professors are very concerned about tenure and may explore alternatives should tenure be denied. Associate professors enjoy the recent promotion and tenure and become more involved in their institution. However, they may go through a "sophomore slump" since they are no longer receiving the attention and help that assistant professors receive. Full professors often go through a transition period or midlife crisis (Levinson et al., 1978; Sloan, 1989). They may feel less enthusiasm for teaching and research and may suffer declines in student ratings and research productivity. In general terms, these professors must choose between stagnation and diversification. As retirement nears, the professor may start to withdraw gradually, possibly become more "mellow," and be very satisfied with service to the department and the profession. Professors need encouragement and help to be most effective in each of these stages.

Faculty development can be accomplished by the individual faculty member, but it is helpful if the department chair or the dean provides some encouragement and modest financial support. Growth or creativity contracts which list what will be done over a three- to five-year period are useful (Boyer, 1990; Simpson and Oggel, 1984). They should be drawn up by the professor. The advantage of a growth contract agreed to by the chair and the dean is that the professor knows that successful completion will be recognized and rewarded. Otherwise, a professor embarking on a new path may find his or her efforts ignored. The growth contract recognizes that universities need faculty with interest and strength in a variety of areas, not just research.

Mentoring is another type of faculty development which can be particularly advantageous to new faculty (Sands et al., 1991). New faculty with mentors often get off to a much faster start in teaching and research (Boice, 1990). Those who receive role-specific modeling in teaching or research receive higher teaching ratings or are more productive in research. However, since people prefer mentors of the same gender, women are at a disadvantage in engineering. Women faculty get less faculty support than men but need more (Gibbons, 1992; Sands et al., 1991).

An obvious area for faculty growth is in teaching (Culver, 1990; Terry et al., 1991). Many professors are acquainted only with the lecture style of teaching, and then only with a noninteractive lecture style. Better teachers know instinctively what works but usually do not know why and cannot explain how someone else can improve. For good teachers a very modest amount of study can have a major impact on their understanding of the teaching process since they already have a rudimentary knowledge structure and are usually motivated to do better. Poor teachers need to read about teaching and observe good teachers. Then they need to experiment, receive feedback and encouragement, and try again. Of course, poor teachers must also have the motivation to improve. Boice (1991) found that new faculty wanted more help with teaching, and he observed that formal teaching development programs worked if the new faculty would enroll in them.

For engineering professors *ASEE Prism* is the most accessible source of teaching information on a monthly basis. The annual meeting of ASEE and the Frontiers in Education Conference cosponsored by ASEE and IEEE are good choices for workshops, symposia, and personal contact. Most universities have in-house teaching programs which can be useful if only as an opportunity to meet other professors who are vitally interested in teaching. There may also be for-credit courses with titles such as "Educational Psychology for College Teachers."

Even if there are no courses, good teachers can be talked to and observed. One possibility is to work with a master teacher (Carpinelli et al., 1989) or mentor (Gibbons, 1992; Sands, et al., 1991). This could be done on campus or while on sabbatical. A word of caution is in order when you observe any professor: Many teachers are good teachers because they have major strengths in the second dimension of good teaching—rapport. The performance (lecture) ability of these professors may just be adequate, but the students respond to the rapport. Thus the observer must watch much more than just lectures. A formal mentoring program where new professors are assigned to teach recitation sections and are expected to attend lectures is also useful. It involves an assistant professor closely with an experienced teacher and encourages informal discussions on teaching methods. In addition, since it is a rare professor who does not prepare for class when he or she knows a colleague will be present, the lectures will be well done.

Once you see, read, or hear about something you think will work for you, try it on a small scale. Students usually interpret change and experiments as interest in teaching, and they respond favorably.

A second major problem teachers have with teaching is content boredom. This is somewhat paradoxical since many professors are professors because they love the discipline, but anyone can become bored with teaching the same material semester after semester. Professors who teach because they love students are much less likely to suffer from boredom since the students change every semester. There are several obvious solutions when content boredom sets in, but all the solutions require a little extra work.

- Teach a new course.
- Team-teach, particularly a multidisciplinary course.
- Teach outside your discipline. Examples include teaching mathematics or physics or another area of engineering.
- Write a textbook.
- Develop courseware.
- Teach the same content but use a radically different teaching method.

The university can help a faculty member develop skill in teaching. Paying for trips to ASEE meetings sends a not-so-subtle message that these meetings are as important as technical society meetings. Modest engineering-wide grants awarded competitively can help professors develop innovative teaching methods. Sabbaticals can be granted for teaching as well as for research reasons. Departments can organize mentoring programs, luncheons to discuss teaching, workshops and seminars. Teaching awards are nice but are most effective if made as a salary increase so that they occur year after year.

Faculty members also need to consider development in research. Research in the same area year after year can become routine. To get past the routine and develop new ideas, a professor can start a totally new research area, though this is very time-consuming and is often easiest to do while on sabbatical. Perhaps one can ease into a new area by joining an interdisciplinary research team. Somewhat less drastic steps to invigorate a research program include going to different research conferences, auditing a graduate-level course in a new area, writing a critical review or a research monograph, serving as an NSF program director on a rotating assignment, and integrating research and teaching by teaching a graduate-level seminar. Particularly for new faculty, it would probably be useful to be mentored in how to serve as a research adviser.

Faculty may also want to have a long-term development plan in engineering practice. For young faculty with no, or very little, practical engineering experience, summer jobs in industry can be helpful. However, the common wisdom is that this should not be done until tenure has been obtained. This seems to be another case where tenure skews the educational system. Industrial sabbaticals can be useful, particularly in research areas where industry is at the forefront. Consulting is also helpful, although the contact is usually too short to get a complete industrial flavor. To a lesser extent, working with other engineers through professional societies can be useful.

Finally, some professors may want to include service or administration in their development plans. This is not really a sign of the onset of senility. One of the duties of faculty is to do their fair share in faculty governance (see Section 17.4). The faculty member may decide to do this by becoming involved in the university senate, the faculty union, the American

Association of University Professors, or heavy university committee duties. An alternative is administrative duties such as assistant department chair, department chair, or assistant dean. It would be nice if universities helped to train professors for these positions, but in the absence of a formal training program the professor can talk to professors who have held these positions in the past, read a few books, and perhaps find a suitable workshop.

A fully functioning department needs faculty who are interested in all areas of research, teaching, engineering practice, service, and administration. Astin (1985) states that many universities suffer from a "Jack Armstrong" syndrome and expect faculty members to excel at everything. Very few professors can be good in all areas simultaneously. A functioning department needs professors who specialize in one or two. The current problem and challenge for the future is that some areas such as research receive many more rewards than the others. A department can find itself with few professors interested in students, service, engineering practice, or administration. The results can include student revolts, a breakdown in service and a lack of curriculum development, difficulty at accreditation time, and a lack of leadership. Balance is needed but is difficult to maintain for long periods.

17.4. PROFESSIONAL ETHICS

The privileges of academic freedom, the latitude given to professors to choose research areas, and the security of tenure must be balanced with self-policed ethical behavior. Engineering professors have fewer constraints than their industrial counterparts and fewer external agencies watching their behavior than medical doctors or lawyers, so ethical behavior must be self-directed. Since ethical behavior must come from within, it is useful to study codes of ethics and to reflect on the applications of these codes. Henninger (1991) has a useful list of references on academic ethics.

Some behavior, upon reflection, will clearly be seen as unethical. Other behavior falls into grey areas where it is arguable whether it is ethical or not. The professor may decide to avoid this behavior so that there is no question of impropriety. Alternatively, she or he may decide that the behavior is ethical, but in order to avoid the appearance of unethical behavior will inform the proper administrative authorities in advance. An example of behavior in a grey area involves a professor who commercializes the results of university research by starting a high-technology company. Since large amounts of money may be involved, some people will question the ethics of almost any arrangement.

A general code of ethics for engineers was introduced and discussed in Table 12-1. Naturally, this code applies to engineering professors as well as other engineers. The ramifications of any ethical code for an individual are often not clear until particular cases are discussed in detail. For example, does teaching when one either does not know how to teach or when one is not a competent teacher violate Canon 2 ("Engineers shall perform services only in areas of their competence.")?

The engineers' code of ethics was not written with the requirements of engineering professors in mind. The professorial aspects of the engineering professor's position are more

TABLE 17-3 SUMMARY OF AAUP STATEMENT ON PROFESSIONAL ETHICS
(Adapted from AAUP, 1969)

The professor recognizes special responsibilities:

1. Seek and state truth in **subject** as he or she sees it. Intellectual honesty must be practiced.

2. Encourage **students** in the pursuit of learning. The professor will respect students, avoid exploiting students and honestly evaluate students.

3. Respect **colleagues** and defend their right of free inquiry. Acknowledge academic debts and accept faculty responsibility for institutional governance.

4. Determine amount and character of outside work with due regard to paramount responsibility within **institution** to be an effective teacher and scholar. Give due notice of intent to leave.

5. As a **citizen** speak as an indivdual bound by the rights and obligations of a citizen.

---

closely related to the statement of professional ethics made by the American Association of University Professors. The AAUP statement of ethics is summarized in Table 17-3 (AAUP, 1969). Engineering professors need to adhere to both the engineering code of ethics and to the AAUP statement.

There are many ramifications of the AAUP statement of ethics. A complete enumeration is obviously impossible, and each case must be looked at individually. As an example, a few of the ramifications of each paragraph of the AAUP statement are delineated below.

I. Intellectual honesty obviously requires that research data be reported accurately. Falsification of data is unethical and illegal. Data which may be questionable can be reported, but all questions about the data must be fully discussed. Prior work must be acknowledged (see also item III).

II. Exploitation of students includes the sexual exploitation of students. It is obviously unethical to exchange grades for sexual favors. Dating a student can inadvertently lead to ethical problems. It is probably better to wait until the person is a former student to begin a romantic relationship.

A grey area of the ethical code involves the ethics of requiring students to purchase your textbook for a course. One solution to this problem is to donate the royalty income from your students to the university.

III. Professors should not let personal differences cloud professional evaluations of the work of colleagues. Accepting a share of institutional governance requires that the professor do his or her fair share of committee duties. This may also mean that the professor should accept her or his turn as a member of the faculty senate or as the departmental chair.

IV. Professors should observe the regulations of the institution as long as they do not compromise academic freedom. (The AAUP is very clear that academic freedom is a higher value than following the institution's regulations.) The professor may constructively criticize and try to change institutional regulations. However, we interpret this as meaning that trying to punish the institution would be unethical. Thus, a professor could ethically sue her or his university, but collecting punitive damages may well be unethical. If there is a conflict between outside work such as consulting and university duties, the university duties should be considered more important.

V. The professor has all the rights and obligations of a citizen. This can be interpreted to mean that outside her or his subject area the professor has no special privilege of academic freedom beyond those of every citizen.

Intellectual honesty and responsibility in research has become a topic of national importance, and the federal government through the Public Health Service has established a policy (Public Health Service, 1989). This policy goes beyond statements of ethical codes since it states that individuals have a responsibility to report on others. Briefly, it states:

1 Academics have a responsibility to report the lack of integrity of others in research and scholarship. "Whistle blowers" should be protected from reprisals.

2 Investigations should not be handled by associates of the person whose work is being investigated.

3 Investigations must be confidential.

4 The person being investigated has the right to communicate with the investigator(s) and should be advised of any decisions.

In actual practice many professors have been very reluctant to accuse others formally of unethical scholarship or cheating on research results. Such allegations can become very time-consuming, and it is widely perceived that whistle blowers often receive reprisals in some form. Clearly informing all students doing research of the ethical standards they are expected to follow can help eliminate the need to report others.

With all this talk of ethics it is useful to insert a healthy note of skepticism. "In all of this, however, we must be on guard against any group which seeks recognition as spokesman for "the profession," and then seeks to impose its narrow definition of engineering ethics on us all" (Florman, 1976, p. 31).

17.5. GUIDEPOSTS FOR ENGINEERING EDUCATION (HOUGEN'S PRINCIPLES)

Olaf Hougen was one of the pioneers in chemical engineering education. In a memoriam, Bird (1986) delineated the principles that Hougen used to guide the development of the Department of Chemical Engineering at the University of Wisconsin. We repeat these principles here since we believe that many of them will prove to be useful guiding principles for all engineering educators. The statement of the principles are quotations from Bird (1986).

1 "The undergraduate program should be practical and conservative, whereas the graduate program should be imaginative and exploratory."

Undergraduate programs are to a large extent training for industry and thus should prepare students for responsible engineering jobs. Graduate research should move boldly into new areas.

2 "There should be a smooth flow of information from graduate research to graduate teaching to undergraduate teaching."

Since the graduate program moves boldly into new areas, it can serve as a testing ground for new material. Once this material has proved its worth, it should be moved into the undergraduate program. Note that this implies that professors are involved in teaching at both the graduate and undergraduate levels, and in research.

3 "If you can't find relevant problems to give the student, then you shouldn't be teaching the material to the students."

If there are no industrial problems currently or in the future which can be solved with a method, then that material should not be part of an engineering curriculum.

4 "Use the best available information from the modern sciences."

Engineering should be based on scientific knowledge, and it should be up-to-date.

5 "Well-founded and well-tested empiricisms are to be preferred over theories that have only a limited range of applicability."

Correlations should be scientifically based, and founded on extensive data. The data should be as comprehensive as possible since graduates will hold responsible industrial positions.

6 "It is vital for engineers to know how to solve problems with limited and incomplete data."

Complete data is a luxury that is often unavailable. Students must be well-versed in estimation methods, particularly for physical properties.

7 "Students are impressionable and learn quickly, and therefore a professor must make certain that he [or she] teaches in a responsible way."

Wild conjectures presented as fact or unethical behavior have no place in teaching.

8 "It is important that the students have a good grounding in the basic fundamentals; there's nothing worse than a student who has a thin veneer of high-powered theory."

The basic ideas need to be stressed. Both undergraduate and graduate students with weak backgrounds should be encouraged to take remedial coursework.

9 "We must always recognize that our students and our teaching assistants are young professionals."

The students and teaching assistants need the challenge and reward of helping to develop the engineering profession.

10 ". . . faculty members have an obligation to assist colleagues in other institutions."

Visitors, particularly those from other countries, should be treated with respect and be provided with whatever information they need. In addition, faculty members have a responsibility to prepare excellent textbooks.

11 "We have, as faculty members in a state-supported institution, a responsibility to serve the taxpayers by performing our job well."

Even though resources might be limited, the faculty needs to perform its assignments as well as possible.

12 "Do not show emotions of bitterness or beratement or belittlement; ascribe the best motives to your associates; say nothing derogatory."

Florman (1987) points out that there is a fine line between useful argument and divisiveness. We must believe that all our associates have the best wishes of the university and the engineering discipline at heart. Hougen's is difficult advice to follow; however, if followed, it will lead to a collegial atmosphere within a department.

17.6. CHAPTER COMMENTS

Many of the topics in this chapter are only indirectly related to teaching in the classroom, yet they can have a major impact on how well a professor teaches. Tenure and promotion are issues of vital interest to potential faculty members. The other topics in this chapter seem to be of more interest to older faculty. Ethical concerns don't suddenly arise when one becomes a professor; courses at all levels should consider ethics. As is often the case, however, the topic is appended awkwardly to the end of a class, with the result that students don't appreciate its relevance. Graduate students are no different in this regard; however, they do find case studies to be of considerable interest. We suggest then that ethics be taught by case studies.

17.7. SUMMARY AND OBJECTIVES

After reading this chapter, you should be able to:

• Explain what tenure is and discuss the usual procedures followed for promotion and tenure at universities.
• Discuss the environment for engineering faculty and ways to improve it.
• Discuss methods for developing faculty and prepare a personal development plan.
• Outline the AAUP ethical standards and discuss case studies to determine if the AAUP guidelines are satisfied.
• Determine the applicability of Hougen's principles in one's own engineering discipline.

HOMEWORK

1 Make a list of ten advantages of tenure. Make a counterlist of ten disadvantages. Develop an alternative to tenure which would retain many of the advantages but have fewer disadvantages.

2 Develop a plan for how you will get promoted to associate professor.

3 Assume that you have just been appointed department chair. At your university the department chairs set raises within very broad guidelines. However, the total dollar pool for raises is a fixed sum which averages to 5 percent of the total faculty salaries. Determine a scenario for how you will reward faculty. Consider the following faculty members:

a R does research. He is nationally known and has a standing offer for a position from another university. His teaching ratings are abysmal.

b T is a wonderful teacher, but he has not done research for ten years. He routinely alternates winning the best teacher award with professor S.

c E is a good teacher, does modest research, and serves the department whenever asked to do so.

d A is the best known professor in your department and is a member of the National Academy of Engineering. He is getting ready to retire in a year or two and is no longer doing research.

e S is the chairman of the undergraduate curriculum committee, does all the departmental advising of undergraduates, is adviser to the student professional society, and is a good teacher. The students talk to him all the time, and he single-handedly prevented a revolt of the seniors in Prof. R's class. He is not doing research.

f D has been an associate professor for the last twenty years. He is the outstanding racquetball player on the faculty, but you cannot think of anything else outstanding about him. He is a member of the organizing committee for a proposed faculty union.

g N is a new assistant professor who has been with the department for one year. She seems to be off to a fast start in her career and already has one research grant.

4 Discuss the following case studies. Is the professor's behavior ethical?

a B is single. She has started dating one of the graduate students at your university. Consider three different sub cases: 1. The graduate student is not in Prof. B's department. 2. The graduate student is in Prof. B's department, but she is not his adviser and he is not taking any courses from her. 3. Professor B is the graduate student's research adviser.

b C is a highly sought-after consultant. He normally teaches Monday, Wednesday and Friday and is often gone on Tuesday or Thursday. He has the opportunity to make a great deal of money consulting for a new client, but would have to miss his Wednesday and Friday classes.

c K is the department chair. He has allowed other professors in the department $1000 for travel to professional meetings. So far this year Prof. K has spent $3000 for travel to professional meetings himself.

REFERENCES

AAUP, "Statement on professional ethics," AAUP Bull., 86 (Spring 1969).

Addy, A. L. and Dutton, J. C., "Clearing the hurdles along the tenure track," *Proceedings ASEE Annual Conference*, ASEE, Washington, DC, 1266, 1987.

AIChE, Guidelines on 'People-Related' Employment Issues, American Institute of Chemical Engineers, New York, 1990.

Anonymous, "Fact file: Recipients of doctorates from U.S. universities, 1989," *Chronicle Higher Educ.*, A13 (March 6, 1991).

Astin, A.W., *Achieving Educational Excellence,* Jossey-Bass, San Francisco, 1985.

Balachandran, S. and Udoka, S., "Approaches for mentoring new engineering educators," *Proceedings ASEE Annual Conference*, ASEE, Washington, DC, 916, 1990.

Barber, E. G., Morgan, R. P., Darby, W. P., Sallman-Smith, L. J., "To pursue or not pursue a graduate engineering degree," *Eng. Educ.,* 550 (July/Aug., 1989).

Beaufait, F., "Planning for tenure and promotion: Be your own coach," *Eng. Educ.*, 31 (Jan/Feb., 1990).

Beaufait, F. and Harris, W., "Enhancing the engineering faculty environment," *Eng. Educ.*, 566 (July/Aug, 1989).

Bird, R. B., "Hougen's principles," *Chem. Eng. Educ.*, 161 (Fall 1986).

Blackmond, D. G., "Graduate education in chemical engineering: A workshop to help students answer the questions What is it? Why go? What comes after?" *Chem. Eng. Educ.,* 20, 174 (Fall 1986).

Boice, R., "New faculty as teachers," *J. Higher Ed.,* 62, 150 (March/April, 1991).

Boyer, E. L., *Scholarship Reconsidered. Priorities of the Professoriate*, Carnegie Foundation for the Advancement of Teaching, Princeton, NJ, 1990.

Bresler, J. B., "Teaching effectiveness and government awards," *Science,* 160, 164 (1968).

Canelos, J. J. and Elliott, C. A., "Further investigations of teaching and course effectiveness evaluation: An ongoing project at Penn State University," *Proceedings ASEE/IEEE Frontiers in Education Conference*, IEEE, New York, 77, 1985.

Carpinelli, J., Dresnack, R., Grow, J., and Hubbi, W., "Experience with the New Jersey master faculty program," *Proceedings ASEE/IEEE Frontiers in Education Conference*, IEEE, New York, 261, 1989.

Cavin, R. K., III, and O'Neal, J. R., Jr., "Is sponsored research detrimental to undergraduate engineering education?" *Proceedings ASEE/IEEE Frontiers in Education Conference*, IEEE, New York, 262, 1991.

Culver, R. S., "Motivating faculty to improve teaching: Four practical suggestions," *Proceedings ASEE Annual Conference*, ASEE, Washington, DC, 1655, 1990.

Eble, K. E., *The Craft of Teaching*, 2nd ed., Jossey Bass, San Francisco, 1988.

Eisenberg, E. and Galanti, A. V., The engineer and academia: An analysis of the changing relationship. *Eng. Educ.*, 232 (Dec. 1982).

Engelken, R. D., "Development, "survival," and retention of young engineering faculty: A front line view," *Proceedings ASEE/IEEE Frontiers in Education Conference*, IEEE, New York, 354, 1986.

Feldman, K. A., "Research productivity and scholarly accomplishment of college teachers as related to their instructional effectiveness: A review and exploration," *Res. Higher Educ.*, 26, 227(1987).

Florman, S. C., *The Existential Pleasures of Engineering*, St. Martin's Press, New York, 1976.

Florman, C. S., *The Civilized Engineer*, St. Martin's Press, New York, 1987.

Friel, L., "A symbiosis," *Proceedings ASEE/IEEE Frontiers in Education Conference*, IEEE, New York, 105, 1985.

Gibbons, A., "Key issue mentoring," *Science*, 255, 1368 (13 March 1992).

Graham, A. R., "The faculty team: Who are the players?" *Proceedings ASEE/IEEE Frontiers in Education Conference*, IEEE, New York, 436, 1986.

Henninger, E. H., "Ethics and professionalism in higher education," *Proceedings ASEE/IEEE Frontiers in Education Conference*, IEEE, New York, 156, 1991.

Howard, G. M., "The importance of teaching from an assistant professor's point of view," *Chem. Eng. Educ.,* 14, 66 (Spring 1980).

Kauffman, D., "Is graduate school worth it? A cost-benefit analysis with some second-order twists," *Chem. Eng. Educ.,* 19, 208 (Fall 1985).

Kuriger, W. L., "Some statistics regarding student-faculty evaluations," *Eng. Educ.*, 211 (Nov. 1978).

Landis, R. B., "An academic career: It could be for you," *Eng. Educ.,* 79(5) (July/Aug. 1989).

Levinson, D. J., Darrow, C. N., Klein, E. B., Levinson, M. H., and McKee, B., *The Seasons of a Man's Life,* Knopf, New York, 1978.

Lowman, J. R., "Myths, facts and the future of U.S. engineering and science education," *Eng. Educ.,* 365 (April 1991).

Matier, M. W., "Recruiting faculty: Complementary tales from two campuses," *Res. Higher Ed., 32, 31* (Feb. 1991).

McKeachie, W. J., *Teaching Tips,* 8th ed., D.C. Heath, Lexington, MA, 1986.

Mooney, C. J., "Professors feel conflict between roles in teaching and research, say students are badly prepared," *Chron. Higher Educ.,* A15 (May 8, 1991).

Murray, H. G., Rushton, J. P. and Paunonen, S. V., "Teacher personality traits and student instructional ratings in six types of university courses," *J. Educ. Psychol.,* 82, 250 (1990).

Newton, T. G. and Scholz, P. D., "Introducing undergraduates to college teaching," *Eng. Educ.,* 179 (Dec. 1987).

Public Health Service, Department of Health and Human Services, "Responsibilities of awardee and applicant institutions for dealing with and reporting possible misconduct in science," *Fed. Regist.,* 54, 151, (August 8, 1989).

Roberds, R. M., "Research: Blessing or blight to engineering education," *Proceedings ASEE/IEEE Frontiers in Education Conference,* IEEE, New York, 2234, 1988.

Rugarcia, A., "The link between teaching and research: Myth or possibility?" *Eng. Educ.,* 20 (Jan./Feb. 1991).

Sands, R. G., Parson, L. A., and Duane, J., "Faculty mentoring faculty in a public university," *J. Higher Educ.,* 62, 174 (March/April 1991).

Segal, R. A., "Do you mean Professor . . . is still at Princeton? An analysis of tenure," *Princeton Alumni Wkly,* 8 (March 6, 1974).

Seldin, P., "Court challenges to tenure, promotion, and retention decisions," Idea Paper No. 12, Center for Faculty Education and Development, Kansas State University, Manhattan, KS, 1984.

Simpson, E. and Oggel, T., "Growth contracting for faculty development," Idea Paper No. 11, Center for Faculty Evaluation and Development, Kansas State University, Manhattan, KS, 1984.

Sisson, L. E., "Faculty evaluation and reward: Does effective teaching matter?" *Eng. Educ.,* 380 (Feb. 1982).

Sloan, E. D., "Extrinsic versus intrinsic motivation in faculty development," *Chem. Eng. Educ.,* 23, 134 (Summer 1989).

Terry, R. E., Durrant, S. O., Hurt, P. K., and Williamson, K., "Implementation of the Kolb learning style theory in a faculty development program at Brigham Young University," *Proceedings ASEE Annual Conference,* ASEE, Washington, DC, 54, 1991.

Tickton, S. G. (Ed.), *1982 Idea Handbook: Attracting and Retaining Highly Qualified Young Faculty Members at Colleges and Universities,* Academy for Educational Development, Washington, DC,1982.

Turns, S. R., "Faculty research and teaching—A view from the trenches," *Eng. Educ.,* 23 (Jan/Feb. 1991).

Watson, K., "Tenure denied: Assessing your options as the axe descends," *Proceedings ASEE Annual Conference,* ASEE, Washington, DC, 912, 1991.

Yao, J. T. P. and Michael, H. L., "Teaching, research, and service—Which ones to emphasize," *Proceedings ASEE/IEEE Frontiers in Education Conference,* IEEE, New York, 1270, 1987.

OBTAINING AN ACADEMIC POSITION

This appendix closely follows the article by Wankat and Oreovicz (1983)—"The graduate student's guide to academic job hunting"—with the addition of three more-recent references. To obtain an academic position, candidates go through a series of steps, with the foremost requirement being that they have something to sell: a solid graduate education and good research. They should also know at least three professors well. This first step serves as the basis for the second step: building a resume.

The resume should be carefully and professionally done. It should include all significant professional activities, and for academic positions should highlight research and teaching. Included should be papers which have been published, are in press (that is, have been accepted by an editor), have been submitted, and perhaps are in preparation. The latter category will be discounted by many examiners of the resume, but it doesn't hurt to include them. If you have had a substantial share in writing a proposal, include it also. List any TA duties, and if you did more than just grade papers, list the activities. The names, addresses, and phone numbers of references should be included on the resume. Since you will need good references from good people, you should have been getting to know these professors professionally over the last three or four years. We suggest including the references on your resume instead of stating that they are available on request since it reduces the barriers for prospective employers. When the resume is finished, ask two people to proofread it carefully. Many professors use the resume and your cover letter as an indication of how well you can communicate in writing.

In some areas of engineering such as industrial engineering, the Department Chairs Organization collects resumes of Ph.D. candidates interested in academic positions. This can supplement your search, but do not assume that this resume book will get you a job offer.

1 While you are preparing your resume, develop a research plan for the next five or so years. This is a separate document which some schools may want submitted at the same time as the resume. As for research, where do you want to be five years from now? Following up on Ph.D. research is fine but be sure also to branch out from the research of your major professor.

2 What equipment needs do you anticipate for getting started in your research? In many engineering fields it is now common to give new professors a start-up package. You need to determine three acceptable start-up packages. The first is a "blue-sky" package which includes everything that you could profitably use in your research. The second is a "middle-of-the road" package which is sufficient to get you well started on research but leaves one or two major items of equipment for later acquisition. The third is an "absolute minimum" package which is the minimum you can accept and still be able to do research in your area. These packages need to be developed thoughtfully.

3 Are there any major new experimental, numerical, or theoretical skills which will aid you in your research? If so, plan on how you will obtain these skills. Now is a good time to consider the appropriateness of a postdoctoral position. It can give you the opportunity to learn new research skills, work with a well-known professor, publish your Ph.D. papers, write some more papers, and think deeply about research. A word of caution, though: If not planned well, the postdoctoral position can also leave you in a holding pattern.

4 While preparing your resume, prescreen openings. Every March the ASEE publishes the *Engineering College Research and Graduate Study Directory*, and in November the *Undergraduate Programs in Engineering and Engineering Technology Directory*. These two compendiums of data can be very useful for comparing schools and for getting an idea of where to apply. Talk to several professors in your department to obtain a qualitative feel for different departments. Be sure to get several opinions because individual biases can be strong. Now is also the appropriate time to become a reader of the academic openings sections of the appropriate journals in your area of engineering. In addition to the specialized journals, read *ASEE Prism*. If you are also interested in a nontraditional position such as general engineering, freshmen engineering, or an interdisciplinary engineering position, then check out the *Chronicle of Higher Education* in your university library.

5 Next, decide on the schools to which you will apply. The prime source of these schools consists of those who have advertised. However, if you are interested in a particular university, send a resume even if you haven't seen an advertisement. Perhaps you missed it.

6 Prepare three generic cover letters on your word processor. One should be for schools which have advertised a position close to your qualifications. Another letter is for schools which have advertised a position which really doesn't fit your qualifications. Since many schools will bend qualifications for strong candidates, it pays to write to these schools. The third letter is for schools which have not advertised. The cover letters should be personalized by naming the school and the position you are applying for. The writing in the cover letter must be impeccable, or your resume may not receive the attention it deserves. Proofread all cover letters to be sure that you name the school the letter is going to and the correct position. Nothing is more damaging to a candidate than a letter which applies for the wrong position. Since academic searches are usually conducted from October to May (Stevens, 1990), you need to plan accordingly in developing your job search.

If you can get your letters sent out a few months before a major professional society meeting, you can use the meeting to further your job search. Mention in your cover letter that you will be presenting a paper at the meeting (this obviously requires advance planning) and that you would be happy to meet with them at this meeting. Many departments use professional society meetings as a chance to screen candidates before inviting them for a campus visit. The department may well send someone to listen to your presentation and may arrange an informal meeting with you. Come to this meeting prepared with extra copies of your research plan and resume. The professional society meeting can also provide an opportunity to meet with professors from schools you haven't yet applied to.

7 Once the letters and resumes have been sent, you sit and wait. Most schools will quickly send you a letter of acknowledgment, but this is likely to be the only thing which is done quickly. It is not unusual for departments to receive several hundred applications for a single position, and processing all of these applications takes time. Unless your obvious superstar status shines through, expect to receive many more negative responses than positive ones. For this reason, you need to apply to a relatively large number of schools.

8 Once you have at least one positive response, you can plan the interview trip. Arrange it at a time which is convenient for both you and the school. If you get several invitations to interview, put the schools you are most interested in third or fourth. The first and second interview trips will be learning experiences, and you will probably wish that you could do them over again. The third or fourth visit is also the best because by this time your seminar and your ability to answer questions have been polished. But don't take too many interview trips. They are tiring and time-consuming, and your interest and effectiveness will wane after some point. The key to the interview trip is preparation. Be prepared to discuss yours and others' research. Get a copy of the school's research report and study it. Find out what research professors at the school are doing. You might even consider reading some of their recent articles. Be prepared to elaborate on your research plans for the upcoming years.

9 Above all, be prepared for your seminar. Many schools use the seminar as a measure of how good your research is and how good a teacher you will be (ASEE, no date). Practice the seminar ahead of time and be sure it fits within the time guidelines. Remember to start fairly slowly with a general introduction that everyone in the audience can follow. Then lead up to the research results which will excite the experts. Rehearse by having your major professor and graduate student colleagues ask you questions so that you can practice answering both friendly and hostile questions. Tips on giving the seminar are given by Wankat and Oreovicz (1983) and ASEE (no date).

10 Observe social amenities. During the visit do your best to shine both professionally and socially. If you are traveling across the country beware of jet lag. Relax and take a nap on the plane so that you will be fresh for dinner. During social occasions follow normal rules of etiquette—in particular, don't drink too much. If the evening starts to get too late, be assertive about your need to sleep. On the interview day be interested in the research of others and be enthusiastic about your own research, making sure that your enthusiasm carries over to your seminar presentation. Avoid becoming defensive during the question period. Some of the questions may be purposefully hostile to see how you perform under pressure. If you don't know the answer to a question, say "I don't know. That is an interesting question and I'll find out the answer when I get back home."

11 Ask questions and determine the school's climate. Ask the department chair about teaching loads, start-up funds, office and laboratory space, travel money, and so forth. Ask other faculty questions which will help you to determine the school's environment for teaching and research: How qualified are the undergraduate and the graduate students? Are secretarial and other services satisfactory? Is the research space you will be assigned adequate, or would extensive remodeling be required? Are the assistant professors generally happy, and do they feel they have been fairly treated? Do the professors in the department work well together, or is there significant bickering and fighting? Since many people will not be bluntly honest, you will have to pay attention to numerous and subtle cues to get a good picture of the department's health.

12 Use the return flight to begin your follow-up. Record your impressions of the department and note any questions that you forgot to ask but will ask if the department makes you a job offer. Make a to-do list of things to follow up on. Send a thank you letter to your host or hostess. If you have promised anyone reprints, send them. Send in your receipts and expenses for the trip. Then, sit back and wait some more.

13 Assuming that you get a job offer, now is the time for you to negotiate. Your ability to negotiate is greatest when you have received, but not yet accepted, a job offer. Although practically anything can be negotiated, a new assistant professor is most likely to want to negotiate the start-up package, teaching load and assignment, and salary. If the offered start-up package is less than your minimum package, then you might be better off refusing the offer. An alternative is to arrange a compromise, such as asking for the necessary amount of money but offering to spread it over two fiscal years. Of course, if you are being unreasonable, you may not find any schools which will provide sufficient funds. To avoid having this occur, discuss your start-up needs with at least two professors at your university before starting the negotiation. Items involving salary, summer salary, and start-up funds should all be obtained in writing to avoid future misunderstanding.

14 To accept or not to accept? Most engineering departments have only one position open at a time. Thus, the department will want to know fairly quickly if you will accept the position. Murphy's law holds here: The timing of offers never works out well. If you can, schedule all interview trips close to each other so that you can at least visit each school before you need to accept an offer. After one school gives you an offer, it is certainly ethical to visit other schools as long as you haven't decided to accept the first offer. An interview trip after you have an offer may well be your best interview trip since some of the pressure of finding a job has been removed. Tell the other schools about your deadline for making a decision. The school making the first offer probably will extend the time for decision if pressed but won't like to do so. Usually, when a candidate keeps pushing the decision back, he or she eventually accepts another offer. Once you have made a decision, accept the offer in writing, inform other schools of your decision, and then get back to work so that you can finish your thesis and report to work on time.

The procedure for professors who want to change jobs can be quite similar (Baldwin, 1990). A professor with a job may want to state on the resume that references will be supplied on request. In this way, the candidate can prescreen possible offers before letting the department chair know that other schools are interested. The professor already has a track record, and often interested universities will call him or her instead of the other way around. It is obviously flattering to be offered an interview trip which you haven't requested, but there may be good reasons to stay where you are. Every department has problems—at least you know what they are at your university. If the potential job offer is too good to turn down, then of course you will go and interview. Although there are arguments both for telling and not telling your current chair about the interview, in most cases it is probably preferable to tell the chair instead of having that person hear about the interview trip through the grapevine.

HOMEWORK

1 Prepare your resume with a focus on academic positions. Prepare a research plan for the next five years as a supplement to the resume.
2 Screen advertisements in appropriate journals for two or three months and develop a list of potential academic employers.
3 Write a cover letter to apply for an academic position.

REFERENCES

ASEE, "Thinking of an Academic Career," American Society for Engineering Education, Washington, DC, no date.

Baldwin, W. L., "A process for seeking a new position in higher education: A faculty perspective," *Proceedings ASEE Annual Conference, ASEE*, Washington, DC, 161, 1990.

Stevens, C. A., "A dean's view of hiring faculty," *Proceedings ASEE/IEEE Frontiers in Education Conference,* IEEE, New York, 159, 1990.

Wankat, P. C. and Oreovicz, F. S., "The graduate student's guide to academic job hunting," *Chem. Eng. Educ.*, 17, 178 (Fall 1983).

SAMPLE TEACHING COURSE OUTLINE

A sample outline for a three credit, one semester teaching course.

CLASS	TOPIC	CHAPTER
1	Housekeeping and introduction	1
2	Designing the first course	3
3	Efficiency	2
4	Efficiency	2
5	Finding an academic job (Panel)	App. A
6	Taxonomy and objectives	4
7	Educational objectives and text selection	4
8	Problem solving	5
9	Problem solving and creativity	5
10	Teaching methods: Lecture	6
11	Teaching methods: Lecture	6
12	Teaching methods: Discussion	7
13	Teaching methods: Cooperative groups	7
14	Teaching methods: Mastery and PSI	7
15	Teaching technology: (A) Computer (demonstration)	8
16	Teaching technology: (B) AudioTutorial [field trip]	8
17	Teaching technology: (C) TV, video	8
18	Communication skills I	
19	First progress reports on project (written and Oral)	
20	Modified debate: Teaching methods	
21	Teaching design	9

CLASS	TOPIC	CHAPTER
22	Teaching laboratory	9
23	One-to-one teaching	10
24	Review (and catchup)	
25	First test	
26	Go over test	
27.	Testing	11
28	Testing, grading and homework	11
29	Cheating and discipline	12
30	Students and teachers: Types and styles (MBTI)	13
31	Students and teachers: Types and styles (MBTI)	13
32	Development models (Piaget)	14
33	Development models (Perry)	14
34	Development models (Perry)	14
35	Learning theories	15
	Second progress report on project due (written)	
36	Learning theories	15
37	Learning theories	15
38	Evaluation of teaching	16
39	Evaluation of teaching	16
40	Professional concerns	17
41	Ethics	17
42	Second test	
43	Go over test/Communications skills II	
44	Student project presentations	
	(All written projects due)	
45	Student project presentations	
Finals	Student project presentations	